Solid-State Physics

Subject References

Solid state, of necessity, draws on many other disciplines. Suggested background reading is listed in this bibliography.

Mechanics

1. Fetter AL and Walecka JD, *Theoretical Mechanics of Particles and Continua*, McGraw-Hill Book Co., New York, 1980. Advanced
2. Goldstein H, *Classical Mechanics*, 2nd edn., Addison-Wesley Publishing Co., Reading, MA 1980. Advanced
3. Marion JB and Thornton ST, *Classical Dynamics of Particles and Systems*, Saunders College Publ. Co., Fort Worth, 1995. Intermediate

Electricity

4. Jackson JD, *Classical Electrodynamics*, John Wiley and Sons, 2nd edn. New York, 1975. Advanced
5. Reitz JD, Milford FJ and Christy RW, *Foundations of Electromagnetic Theory*, Addison-Wesley Publishing Co., Reading, MA, 1993. Intermediate

Optics

6. Guenther RD, *Modern Optics*, John Wiley and Sons, New York, 1990. Intermediate
7. Klein MV and Furtak TE, *Optics*, 2nd edn, John Wiley and Sons, New York, 1986. Intermediate

Thermodynamics

8. Espinosa TP, *Introduction to Thermophysics*, W. C. Brown, Dubuque, IA, 1994. Intermediate
9. Callen HB, *Thermodynamics and an Introduction to Thermostatics*, John Wiley and Sons, New York, 1985. Intermediate to Advanced

Statistical Mechanics

10. Kittel C and Kroemer H, *Thermal Physics*, 2nd edn, W. H. Freeman and Co., San Francisco, 1980. Intermediate
11. Huang K, *Statistical Physics*, 2nd edn, John Wiley and Sons, New York, 1987. Advanced

Critical Phenomena

12. Binney JJ, Dowrick NJ, Fisher AJ, and Newman MEJ, *The Theory of Critical Phenomena*, Clarendon Press, Oxford, 1992. Advanced

James D. Patterson
Bernard C. Bailey

Solid-State Physics

Introduction to the Theory

With 202 Figures

Springer

Professor Emeritus, James D. Patterson, Ph.D.
3504 Parkview Drive
Rapid City, SD 57701
USA
e-mail: *marluce@rap.midco.net*

Dr. Bernard C. Bailey, Ph.D.
310 Taylor Ave. # C18
Cape Canaveral, FL 32920
USA
e-mail: *bbailey@brevard.net*

ISBN-10 3-540-24115-9 Springer Berlin Heidelberg New York
ISBN-13 978-3-540-24115-7 Springer Berlin Heidelberg New York

Library of Congress Control Number: 2006938043

This work is subject to copyright. All rights are reserved, whether the whole or part of the material is concerned, specifically the rights of translation, reprinting, reuse of illustrations, recitation, broadcasting, reproduction on microfilm or in any other way, and storage in data banks. Duplication of this publication or parts thereof is permitted only under the provisions of the German Copyright Law of September 9, 1965, in its current version, and permission for use must always be obtained from Springer. Violations are liable to prosecution under the German Copyright Law.

Springer is part of Springer Science+Business Media

springer.com

© Springer-Verlag Berlin Heidelberg 2007

The use of general descriptive names, registered names, trademarks, etc. in this publication does not imply, even in the absence of a specific statement, that such names are exempt from the relevant protective laws and regulations and therefore free for general use.

This reprint has been authorized by Springer-Verlag (Berlin/Heidelberg/New York) for sale in the Mainland China only and not for export therefrom.

Preface

Learning solid-state physics requires a certain degree of maturity, since it involves tying together diverse concepts from many areas of physics. The objective is to understand, in a basic way, how solid materials behave. To do this one needs both a good physical and mathematical background. One definition of solid-state physics is that it is the study of the physical (e.g. the electrical, dielectric, magnetic, elastic, and thermal) properties of solids in terms of basic physical laws. In one sense, solid-state physics is more like chemistry than some other branches of physics because it focuses on common properties of large classes of materials. It is typical that solid-state physics emphasizes how physical properties link to the electronic structure. In this book we will emphasize crystalline solids (which are periodic 3D arrays of atoms).

We have retained the term solid-state physics, even though condensed-matter physics is more commonly used. Condensed-matter physics includes liquids and non-crystalline solids such as glass, about which we have little to say. We have also included only a little material concerning soft condensed matter (which includes polymers, membranes and liquid crystals – it also includes wood and gelatins).

Modern solid-state physics came of age in the late 1930s and early 1940s (see Seitz [82]), and had its most extensive expansion with the development of the transistor, integrated circuits, and microelectronics. Most of microelectronics, however, is limited to the properties of inhomogeneously doped semiconductors. Solid-state physics includes many other areas of course; among the largest of these are ferromagnetic materials, and superconductors. Just a little less than half of all working physicists are engaged in condensed matter work, including solid-state.

One earlier version of this book was first published 30 years ago (J.D. Patterson, *Introduction to the Theory of Solid State Physics*, Addison-Wesley Publishing Company, Reading, Massachusetts, 1971, copyright reassigned to JDP 13 December, 1977), and bringing out a new modernized and expanded version has been a prodigious task. Sticking to the original idea of presenting basics has meant that the early parts are relatively unchanged (although they contain new and reworked material), dealing as they do with structure (Chap. 1), phonons (2), electrons (3), and interactions (4). Of course, the scope of solid-state physics has greatly expanded during the past 30 years. Consequently, separate chapters are now devoted to metals and the Fermi surface (5), semiconductors (6), magnetism (7, expanded and reorganized), superconductors (8), dielectrics and ferroelectrics (9), optical properties (10), defects (11), and a final chapter (12) that includes surfaces, and brief mention of modern topics (nanostructures, the quantum Hall effect, carbon nanotubes, amorphous materials, and soft condensed matter). The reference list has been brought up to date, and several relevant topics are further discussed in the appendices. The table of contents can be consulted for a full list of what is now included.

The fact that one of us (JDP) has taught solid-state physics over the course of this 30 years has helped define the scope of this book, which is intended as a textbook. Like golf, teaching is a humbling experience. One finds not only that the students don't understand as much as one hopes, but one constantly discovers limits to his own understanding. We hope this book will help students to begin a life-long learning experience, for only in that way can they gain a deep understanding of solid-state physics.

Discoveries continue in solid-state physics. Some of the more obvious ones during the last thirty years are: quasicrystals, the quantum Hall effect (both integer and fractional – where one must finally confront new aspects of electron–electron interactions), high-temperature superconductivity, and heavy fermions. We have included these, at least to some extent, as well as several others. New experimental techniques, such as scanning probe microscopy, LEED, and EXAFS, among others have revolutionized the study of solids. Since this is an introductory book on solid-state theory, we have only included brief summaries of these techniques. New ways of growing crystals and new "designer" materials on the nanophysics scale (superlattices, quantum dots, etc.) have also kept solid-state physics vibrant, and we have introduced these topics. There have also been numerous areas in which applications have played a driving role. These include semiconductor technology, spin-polarized tunneling, and giant magnetoresistance (GMR). We have at least briefly discussed these as well as other topics.

Greatly increased computing power has allowed many ab initio methods of calculations to become practical. Most of these require specialized discussions beyond the scope of this book. However, we continue to discuss pseudopotentials, and have added a Section on density functional techniques.

Problems are given at the end of each chapter (many new problems have been added). Occasionally they are quite long and have different approximate solutions. This may be frustrating, but it appears to be necessary to work problems in solid-state physics in order to gain a physical feeling for the subject. In this respect, solid-state physics is no different from many other branches of physics.

We should discuss what level of students for which this book is intended. One could perhaps more appropriately ask what degree of maturity of the students is assumed? Obviously, some introduction to quantum mechanics, solid-state physics, thermodynamics, statistical mechanics, mathematical physics, as well as basic mechanics and electrodynamics is necessary. In our experience, this is most commonly encountered in graduate students, although certain mature undergraduates will be able to handle much of the material in this book.

Although it is well to briefly mention a wide variety of topics, so that students will not be "blind sided" later, and we have done this in places, in general it is better to understand one topic relatively completely, than to scan over several. We caution professors to be realistic as to what their students can really grasp. If the students have a good start, they have their whole careers to fill in the details.

The method of presentation of the topics draws heavily on many other solid-state books listed in the bibliography. Acknowledgment due the authors of these books is made here. The selection of topics was also influenced by discussion with colleagues and former teachers, some of whom are mentioned later.

We think that solid-state physics abundantly proves that more is different, as has been attributed to P. W. Anderson. There really are emergent properties at higher levels of complexity. Seeking them, including applications, is what keeps solid-state physics alive.

In this day and age, no one book can hope to cover all of solid-state physics. We would like to particularly single out the following books for reference and or further study. Terms in brackets refer to references listed in the Bibliography.

1. Kittel – 7th edition – remains unsurpassed for what it does [23, 1996]. Also Kittel's book on advanced solid-state physics [60, 1963] is very good.

2. Ashcroft and Mermin, *Solid State Physics* – has some of the best explanations of many topics I have found anywhere [21, 1976].

3. Jones and March – a comprehensive two-volume work [22, 1973].

4. J.M. Ziman – many extremely clear physical explanation [25, 1972], see also Ziman's classic *Electrons and Phonons* [99, 1960].

5. O. Madelung, *Introduction to Solid-State Theory* – Complete with a very transparent and physical presentation [4.25].

6. M.P. Marder, *Condensed Matter Physics* – A modern presentation, including modern density functional methods with references [3.29].

7. P. Phillips, *Advanced Solid State Physics* – A modern Frontiers in Physics book, bearing the imprimatur of David Pines [A.20].

8. Dalven – a good start on applied solid-state physics [32, 1990].

9. Also Oxford University Press has recently put out a "Master Series in Condensed Matter Physics." There are six books which we recommend.
   a) Martin T. Dove, *Structure and Dynamics* – An atomic view of Materials [2.14].
   b) John Singleton, *Band Theory and Electronic Properties of Solids* [3.46].
   c) Mark Fox, *Optical Properties of Solids* [10.12].
   d) Stephen Blundell, *Magnetism in Condensed Matter* [7.9].
   e) James F. Annett, *Superconductivity, Superfluids, and Condensates* [8.3].
   f) Richard A. L. Jones, *Soft Condensed Matter* [12.30].

A word about notation is in order. We have mostly used SI units (although gaussian is occasionally used when convenient); thus E is the electric field, D is the electric displacement vector, P is the polarization vector, H is the magnetic field, B is the magnetic induction, and M is the magnetization. Note that the above quantities are in boldface. The boldface notation is used to indicate a vector. The magnitude of a vector V is denoted by V. In the SI system μ is the permeability (μ also represents other quantities). μ_0 is the permeability of free space, ε is the permittivity, and ε_0 the permittivity of free space. In this notation μ_0 should not be confused with μ_B, which is the Bohr magneton [$= |e| \hbar/2m$, where e = magnitude of electronic charge (i.e. e means $+|e|$ unless otherwise noted), $\hbar$ = Planck's constant divided by 2π, and m = electronic mass]. We generally prefer to write $\int A d^3 r$ or $\int A d\mathbf{r}$ instead of $\int A\, dx\, dy\, dz$, but they all mean the same thing. Both $\langle i|H|j\rangle$ and $(i|H|j)$ are used for the matrix elements of an operator H. Both mean $\int \psi^* H \psi d\tau$ where the integral over τ means to integrate over whatever space is appropriate (e.g., it could mean an integral over real space and a sum over spin space). By $\sum$ a summation is indicated and by $\prod$ a product. The Kronecker delta δ_{ij} is 1 when $i = j$ and zero when $i \neq j$. We have not used covariant and contravariant spaces; thus δ_{ij} and δ_i^j, for example, mean the same thing. We have labeled sections by A for advanced, B for basic, and EE for material that might be especially interesting for electrical engineers, and similarly MS for materials science, and MET for metallurgy. Also by [number], we refer to a reference at the end of the book.

There are too many colleagues to thank, to include a complete list. JDP wishes to specifically thank several. A beautifully prepared solid-state course by Professor W. R Wright at the University of Kansas gave him his first exposure to a logical presentation of solid-state physics, while also at Kansas, Dr. R.J. Friauf, was very helpful in introducing JDP to the solid-state. Discussions with Dr. R.D. Redin, Dr. R.G. Morris, Dr. D.C. Hopkins, Dr. J. Weyland, Dr. R.C. Weger and others who were at the South Dakota School of Mines and Technology were always useful. Sabbaticals were spent at Notre Dame and the University of Nebraska, where working with Dr. G.L. Jones (Notre Dame) and D.J. Sellmyer (Nebraska) deepened JDP's understanding. At the Florida Institute of Technology, Drs. J. Burns, and J. Mantovani have read parts of this book, and discussions with Dr. R. Raffaelle and Dr. J. Blatt were useful. Over the course of JDP's career, a variety of summer jobs were held that bore on solid-state physics; these included positions at Hughes Semiconductor Laboratory, North American Science Center, Argonne National Laboratory, Ames Laboratory of Iowa State University, the Federal University of Pernambuco in Recife, Brazil, Sandia National Laboratory, and the Marshal Space Flight Center. Dr. P. Richards of Sandia, and Dr. S.L. Lehoczky of Marshall, were particularly helpful to JDP. Brief, but very pithy conversations of JDP with Dr. M. L. Cohen of the University of Califonia/ Berkeley, over the years, have also been uncommonly useful.

Dr. B.C. Bailey would like particularly to thank Drs. J. Burns and J. Blatt for the many years of academic preparation, mentorship, and care they provided at Florida Institute of Technology. A special thanks to Dr. J.D. Patterson who, while Physics Department Head at Florida Institute of Technology, made a conscious decision to take on a coauthor for this extraordinary project.

All mistakes, misconceptions and failures to communicate ideas are our own. No doubt some sign errors, misprints, incorrect shading of meanings, and perhaps more serious errors have crept in, but hopefully their frequency decreases with their gravity.

Most of the figures, for the first version of this book, were prepared in preliminary form by Mr. R.F. Thomas. However, for this book, the figures are either new or reworked by the coauthor (BCB).

We gratefully acknowledge the cooperation and kind support of Dr. C. Asheron, Ms. E. Sauer, and Ms. A. Duhm of Springer. Finally, and most importantly, JDP would like to note that without the constant encouragement and patience of his wife Marluce, this book would never have been completed.

October 2005

J.D. Patterson, Rapid City, South Dakota
B.C. Bailey, Cape Canaveral, Florida

Contents

1 Crystal Binding and Structure 1
- 1.1 Classification of Solids by Binding Forces (B) 2
  - 1.1.1 Molecular Crystals and the van der Waals Forces (B) 2
  - 1.1.2 Ionic Crystals and Born–Mayer Theory (B) 6
  - 1.1.3 Metals and Wigner–Seitz Theory (B) 9
  - 1.1.4 Valence Crystals and Heitler–London Theory (B) 11
  - 1.1.5 Comment on Hydrogen-Bonded Crystals (B) 12
- 1.2 Group Theory and Crystallography 13
  - 1.2.1 Definition and Simple Properties of Groups (AB) 14
  - 1.2.2 Examples of Solid-State Symmetry Properties (B) 17
  - 1.2.3 Theorem: No Five-fold Symmetry (B) 20
  - 1.2.4 Some Crystal Structure Terms and Nonderived Facts (B) 22
  - 1.2.5 List of Crystal Systems and Bravais Lattices (B) 23
  - 1.2.6 Schoenflies and International Notation for Point Groups (A) 26
  - 1.2.7 Some Typical Crystal Structures (B) 28
  - 1.2.8 Miller Indices (B) 31
  - 1.2.9 Bragg and von Laue Diffraction (AB) 31
- Problems 38

2 Lattice Vibrations and Thermal Properties 41
- 2.1 The Born–Oppenheimer Approximation (A) 42
- 2.2 One-Dimensional Lattices (B) 47
  - 2.2.1 Classical Two-Atom Lattice with Periodic Boundary Conditions (B) 48
  - 2.2.2 Classical, Large, Perfect Monatomic Lattice, and Introduction to Brillouin Zones (B) 51
  - 2.2.3 Specific Heat of Linear Lattice (B) 61
  - 2.2.4 Classical Diatomic Lattices: Optic and Acoustic Modes (B) 64
  - 2.2.5 Classical Lattice with Defects (B) 69
  - 2.2.6 Quantum-Mechanical Linear Lattice (B) 75

XII Contents

    2.3   Three-Dimensional Lattices 84
           2.3.1   Direct and Reciprocal Lattices and Pertinent
                     Relations (B) 84
           2.3.2   Quantum-Mechanical Treatment and Classical
                     Calculation of the Dispersion Relation (B) 86
           2.3.3   The Debye Theory of Specific Heat (B) 91
           2.3.4   Anharmonic Terms in The Potential /
                     The Gruneisen Parameter (A) 99
           2.3.5   Wave Propagation in an Elastic Crystalline Continuum
                     (MET, MS) 102
    Problems 108

3 Electrons in Periodic Potentials 113
    3.1   Reduction to One-Electron Problem 114
           3.1.1   The Variational Principle (B) 114
           3.1.2   The Hartree Approximation (B) 115
           3.1.3   The Hartree–Fock Approximation (A) 119
           3.1.4   Coulomb Correlations and the Many-Electron
                     Problem (A) 135
           3.1.5   Density Functional Approximation (A) 137
    3.2   One-Electron Models 148
           3.2.1   The Kronig–Penney Model (B) 148
           3.2.2   The Free-Electron or Quasifree-Electron
                     Approximation (B) 158
           3.2.3   The Problem of One Electron in a Three-Dimensional
                     Periodic Potential 173
           3.2.4   Effect of Lattice Defects on Electronic States
                     in Crystals (A) 205
    Problems 209

4 The Interaction of Electrons and Lattice Vibrations 213
    4.1   Particles and Interactions of Solid-state Physics (B) 213
    4.2   The Phonon–Phonon Interaction (B) 219
           4.2.1   Anharmonic Terms in the Hamiltonian (B) 219
           4.2.2   Normal and Umklapp Processes (B) 220
           4.2.3   Comment on Thermal Conductivity (B) 223
    4.3   The Electron–Phonon Interaction 225
           4.3.1   Form of the Hamiltonian (B) 225
           4.3.2   Rigid-Ion Approximation (B) 229
           4.3.3   The Polaron as a Prototype Quasiparticle (A) 232
    4.4   Brief Comments on Electron–Electron Interactions (B) 242

		Contents	XIII

- 4.5 The Boltzmann Equation and Electrical Conductivity 244
  - 4.5.1 Derivation of the Boltzmann Differential Equation (B) 244
  - 4.5.2 Motivation for Solving the Boltzmann Differential Equation (B) 246
  - 4.5.3 Scattering Processes and Q Details (B) 247
  - 4.5.4 The Relaxation-Time Approximate Solution of the Boltzmann Equation for Metals (B) 251
- 4.6 Transport Coefficients 253
  - 4.6.1 The Electrical Conductivity (B) 254
  - 4.6.2 The Peltier Coefficient (B) 254
  - 4.6.3 The Thermal Conductivity (B) 254
  - 4.6.4 The Thermoelectric Power (B) 255
  - 4.6.5 Kelvin's Theorem (B) 255
  - 4.6.6 Transport and Material Properties in Composites (MET, MS) 256
- Problems 263

5 Metals, Alloys, and the Fermi Surface 265
- 5.1 Fermi Surface (B) 265
  - 5.1.1 Empty Lattice (B) 266
  - 5.1.2 Exercises (B) 267
- 5.2 The Fermi Surface in Real Metals (B) 271
  - 5.2.1 The Alkali Metals (B) 271
  - 5.2.2 Hydrogen Metal (B) 271
  - 5.2.3 The Alkaline Earth Metals (B) 271
  - 5.2.4 The Noble Metals (B) 271
- 5.3 Experiments Related to the Fermi Surface (B) 273
- 5.4 The de Haas–van Alphen effect (B) 274
- 5.5 Eutectics (MS, ME) 278
- 5.6 Peierls Instability of Linear Metals (B) 279
  - 5.6.1 Relation to Charge Density Waves (A) 282
  - 5.6.2 Spin Density Waves (A) 283
- 5.7 Heavy Fermion Systems (A) 283
- 5.8 Electromigration (EE, MS) 284
- 5.9 White Dwarfs and Chandrasekhar's Limit (A) 286
  - 5.9.1 Gravitational Self-Energy (A) 287
  - 5.9.2 Idealized Model of a White Dwarf (A) 287
- 5.10 Some Famous Metals and Alloys (B, MET) 290
- Problems 291

XIV Contents

6 Semiconductors ... 293
6.1 Electron Motion ... 296
- 6.1.1 Calculation of Electron and Hole Concentration (B) ... 296
- 6.1.2 Equation of Motion of Electrons in Energy Bands (B) ... 302
- 6.1.3 Concept of Hole Conduction (B) ... 305
- 6.1.4 Conductivity and Mobility in Semiconductors (B) ... 307
- 6.1.5 Drift of Carriers in Electric and Magnetic Fields: The Hall Effect (B) ... 309
- 6.1.6 Cyclotron Resonance (A) ... 311

6.2 Examples of Semiconductors ... 319
- 6.2.1 Models of Band Structure for Si, Ge and II-VI and III-V Materials (A) ... 319
- 6.2.2 Comments about GaN (A) ... 324

6.3 Semiconductor Device Physics ... 325
- 6.3.1 Crystal Growth of Semiconductors (EE, MET, MS) ... 325
- 6.3.2 Gunn Effect (EE) ... 326
- 6.3.3 *pn*-Junctions (EE) ... 328
- 6.3.4 Depletion Width, Varactors, and Graded Junctions (EE) ... 331
- 6.3.5 Metal Semiconductor Junctions — the Schottky Barrier (EE) ... 334
- 6.3.6 Semiconductor Surface States and Passivation (EE) ... 335
- 6.3.7 Surfaces Under Bias Voltage (EE) ... 337
- 6.3.8 Inhomogeneous Semiconductors Not in Equilibrium (EE) ... 338
- 6.3.9 Solar Cells (EE) ... 344
- 6.3.10 Transistors (EE) ... 350
- 6.3.11 Charge-Coupled Devices (CCD) (EE) ... 350

Problems ... 351

7 Magnetism, Magnons, and Magnetic Resonance ... 353
7.1 Types of Magnetism ... 354
- 7.1.1 Diamagnetism of the Core Electrons (B) ... 354
- 7.1.2 Paramagnetism of Valence Electrons (B) ... 355
- 7.1.3 Ordered Magnetic Systems (B) ... 358

7.2 Origin and Consequences of Magnetic Order ... 371
- 7.2.1 Heisenberg Hamiltonian ... 371
- 7.2.2 Magnetic Anisotropy and Magnetostatic Interactions (A) ... 383
- 7.2.3 Spin Waves and Magnons (B) ... 388
- 7.2.4 Band Ferromagnetism (B) ... 405
- 7.2.5 Magnetic Phase Transitions (A) ... 414

7.3 Magnetic Domains and Magnetic Materials (B) ... 420
- 7.3.1 Origin of Domains and General Comments (B) ... 420
- 7.3.2 Magnetic Materials (EE, MS) ... 430

	7.4	Magnetic Resonance and Crystal Field Theory	432
		7.4.1 Simple Ideas About Magnetic Resonance (B)	432
		7.4.2 A Classical Picture of Resonance (B)	433
		7.4.3 The Bloch Equations and Magnetic Resonance (B)	436
		7.4.4 Crystal Field Theory and Related Topics (B)	442
	7.5	Brief Mention of Other Topics	450
		7.5.1 Spintronics or Magnetoelectronics (EE)	450
		7.5.2 The Kondo Effect (A)	453
		7.5.3 Spin Glass (A)	454
		7.5.4 Solitons (A, EE)	456
	Problems		457

8 Superconductivity ... 459

	8.1	Introduction and Some Experiments (B)	459
		8.1.1 Ultrasonic Attenuation (B)	463
		8.1.2 Electron Tunneling (B)	463
		8.1.3 Infrared Absorption (B)	463
		8.1.4 Flux Quantization (B)	463
		8.1.5 Nuclear Spin Relaxation (B)	463
		8.1.6 Thermal Conductivity (B)	464
	8.2	The London and Ginzburg–Landau Equations (B)	465
		8.2.1 The Coherence Length (B)	467
		8.2.2 Flux Quantization and Fluxoids (B)	471
		8.2.3 Order of Magnitude for Coherence Length (B)	472
	8.3	Tunneling (B, EE)	473
		8.3.1 Single-Particle or Giaever Tunneling	473
		8.3.2 Josephson Junction Tunneling	475
	8.4	SQUID: Superconducting Quantum Interference (EE)	479
		8.4.1 Questions and Answers (B)	481
	8.5	The Theory of Superconductivity (A)	482
		8.5.1 Assumed Second Quantized Hamiltonian for Electrons and Phonons in Interaction (A)	482
		8.5.2 Elimination of Phonon Variables and Separation of Electron–Electron Attraction Term Due to Virtual Exchange of Phonons (A)	486
		8.5.3 Cooper Pairs and the BCS Hamiltonian (A)	489
		8.5.4 Remarks on the Nambu Formalism and Strong Coupling Superconductivity (A)	500
	8.6	Magnesium Diboride (EE, MS, MET)	501
	8.7	Heavy-Electron Superconductors (EE, MS, MET)	501
	8.8	High-Temperature Superconductors (EE, MS, MET)	502
	8.9	Summary Comments on Superconductivity (B)	504
	Problems		507

9 Dielectrics and Ferroelectrics 509
- 9.1 The Four Types of Dielectric Behavior (B) 509
- 9.2 Electronic Polarization and the Dielectric Constant (B) 510
- 9.3 Ferroelectric Crystals (B) 516
  - 9.3.1 Thermodynamics of Ferroelectricity by Landau Theory (B) 518
  - 9.3.2 Further Comment on the Ferroelectric Transition (B, ME) 520
  - 9.3.3 One-Dimensional Model of the Soft Mode of Ferroelectric Transitions (A) 521
- 9.4 Dielectric Screening and Plasma Oscillations (B) 525
  - 9.4.1 Helicons (EE) 527
  - 9.4.2 Alfvén Waves (EE) 529
- 9.5 Free-Electron Screening 531
  - 9.5.1 Introduction (B) 531
  - 9.5.2 The Thomas–Fermi and Debye–Huckel Methods (A, EE) 531
  - 9.5.3 The Lindhard Theory of Screening (A) 535
- Problems 540

10 Optical Properties of Solids 543
- 10.1 Introduction (B) 543
- 10.2 Macroscopic Properties (B) 544
  - 10.2.1 Kronig–Kramers Relations (A) 548
- 10.3 Absorption of Electromagnetic Radiation–General (B) 550
- 10.4 Direct and Indirect Absorption Coefficients (B) 551
- 10.5 Oscillator Strengths and Sum Rules (A) 558
- 10.6 Critical Points and Joint Density of States (A) 559
- 10.7 Exciton Absorption (A) 560
- 10.8 Imperfections (B, MS, MET) 561
- 10.9 Optical Properties of Metals (B, EE, MS) 563
- 10.10 Lattice Absorption, Restrahlen, and Polaritons (B) 569
  - 10.10.1 General Results (A) 569
  - 10.10.2 Summary of the Properties of $\varepsilon(q, \omega)$ (B) 576
  - 10.10.3 Summary of Absorption Processes: General Equations (B) 577
- 10.11 Optical Emission, Optical Scattering and Photoemission (B) 578
  - 10.11.1 Emission (B) 578
  - 10.11.2 Einstein A and B Coefficients (B, EE, MS) 579
  - 10.11.3 Raman and Brillouin Scattering (B, MS) 580
- 10.12 Magneto-Optic Effects: The Faraday Effect (B, EE, MS) 582
- Problems 585

11 Defects in Solids 587
11.1 Summary About Important Defects (B) 587
11.2 Shallow and Deep Impurity Levels in Semiconductors (EE) 590
11.3 Effective Mass Theory, Shallow Defects, and Superlattices (A) ... 591
11.3.1 Envelope Functions (A) 591
11.3.2 First Approximation (A) 592
11.3.3 Second Approximation (A) 593
11.4 Color Centers (B) 596
11.5 Diffusion (MET, MS) 598
11.6 Edge and Screw Dislocation (MET, MS) 599
11.7 Thermionic Emission (B) 601
11.8 Cold-Field Emission (B) 604
11.9 Microgravity (MS) 606
Problems 607

12 Current Topics in Solid Condensed–Matter Physics 609
12.1 Surface Reconstruction (MET, MS) 610
12.2 Some Surface Characterization Techniques (MET, MS, EE) ... 611
12.3 Molecular Beam Epitaxy (MET, MS) 613
12.4 Heterostructures and Quantum Wells 614
12.5 Quantum Structures and Single-Electron Devices (EE) ... 615
12.5.1 Coulomb Blockade (EE) 616
12.5.2 Tunneling and the Landauer Equation (EE) 619
12.6 Superlattices, Bloch Oscillators, Stark–Wannier Ladders ... 622
12.6.1 Applications of Superlattices and Related Nanostructures (EE) 625
12.7 Classical and Quantum Hall Effect (A) 627
12.7.1 Classical Hall Effect – CHE (A) 627
12.7.2 The Quantum Mechanics of Electrons in a Magnetic Field: The Landau Gauge (A) 630
12.7.3 Quantum Hall Effect: General Comments (A) 632
12.8 Carbon – Nanotubes and Fullerene Nanotechnology (EE) ... 636
12.9 Amorphous Semiconductors and the Mobility Edge (EE) 637
12.9.1 Hopping Conductivity (EE) 638
12.10 Amorphous Magnets (MET, MS) 639
12.11 Soft Condensed Matter (MET, MS) 640
12.11.1 General Comments 640
12.11.2 Liquid Crystals (MET, MS) 640
12.11.3 Polymers and Rubbers (MET, MS) 641
Problems 644

Appendices 647
- A Units 647
- B Normal Coordinates 649
- C Derivations of Bloch's Theorem 652
  - C.1 Simple One-Dimensional Derivation 652
  - C.2 Simple Derivation in Three Dimensions 655
  - C.3 Derivation of Bloch's Theorem by Group Theory 656
- D Density Matrices and Thermodynamics 657
- E Time-Dependent Perturbation Theory 658
- F Derivation of The Spin-Orbit Term From Dirac's Equation 660
- G The Second Quantization Notation for Fermions and Bosons 662
  - G.1 Bose Particles 662
  - G.2 Fermi Particles 663
- H The Many-Body Problem 665
  - H.1 Propagators 666
  - H.2 Green Functions 666
  - H.3 Feynman Diagrams 667
  - H.4 Definitions 667
  - H.5 Diagrams and the Hartree and Hartree–Fock Approximations 668
  - H.6 The Dyson Equation 671

Bibliography 673
- Chapter 1 673
- Chapter 2 674
- Chapter 3 676
- Chapter 4 678
- Chapter 5 679
- Chapter 6 681
- Chapter 7 683
- Chapter 8 685
- Chapter 9 687
- Chapter 10 688
- Chapter 11 689
- Chapter 12 690
- Appendices 694
- Subject References 695
- Further Reading 698

Index 703

7 Magnetism, Magnons, and Magnetic Resonance

The first chapter was devoted to the solid-state medium (i.e. its crystal structure and binding). The next two chapters concerned the two most important types of energy excitations in a solid (the electronic excitations and the phonons). *Magnons* are another important type of energy excitation and they occur in magnetically ordered solids. However, it is not possible to discuss magnons without laying some groundwork for them by discussing the more elementary parts of magnetic phenomena. Also, there are many magnetic properties that cannot be discussed by using the concept of magnons. In fact, the study of magnetism is probably the first solid-state property that was seriously studied, relating as it does to lodestone and compass needles.

Nearly all the magnetic effects in solids arise from electronic phenomena, and so it might be thought that we have already covered at least the fundamental principles of magnetism. However, we have not yet discussed in detail the electron's spin degree of freedom, and it is this, as well as the orbital angular moment that together produce magnetic moments and thus are responsible for most magnetic effects in solids. When all is said and done, because of the richness of this subject, we will end up with a rather large chapter devoted to magnetism.

We will begin by briefly surveying some of the larger-scale phenomena associated with magnetism (diamagnetism, paramagnetism, ferromagnetism, and allied topics). These are of great technical importance. We will then show how to understand the origin of ordered magnetic structures from a quantum-mechanical viewpoint (in fact, strictly speaking this is the only way to understand it). This will lead to a discussion of the Heisenberg Hamiltonian, mean field theory, spin waves and magnons (the quanta of spin waves). We will also discuss the behavior of ordered magnetic systems near their critical temperature, which turns out also to be incredibly rich in ideas.

Following this we will discuss magnetic domains and related topics. This is of great practical importance.

Some of the simpler aspects of magnetic resonance will then be discussed as it not only has important applications, but magnetic resonance experiments provide direct measurements of the very small energy differences between magnetic sublevels in solids, and so they can be very sensitive probes into the inner details of magnetic solids.

We will end the chapter with some brief discussion of recent topics: the Kondo effect, spin glasses, magnetoelectronics, and solitons.

7.1 Types of Magnetism

7.1.1 Diamagnetism of the Core Electrons (B)

All matter shows diamagnetic effects, although these effects are often obscured by other stronger types of magnetism. In a solid in which the diamagnetic effect predominates, the solid has an induced magnetic moment that is in the opposite direction to an external applied magnetic field.

Since the diamagnetism of conduction electrons (Landau diamagnetism) has already been discussed (Sect. 3.2.2), this Section will concern itself only with the diamagnetism of the core electrons.

For an external magnetic field H in the z direction, the Hamiltonian (SI, $e > 0$) is given by

$$\mathcal{H} = \frac{p^2}{2m} + V(r) + \frac{e\hbar\mu_0 H}{2mi}\left(x\frac{\partial}{\partial y} - y\frac{\partial}{\partial x}\right) + \frac{e^2\mu_0^2 H^2}{8m}(x^2 + y^2).$$

For purely diamagnetic atoms with zero total angular momentum, the term involving first derivatives has zero matrix elements and so will be neglected. Thus, with a spherically symmetric potential $V(r)$, the one-electron Hamiltonian is

$$\mathcal{H} = \frac{p^2}{2m} + V(r) + \frac{e^2\mu_0^2 H^2}{8m}(x^2 + y^2). \tag{7.1}$$

Let us evaluate the susceptibility of such a diamagnetic substance. It will be assumed that the eigenvalues of (7.1) (with $H = 0$) and the eigenkets $|n\rangle$ are precisely known. Then by first-order perturbation theory, the energy change in state n due to the external magnetic field is

$$E' = \frac{e^2\mu_0^2 H^2}{8m}\langle n|x^2 + y^2|n\rangle. \tag{7.2}$$

For simplicity, it will be assumed that $|n\rangle$ is spherically symmetric. In this case

$$\langle n|x^2 + y^2|n\rangle = \tfrac{2}{3}\langle n|r^2|n\rangle. \tag{7.3}$$

The induced magnetic moment μ can now be readily evaluated:

$$\mu = -\frac{\partial E'}{\partial(\mu_0 H)} = -\frac{e^2\mu_0 H}{6m}\langle n|r^2|n\rangle. \tag{7.4}$$

If N is the number of atoms per unit volume, and Z is the number of core electrons, then the magnetization M is $ZN\mu$, and the magnetic susceptibility χ is

$$\chi = \frac{\partial M}{\partial H} = -\frac{ZNe^2\mu_0}{6m}\langle n|r^2|n\rangle. \tag{7.5}$$

If we make an obvious reinterpretation of $\langle n|r^2|n\rangle$, then this result agrees with the classical result [7.39 p. 418]. The derivation of (7.5) assumes that the core electrons do not interact and that they are all in the same state $|n\rangle$. For core electrons on different atoms noninteraction would appear to be reasonable. However, it is not clear that this would lead to reasonable results for core electrons on the same atom. A generalization to core atoms in different states is fairly obvious.

A measurement of the diamagnetic susceptibility, when combined with theory (similar to the above), can sometimes provide a good test for any proposed forms for the core wave functions. However, if paramagnetic or other effects are present, they must first be subtracted out, and this procedure can lead to uncertainty in interpretation.

In summary, we can make the following statements about diamagnetism:

1. Every solid has diamagnetism although it may be masked by other magnetic effects.
2. The diamagnetic susceptibility (which is negative) is temperature independent (assuming we can regard $\langle n|r^2|n\rangle$ as temperature independent).

7.1.2 Paramagnetism of Valence Electrons (B)

This Section is begun by making several comments about paramagnetism:

1. One form of paramagnetism has already been studied. This is the Pauli paramagnetism of the free electrons (Sect. 3.2.2).
2. When discussing paramagnetic effects, in general both the orbital and intrinsic spin properties of the electrons must be considered.
3. A paramagnetic substance has an induced magnetic moment in the same direction as the applied magnetic field.
4. When paramagnetic effects are present, they generally are much larger than the diamagnetic effects.
5. At high enough temperatures, all substances appear to behave in either a paramagnetic fashion or a diamagnetic fashion (even ferromagnetic solids, as we will discuss, become paramagnetic above a certain temperature).
6. The calculation of the paramagnetic susceptibility is a statistical problem, but the general reason for paramagnetism is unpaired electrons in unfilled shells of electrons.
7. The study of paramagnetism provides a natural first step for understanding ferromagnetism.

The calculation of a paramagnetic susceptibility will only be outlined. The perturbing part of the Hamiltonian is of the form [94], $e > 0$,

$$\mathcal{H}' = \frac{e\mu_0 H}{2m} \cdot (L + 2S), \tag{7.6}$$

where L is the total orbital angular momentum operator, and S is the total spin operator. Using a canonical ensemble, we find the magnetization of a sample to be given by

$$\langle M \rangle = N\text{Tr}\left[\mu \exp\left(\frac{F - \mathcal{H}'}{kT}\right)\right], \tag{7.7}$$

where N is the number of atoms per unit volume, μ is the magnetic moment operator proportional to $(L + 2S)$, and F is the Helmholtz free energy.

Once (7.7) has been computed, the magnetic susceptibility is easily evaluated by means of

$$\chi \equiv \frac{\partial \langle M \rangle}{\partial H}. \tag{7.8}$$

Equations (7.7) and (7.8) are always appropriate for evaluating χ, but the form of the Hamiltonian is modified if one wants to include complicated interaction effects.

At lower temperatures we expect that interactions such as crystal-field effects will become important. Properly including these effects for a specific problem is usually a research problem. The effects of crystal fields will be discussed later in the chapter.

Let us consider a particularly simple case of paramagnetism. This is the case of a particle with spin S (and no other angular momentum). For a magnetic field in the z-direction we can write the Hamiltonian as (charge on electron is $e > 0$)

$$\mathcal{H}' = \frac{e\mu_0 H}{2m} \cdot 2S_z. \tag{7.9}$$

Let us define $g\mu_B$ in such a way that the eigenvalues of (7.9) are

$$E = g\mu_B \mu_0 H M_S, \tag{7.10}$$

where $\mu_B = e\hbar/2m$ is the Bohr magneton, and g is sometimes called simply the g-factor. The use of a g-factor allows our formalism to include orbital effects if necessary. In (7.10) $g = 2$ (spin only).

If N is the number of particles per unit volume, then the average magnetization can be written as[1]

$$\langle M \rangle = N \frac{\sum_{M_S=-S}^{S} M_S g\mu_B \exp(M_S g\mu_B \mu_0 H / kT)}{\sum_{M_S=-S}^{S} \exp(M_S g\mu_B \mu_0 H / kT)}. \tag{7.11}$$

---

[1] Note that μ_B has absorbed the $\hbar$ so M_S and S are either integers or half-integers. Also note (7.11) is invariant to a change of the dummy summation variable from M_S to $-M_S$.

For high temperatures (and/or weak magnetic fields, so only the first two terms of the expansion of the exponential need be retained) we can write

$$\langle M \rangle \cong Ng\mu_B \frac{\sum_{M_S=-S}^{S} M_S(1 + M_S g\mu_B \mu_0 H/kT)}{\sum_{M_S=-S}^{S}(1 + M_S g\mu_B \mu_0 H/kT)},$$

which, after some manipulation, becomes to order H

$$\langle M \rangle = g^2 S(S+1)\frac{N\mu_B^2 \mu_0 H}{3kT},$$

or

$$\chi \equiv \frac{\partial \langle M \rangle}{\partial H} = \mu_0 \frac{N p_{\text{eff}}^2 \mu_B^2}{3kT}, \qquad (7.12)$$

[2] where $p_{\text{eff}} = g[S(S+1)]^{1/2}$ is called the *effective magneton number*. Equation (7.12) is the *Curie law*. It expresses the $(1/T)$ dependence of the magnetic susceptibility at high temperature. Note that when $H \to 0$, (7.12) is an exact consequence of (7.11).

It is convenient to have an expression for the magnetization of paramagnets that is valid at all temperatures and magnetic fields.

If we define

$$X = \frac{g\mu_B \mu_0 H}{kT}, \qquad (7.13)$$

then

$$\langle M \rangle = Ng\mu_B \frac{\sum_{M_S=-S}^{S} M_S e^{M_S X}}{\sum_{M_S=-S}^{S} e^{M_S X}}. \qquad (7.14)$$

With a little elementary manipulation, it is possible to perform the sums indicated in (7.14):

$$\langle M \rangle = Ng\mu_B \frac{d}{dX}\left[\ln\left(\frac{\sinh[(S+\frac{1}{2})X]}{\sinh(X/2)}\right)\right],$$

or

$$\langle M \rangle = Ng\mu_B S\left[\frac{2S+1}{2S}\coth\left(\frac{2S+1}{2S}SX\right) - \frac{1}{2S}\coth\left(\frac{SX}{2S}\right)\right]. \qquad (7.15)$$

---

[2] A temperature-independent contribution known as van Vleck paramagnetism may also be important for some materials at low temperature. It may occur due to the effect of excited states that can be treated by second-order perturbation theory. It is commonly important when first-order terms vanish. See Ashcroft and Mermin [7.2 p. 653].

Defining the Brillouin function $B_J(y)$ as

$$B_J(y) = \frac{2J+1}{2J}\coth\left(\frac{2J+1}{2J}y\right) - \frac{1}{2J}\coth\frac{y}{2J}, \qquad (7.16)$$

we can write the magnetization $\langle M \rangle$ as

$$\langle M \rangle = NgS\mu_B B_S(SX). \qquad (7.17)$$

It is easy to recover the high-temperature results (7.12) from (7.17). All we have to do is use

$$B_J(y) = \frac{J+1}{3J}y \quad \text{if} \quad y \ll 1. \qquad (7.18)$$

Then

$$\langle M \rangle \to NgS\mu_B \frac{S(S+1)}{3S}SX,$$

or using (7.13),

$$\langle M \rangle = \frac{Ng^2\mu_B^2 S(S+1)\mu_0 H}{3kT}.$$

7.1.3 Ordered Magnetic Systems (B)

Ferromagnetism and the Weiss Mean Field Theory (B)

Ferromagnetism refers to solids that are magnetized without an applied magnetic field. These solids are said to be spontaneously magnetized. Ferromagnetism occurs when paramagnetic ions in a solid "lock" together in such a way that their magnetic moments all point (on the average) in the same direction. At high enough temperatures, this "locking" breaks down and ferromagnetic materials become paramagnetic. The temperature at which this transition occurs is called the *Curie temperature*.

There are two aspects of ferromagnetism. One of these is the description of what goes on inside a single magnetized *domain* (where the magnetic moments are all aligned). The other is the description of how domains interact to produce the observed magnetic effects such as hysteresis. Domains will be briefly discussed later (Sect. 7.3).

We start by considering various magnetic structures without the complication of domains. Ferromagnetism, especially ferromagnetism in metals, is still not quantitatively and completely understood in all magnetic materials. We will turn to a more detailed study of the fundamental origin of ferromagnetism in Sect. 7.2. Our aim in this Section is to give a brief survey of the phenomena and of some phenomenological ideas.

In the ferromagnetic state at low temperatures, the spins on the various atoms are aligned parallel. There are several other types of ordered magnetic structures. These structures order for the same physical reason that ferromagnetic structures do (i.e. because of exchange coupling between the spins as we will discuss in Sect. 7.2). They also have more complex domain effects that will not be discussed.

Examples of elements that show spontaneous magnetism or ferromagnetism are (1) transition or iron group elements (e.g. Fe, Ni, Co), (2) rare earth group elements (e.g. Gd or Dy), and (3) many compounds and alloys. Further examples are given in Sect. 7.3.2.

The Weiss theory is a mean field theory and is perhaps the simplest way of discussing the appearance of the ferromagnetic state. First, what is mean field theory? Basically, mean field theory is a linearized theory in which the Hamiltonian products of operators representing dynamical observables are approximated by replacing these products by a dynamical observable times the mean or average value of a dynamic observable. The average value is then calculated self-consistently from this approximated Hamiltonian. The nature of this approximation is such that thermodynamic fluctuations are ignored. Mean field theory is often used to get an idea as to what structures or phases are present as the temperature and other parameters are varied. It is almost universally used as a first approximation, although, as discussed below, it can even be qualitatively wrong (in, for example, predicting a phase transition where there is none).

The Weiss mean field theory does the main thing that we want a theory of the magnetic state to do. It predicts a phase transition. Unfortunately, the quantitative details of real phase transitions are typically not what the Weiss theory says they should be. Still, it has several advantages:

1. It provides a comprehensive if at times only qualitative description of most magnetic materials. The Weiss theory (augmented with the concept of domains) is still the most important theory for a practical discussion of many types of magnetic behavior. Many experimental results are still presented within the context of this theory, and so in order to read the experimental papers it is necessary to understand Weiss theory.

2. It is rigorous for infinite-range interactions between spins (which never occur in practice).

3. The Weiss theory originally postulated a mysterious molecular field that was the "real" cause of the ordered magnetic state. This molecular field was later given an explanation based on the exchange effects described by the Heisenberg Hamiltonian (see Sect. 7.2). The Weiss theory gives a very simple way of relating the occurrence of a phase transition to the description of a magnetic system by the Heisenberg Hamiltonian. Of course, the way it relates these two is only qualitatively correct. However, it is a good starting place for more general theories that come closer to describing the behavior of the actual magnetic systems.[3]

---

[3] Perhaps the best simple discussion of the Weiss and related theories is contained in the book by J. S. Smart [92], which can be consulted for further details. By using two sublattices, it is possible to give a similar (to that below) description of antiferromagnetism. See Sect. 7.1.3.

For the case of a simple paramagnet, we have already derived that (see Sect. 7.1.2)

$$M = NgS\mu_B B_S(a), \qquad (7.19)$$

[4] where B_S is defined by (7.16) and

$$a \equiv \frac{Sg\mu_B \mu_0 H}{kT}. \qquad (7.20)$$

Recall also high-temperature (7.18) for $B_S(a)$ can be used.

Following a modern version of the original Weiss theory, we will give a qualitative description of the occurrence of spontaneous magnetization. Based on the concept of the mean or molecular field the spontaneous magnetization must be caused by some sort of atomic interaction. Whatever the physical origin of this interaction, it tends to bring about an ordering of the spins. Weiss did not attempt to derive the origin of this interaction. In fact, all he did was to postulate the existence of a molecular field that would tend to align the spins. His basic assumption was that the interaction would be taken account of if H (the applied magnetic field) were replaced by $H + \gamma M$, where γM is the molecular field. (γ is called the molecular field constant, sometimes the Weiss constant, and has nothing to do with the gyromagnetic ratio γ that will be discussed later.)

Thus the basic equation for ferromagnetic materials is

$$M = Ng\mu_B S B_S(a'), \qquad (7.21)$$

where

$$a' = \frac{\mu_0 Sg\mu_B}{kT}(H + \gamma M). \qquad (7.22)$$

That is, the basic equations of the molecular field theory are the same as the paramagnetic case plus the $H \to H + \gamma M$ replacement. Equations (7.21) and (7.22) are really all there is to the molecular field model. We shall derive other results from these equations, but already the basic ideas of the theory have been covered.

Let us now indicate how this predicts a phase transition. By a phase transition, we mean that spontaneous magnetization ($M \neq 0$ with $H = 0$) will occur for all temperatures below a certain temperature T_c called the *ferromagnetic Curie temperature*.

At the Curie temperature, for a consistent solution of (7.21) and (7.22) we require that the following two equations shall be identical as $a' \to 0$ and $H = 0$:

$$M_1 = Ng\mu_B S B_S(a'), \quad ((7.21) \text{ again})$$

$$M_2 = \frac{kTa'}{Sg\mu_B \gamma \mu_0}, \quad ((7.22) \text{ with } H \to 0).$$

---

[4] Here e can be treated as $|e|$ and so as usual, $\mu_B = |e|\hbar/2m$.

If these equations are identical, then they must have the same slope as $a' \to 0$. That is, we require

$$\left(\frac{dM_1}{da'}\right)_{a' \to 0} = \left(\frac{dM_2}{da'}\right)_{a' \to 0}. \tag{7.23}$$

Using the known behavior of $B_S(a')$ as $a' \to 0$, we find that condition (7.23) gives

$$T_c = \frac{\mu_0 N g^2 S(S+1) \mu_B^2}{3k} \gamma. \tag{7.24}$$

Equation (7.24) provides the relationship between the Curie constant T_c and the Weiss molecular field constant γ. Note that, as expected, if $\gamma = 0$, then $T_c = 0$ (i.e. if $\gamma \to 0$, there is no phase transition). Further, numerical evaluation shows that if $T > T_c$, (7.21) and (7.22) with $H = 0$ have a common solution for M only if $M = 0$. However, for $T < T_c$, numerical evaluation shows that they have a common solution $M \neq 0$, corresponding to the spontaneous magnetization that occurs when the molecular field overwhelms thermal effects.

There is another Curie temperature besides T_c. This is the so-called *paramagnetic Curie temperature* θ that enters into the equation for the high-temperature behavior of the magnetic susceptibility. Within the context of the Weiss theory, these two temperatures turn out to be the same. However, if one makes an experimental determination of T_c (from the transition temperature) and of θ from the high-temperature magnetic susceptibility, θ and T_c do not necessarily turn out to be identical (See Fig. 7.1). We obtain an explicit expression for θ below.

For $\mu_0 H S g \mu_B / kT \ll 1$ we have (by (7.17) and (7.18))

$$M = \frac{\mu_0 N g^2 \mu_B^2 S(S+1)}{3kT} h = C'h. \tag{7.25}$$

For ferromagnetic materials we need to make the replacement $H \to H + \gamma M$ so that $M = C'H + C'\gamma M$ or

$$M = \frac{C'H}{1 - C'\gamma}. \tag{7.26}$$

Substituting the definition of C', we find that (7.26) gives for the susceptibility

$$\chi = \frac{M}{H} = \frac{C}{T - \theta}, \tag{7.27}$$

where

$$C \equiv \text{the Curie–Weiss} = \frac{\mu_0 N g^2 \mu_B^2 S(S+1)}{3k},$$

$$\theta \equiv \text{the paramagnetic Curie temperature} = \frac{\mu_0 N g^2 S(S+1)}{3k} \mu_B^2 \gamma.$$

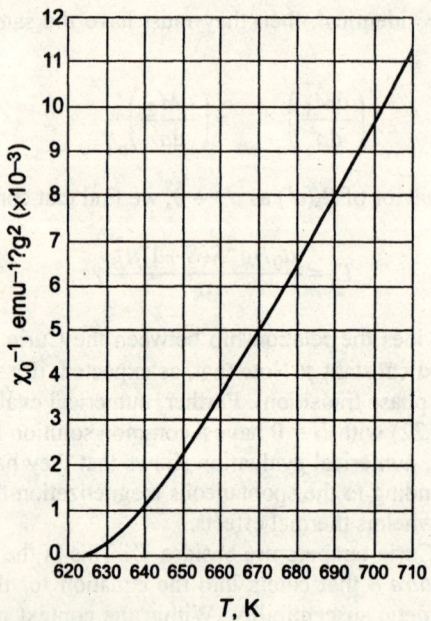

Fig. 7.1. Inverse susceptibility χ_0^{-1} of Ni. [Reprinted with permission from Kouvel JS and Fisher ME, *Phys Rev* **136**, A1626 (1964). Copyright 1964 by the American Physical Society. Original data from Weiss P and Forrer R, *Annales de Physique (Paris)*, **5**, 153 (1926).]

The Weiss theory gives the same result:

$$C\gamma = \theta = T_c = \frac{N\mu_B^2}{3k}(p_{\text{eff}})^2 \mu_0 \gamma, \tag{7.28}$$

where $p_{\text{eff}} = g[S(S+1)]^{1/2}$ is the effective magnetic moment in units of the Bohr magneton. Equation (7.27) is valid experimentally only if $T \gg \theta$. See Fig. 7.1.

It may not be apparent that the above discussion has limited validity. We have predicted a phase transition, and of course γ can be chosen so that the predicted T_c is exactly the experimental T_c. The Weiss prediction of the $(T-\theta)^{-1}$ behavior for χ also fits experiment at high enough temperatures.

However, we shall see that when we begin to look at further details, the Weiss theory begins to break down. In order to keep the algebra fairly simple it is convenient to absorb some of the constants into the variables and thus define new variables. Let us define

$$b \equiv \frac{\mu_0 g \mu_B}{kT}(H + \gamma M), \tag{7.29}$$

and

$$m \equiv \frac{M}{Ng\mu_B S} \equiv B_S(bS), \tag{7.30}$$

which should not be confused with the magnetic moment.

It is also convenient to define a quantity J_{ex} by

$$\gamma = \frac{2ZJ_{ex}}{\mu_0 N g^2 \mu_B^2} \hbar^2, \qquad (7.31)$$

where Z is the number of nearest neighbors in the lattice of interest, and J_{ex} is the *exchange integral*. Compare this to (7.95), which is the same. That is, we will see that (7.31) makes sense from the discussion of the physical origin of the molecular field.

Finally, let us define

$$b_0 = \frac{g\mu_B}{kT} \mu_0 H, \qquad (7.32)$$

and

$$\tau = T/T_c.$$

With these definitions, a little manipulation shows that (7.29) is

$$bS = b_0 S + \frac{3S}{S+1} \frac{m}{\tau}. \qquad (7.33)$$

Equations (7.30) and (7.33) can be solved simultaneously for m (which is proportional to the magnetization). With b_0 equal to zero (i.e. $H = 0$) we combine (7.30) and (7.33) to give a single equation that determines the spontaneous magnetization:

$$m = B_S\left(\frac{3S}{S+1} \frac{m}{\tau}\right). \qquad (7.34)$$

A plot similar to that yielded by (7.34) is shown in Fig. 7.16 ($H = 0$). The fit to experiment of the molecular field model is at least qualitative. Some classic results for Ni by Weiss and Forrer as quoted by Kittel [7.39 p. 448] yield a reasonably good fit.

We have reached the point where we can look at sufficiently fine details to see how the molecular field theory gives predictions that do not agree with experiment. We can see this by looking at the solutions of (7.34) as $\tau \to 0$ (i.e. $T \ll T_c$) and as $\tau \to 1$ (i.e. $T \to T_c$).

We know that for any y that $B_S(y)$ is given by (7.16). We also know that

$$\coth X = \frac{1+e^{-2X}}{1-e^{-2X}}. \qquad (7.35)$$

Since for large X

$$\coth X \cong 1 + 2e^{-2X},$$

we can say that for large y

$$B_S(y) \cong 1 + \frac{2S+1}{S}\exp\left(-\frac{2S+1}{S}y\right) - \frac{1}{S}\exp\left(-\frac{y}{S}\right). \quad (7.36)$$

Therefore by (7.34), m can be written for $T \to 0$ as

$$m \cong 1 + \left(\frac{2S+1}{S}\right)\exp\left[-\frac{3(2S+1)m}{(S+1)\tau}\right] - \frac{1}{S}\exp\left[-\frac{3m}{(S+1)\tau}\right]. \quad (7.37)$$

By iteration, it is clear that $m = 1$ can be used in the exponentials. Further,

$$\exp\left[-2\frac{3}{(S+1)\tau}\right] \ll \exp\left[-\frac{3}{(S+1)\tau}\right],$$

so that the second term can be neglected for all $S \neq 0$ (for $S = 0$ we do not have ferromagnetism anyway). Thus at lower temperature, we finally find

$$m \cong 1 - \frac{1}{S}\exp\left(-\frac{3}{S+1}\frac{T_c}{T}\right). \quad (7.38)$$

Experiment does not agree well with (7.38). For many materials, experiment agrees with

$$m \cong 1 - CT^{3/2}, \quad (7.39)$$

where C is a constant. As we will see in Sect. 7.2, (7.39) is correctly predicted by spin wave theory.

It also turns out that the Weiss molecular field theory disagrees with experiment at temperatures just below the Curie temperature. By making a Taylor series expansion, one can show that for $y \ll 1$,

$$B_S(y) \cong \frac{(2S+1)^2 - 1}{(2S)^2} \cdot \frac{y}{3} - \frac{(2S+1)^4 - 1}{(2S)^4} \cdot \frac{y^3}{45}. \quad (7.40)$$

Combining (7.40) with (7.34), we find that

$$m = K(T_c - T)^{1/2}, \quad (7.41)$$

and

$$\frac{dm^2}{dT} = -K^2 \quad \text{as} \quad T \to T_c^-. \quad (7.42)$$

Equations (7.41) and (7.42) agree only qualitatively with experiment. For many materials, experiment predicts that just below the Curie temperature

$$m \cong A(T_c - T)^{1/3}. \quad (7.43)$$

Perhaps the most dramatic failure of the Weiss molecular field theory occurs when we consider the specific heat. As we will see, the Weiss theory flatly predicts that the specific heat (with no external field) should vanish for temperatures above the Curie temperature. Experiment, however, says nothing of the sort. There is a small residual specific heat above the Curie temperature. This specific heat drops off with temperature. The reason for this failure of the Weiss theory is the neglect of short-range ordering above the Curie temperature.

Let us now look at the behavior of the Weiss predictions for the magnetic specific heat in a little more detail. The energy of a spin in a γM field in the z direction due to the molecular field is

$$E_i = \frac{\mu_0 g \mu_B}{\hbar} S_{iz} \gamma M . \tag{7.44}$$

Thus the internal energy U obtained by averaging E_i for N spins is,

$$U = \mu_0 \frac{N}{2} \frac{g \mu_B}{\hbar} \gamma M \langle S_{iz} \rangle = -\tfrac{1}{2} \mu_0 \gamma M^2 , \tag{7.45}$$

where the factor 1/2 comes from the fact that we do not want to count bonds twice, and $M = -N g \mu_B \langle S_{iz} \rangle / \hbar$ has been used.

The specific heat in zero magnetic field is then given by

$$C_0 = \frac{\partial U}{\partial T} = -\tfrac{1}{2} \mu_0 \gamma \frac{dM^2}{dT} . \tag{7.46}$$

For $T > T_c$, $M = 0$ (with no external magnetic field) and so the specific heat vanishes, which contradicts experiment.

The precise behavior of the magnetic specific heat just above the Curie temperature is of more than passing interest. Experimental results suggest that the specific heat should exhibit a logarithmic singularity or near logarithmic singularity as $T \to T_c$. The Weiss theory is inadequate even to begin attacking this problem.

Antiferromagnetism, Ferrimagnetism, and Other Types of Magnetic Order (B)

Antiferromagnetism is similar to ferromagnetism except that the lowest-energy state involves adjacent spins that are antiparallel rather than parallel (but see the end of this section). As we will see, the reason for this is a change in sign (compared to ferromagnetism) for the coupling parameter or *exchange integral*.

Ferrimagnetism is similar to antiferromagnetism except that the paired spins do not cancel and thus the lowest-energy state has a net spin.

Examples of antiferromagnetic substances are FeO and MnO. Further examples are given in Sect. 7.3.2. The temperature at which an antiferromagnetic substance becomes paramagnetic is known as the *Néel* temperature.

Examples of ferrimagnetism are $MnFe_2O_4$ and $NiFe_2O_7$. Further examples are also given in Sect. 7.3.2.

We now discuss these in more detail by use of mean field theory.[5] We assume near-neighbor and next-nearest-neighbor coupling as shown schematically in Fig. 7.2. The figure is drawn for an assumed ferrimagnetic order below the transition temperature. A and B represent two sublattices with spins S_A and S_B. The coupling is represented by the exchange integrals J (we assume $J_{BA} = J_{AB} < 0$ and these J dominate J_{AA} and $J_{BB} > 0$). Thus we assume the effective field between A and B has a negative sign. For the effective field we write:

$$B_A = -\omega\mu_0 M_B + \alpha_A\mu_0 M_A + B, \tag{7.47}$$

$$B_B = -\omega\mu_0 M_A + \beta_B\mu_0 M_B + B, \tag{7.48}$$

where $\omega > 0$ is a constant proportional to $|J_{AB}| = |J_{BA}|$, while α_A and β_B are constants proportional to J_{AA} and J_{BB}. The M represent magnetization and B is the external field (that is the magnetic induction $B = \mu_0 H_{\text{external}}$).

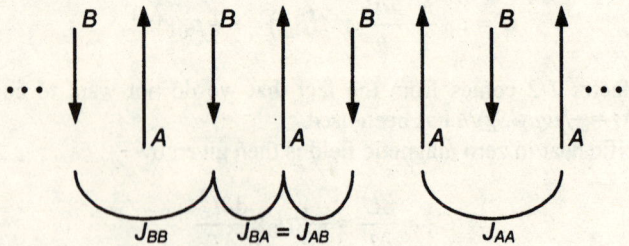

Fig. 7.2. Schematic to represent ferrimagnets

By the mean field approximation with B_{SA} and B_{SB} being the appropriate Brillouin functions (defined by (7.16)):

$$M_A = N_A g_A S_A \mu_B B_{S_A}(\beta g_A \mu_B S_A B_A), \tag{7.49}$$

$$M_B = N_B g_B S_B \mu_B B_{S_B}(\beta g_B \mu_B S_B B_B). \tag{7.50}$$

The S_A, S_B are quantum numbers (e.g. 1, 3/2, etc., labeling the spin). We also will use the result (7.40) for $B_S(x)$ with $x \ll 1$. In the above, N_i is the number of ions of type i per unit volume, g_A and g_B are the *Lande g-factors* (note we are using B not $\mu_0 H$), μ_B is the Bohr magneton and $\beta = 1/(k_B T)$.

Defining the Curie constants

$$C_A = \frac{N_A S_A (S_A + 1) g_A^2 \mu_B^2}{3k}, \tag{7.51}$$

$$C_B = \frac{N_B S_B (S_B + 1) g_B^2 \mu_B^2}{3k}, \tag{7.52}$$

---

[5] See also, e.g., Kittel [7.39 p458ff].

7.1 Types of Magnetism

we have if B_A/T and B_B/T are small:

$$M_A = \frac{C_A B_A}{T}, \tag{7.53}$$

$$M_B = \frac{C_B B_B}{T}. \tag{7.54}$$

This holds above the ordering temperature when $B \to 0$ and even just below the ordering temperature provided $B \to 0$ and M_A, M_B are very small. Thus the equations determining the magnetization become:

$$(T - \alpha_A \mu_0 C_A) M_A + \omega \mu_0 C_A M_B = C_A B, \tag{7.55}$$

$$\omega \mu_0 C_B M_A + (T - \beta_B \mu_0 C_B) M_B = C_B B. \tag{7.56}$$

If the external field $B \to 0$, we can have nonzero (but very small) solutions for M_A, M_B provided

$$(T - \alpha_A \mu_0 C_A)(T - \beta_B \mu_0 C_B) = \omega^2 \mu_0^2 C_A C_B. \tag{7.57}$$

So

$$T_c^{\pm} = \frac{\mu_0}{2}\left(\alpha_A C_A + \beta_B C_B \pm \sqrt{4\omega^2 C_A C_B + (\alpha_A C_A - \beta_B C_B)^2}\right). \tag{7.58}$$

The critical temperature is chosen so $T_c = \omega \mu_0 (C_A C_B)^{1/2}$ when $\alpha_A \to \beta_B \to 0$, and so $T_c = T_c^+$. Above T_c for $B \neq 0$ (and small) with

$$D \equiv (T - T_c^+)(T - T_c^-),$$

$$M_A = D^{-1}[(T - \beta_B \mu_0 C_B) C_A - \omega \mu_0 C_A C_B] B,$$

$$M_B = D^{-1}[(T - \alpha_A \mu_0 C_A) C_B - \omega \mu_0 C_A C_B] B.$$

The reciprocal magnetic susceptibility is then given by

$$\frac{1}{\chi} = \frac{B}{\mu_0 (M_A + M_B)} = \frac{D}{\mu_0 \{T(C_A + C_B) - [(\alpha_A + \beta_B) + 2\omega]\mu_0 C_A C_B\}}. \tag{7.59}$$

Since D is quadratic in T, $1/\chi$ is linear in T only at high temperatures (ferrimagnetism). Also note

$$\frac{1}{\chi} = 0 \quad \text{at} \quad T = T_c^+ = T_c.$$

In the special case where two sublattices are identical (and $\omega > 0$), since $C_A = C_B \equiv C_1$ and $\alpha_A = \beta_B \equiv \alpha_1$,

$$T_c^+ = (\alpha_1 + \omega) C_1 \mu_0, \tag{7.60}$$

and after canceling,

$$\chi^{-1} = \frac{[T - C_1\mu_0(\alpha_1 - \omega)]}{2C_1\mu_0}, \quad (7.61)$$

which is linear in T (antiferromagnetism).

This equation is valid for $T > T_c^+ = \mu_0(\alpha_1+\omega)C_1 \equiv T_N$, the Néel temperature. Thus, if we define

$$\theta \equiv C_1(\omega - \alpha_1)\mu_0,$$

$$\chi_{AF} = \frac{2\mu_0 C_1}{T + \theta}. \quad (7.62)$$

Note:

$$\frac{\theta}{T_N} = \frac{\omega - \alpha_1}{\omega + \alpha_1}.$$

We can also easily derive results for the ferromagnetic case. We choose to drop out one sublattice and in effect double the effect of the other to be consistent with previous work.

$$C_A = C_A^F \equiv 2C_1, \quad \beta_B = 0, \quad C_B = 0,$$

so

$$T_c = \mu_0 \alpha_A^F C_A^F = 2C_1\mu_0\alpha_1 \quad (\text{if } \alpha_1 \equiv \alpha_A^F).$$

Then,[6]

$$\chi = \frac{\mu_0 M_A}{B} = \frac{\mu_0 T(2C_1)}{T(T - 2C_1\mu_0\alpha_1)} = \frac{2C_1\mu_0}{T - 2C_1\mu_0\alpha_1}. \quad (7.63)$$

The paramagnetic case is obtained from neglecting the coupling so

$$\chi = \frac{2C_1\mu_0}{T}. \quad (7.64)$$

The reality of antiferromagnetism has been absolutely determined by neutron diffraction that shows the appearance of magnetic order below the critical temperature. See Fig. 7.3 and Fig. 7.4. Figure 7.5 summarizes our results.

---

[6] $2C_1\mu_0 = C$ of (7.27).

7.1 Types of Magnetism

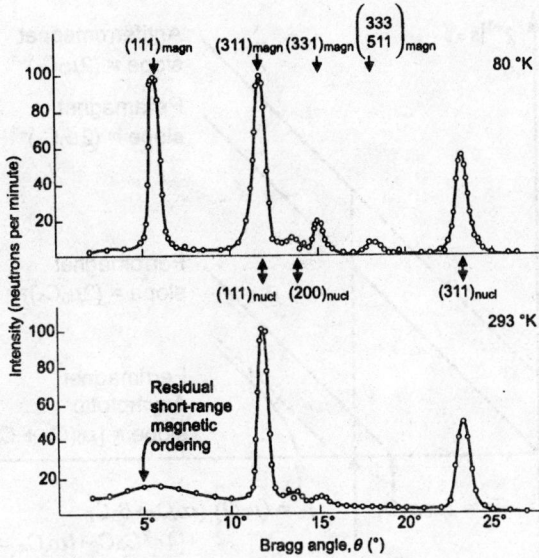

Fig. 7.3. Neutron-diffraction patterns of MnO at 80 K and 293 K. The Curie temperature is 120 K. The low-temperature pattern has extra antiferromagnetic reflections for a magnetic unit twice that of the chemical unit cell. From Bacon GE, *Neutron Diffraction*, Oxford at the Clarendon Press, London, 1962 2nd edn, Fig. 142 p.297. By permission of Oxford University Press. Original data from Shull CG and Smart JS, *Phys Rev*, **76**, 1256 (1949)

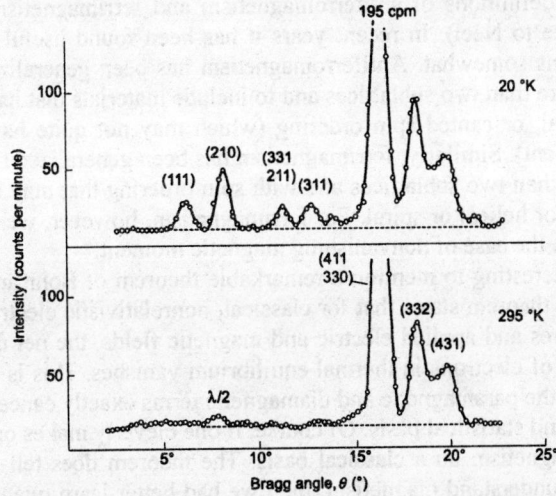

Fig. 7.4. Neutron-diffraction patterns for α-manganese at 20 K and 295 K. Note the antiferromagnetic reflections at the lower temperature. From Bacon GE, *Neutron Diffraction*, Oxford at the Clarendon Press, London, 1962 2nd edn, Fig. 129 p.277. By permission of Oxford University Press. Original data from Shull CG and Wilkinson MK, *Rev Mod Phys*, **25**, 100 (1953)

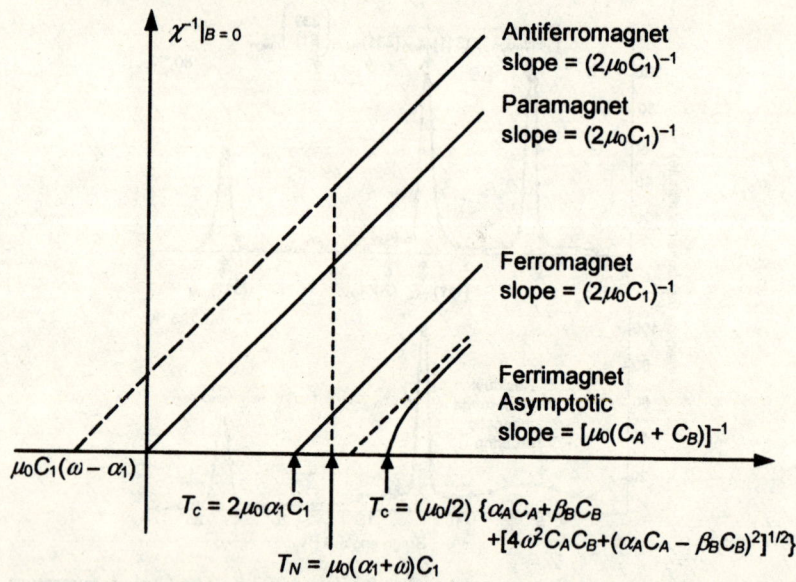

Fig. 7.5. Schematic plot of reciprocal magnetic susceptibility. Note the constants for the various cases can vary. For example α_1 could be negative for the antiferromagnetic case and α_A, β_B could be negative for the ferrimagnetic case. This would shift the zero of χ^{-1}.

The above definitions of antiferromagnetism and ferrimagnetism are the old definitions (due to Néel). In recent years it has been found useful to generalize these definitions somewhat. Antiferromagnetism has been generalized to include solids with more than two sublattices and to include materials that have triangular, helical or spiral, or canted spin ordering (which may not quite have a net zero magnetic moment). Similarly, ferrimagnetism has been generalized to include solids with more than two sublattices and with spin ordering that may be, for example, triangular or helical or spiral. For ferrimagnetism, however, we are definitely concerned with the case of nonvanishing magnetic moment.

It is also interesting to mention a remarkable theorem of Bohr and Van Leeuwen [94]. This theorem states that for classical, nonrelativistic electrons for all finite temperatures and applied electric and magnetic fields, the net magnetization of a collection of electrons in thermal equilibrium vanishes. This is basically due to the fact that the paramagnetic and diamagnetic terms exactly cancel one another on a classical and statistical basis. Of course, if one cleverly makes omissions, one can discuss magnetism on a classical basis. The theorem does tell us that if we really want to understand magnetism, then we had better learn quantum mechanics. See Problem 7.17.

It might be well to learn relativity also. Relativity tells us that the distinction between electric and magnetic fields is just a distinction between reference frames.

7.2 Origin and Consequences of Magnetic Order

7.2.1 Heisenberg Hamiltonian

The Heitler–London Method (B)

In this Section we develop the Heisenberg Hamiltonian and then relate our results to various aspects of the magnetic state. The first method that will be discussed is the Heitler–London method. This discussion will have at least two applications. First, it helps us to understand the covalent bond, and so relates to our previous discussion of valence crystals. Second, the discussion gives us a qualitative understanding of the Heisenberg Hamiltonian. This Hamiltonian is often used to explain the properties of coupled spin systems. The Heisenberg Hamiltonian will be used in the discussion of magnons. Finally, as we will show, the Heisenberg Hamiltonian is useful in showing how an electrostatic exchange interaction approximately predicts the existence of a molecular field and hence gives a fundamental qualitative explanation of the existence of ferromagnetism.

Let a and b label two hydrogen atoms separated by R (see Fig. 7.6). Let the separated ($R \to \infty$) hydrogen atoms be described by the Hamiltonians

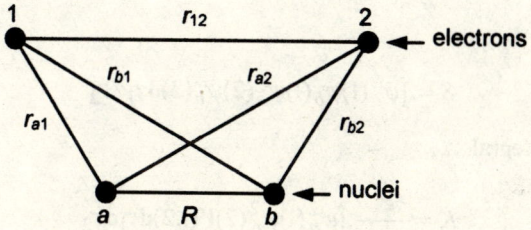

Fig. 7.6. Model for two hydrogen atoms

$$\mathcal{H}_0^a(1) = -\frac{\hbar^2}{2m}\nabla_1^2 - \frac{e^2}{4\pi\varepsilon_0 r_{a1}}, \tag{7.65}$$

and

$$\mathcal{H}_0^b(2) = -\frac{\hbar^2}{2m}\nabla_2^2 - \frac{e^2}{4\pi\varepsilon_0 r_{b2}}. \tag{7.66}$$

Let $\psi_a(1)$ and $\psi_b(2)$ be the spatial ground-state wave functions, that is

$$\mathcal{H}_0^a\psi_a(1) = E_0\psi_a(1), \tag{7.67}$$

or

$$\mathcal{H}_0^b\psi_b(2) = E_0\psi_b(2),$$

where E_0 is the ground-state energy of the hydrogen atom. The zeroth-order hydrogen molecular wave functions may be written

$$\psi_\pm = \psi_a(1)\psi_b(2) \pm \psi_a(2)\psi_b(1). \tag{7.68}$$

In the Heitler–London approximation for un-normalized wave functions

$$E_\pm \cong \frac{\int \psi_\pm \mathcal{H} \psi_\pm d\tau_1 d\tau_2}{\int \psi_\pm^2 d\tau_1 d\tau_2}, \tag{7.69}$$

where $d\tau_i = dx_i dy_i dz_i$ and we have used that wave functions for stationary states can be chosen to be real. In (7.69),

$$\mathcal{H} = \mathcal{H}_0^a(1) + \mathcal{H}_0^b(2) - \frac{e^2}{4\pi\varepsilon_0}\left(\frac{1}{r_{a2}} + \frac{1}{r_{b1}} - \frac{1}{r_{12}} - \frac{1}{R}\right). \tag{7.70}$$

Working out the details when (7.68) is put into (7.69) and assuming $\psi_a(1)$ and $\psi_b(2)$ are normalized we find

$$E_\pm = 2E_0 + \frac{e^2}{4\pi\varepsilon_0 R} + \frac{K \pm J_E}{1 \pm S}, \tag{7.71}$$

where

$$S = \int \psi_a(1)\psi_b(1)\psi_a(2)\psi_b(2) d\tau_1 d\tau_2 \tag{7.72}$$

is the overlap integral,

$$K = \frac{e^2}{4\pi\varepsilon_0} \int \psi_a^2(1)\psi_b^2(2) V(1,2) d\tau_1 d\tau_2 \tag{7.73}$$

is the Coulomb energy of interaction, and

$$J_E = \frac{e^2}{4\pi\varepsilon_0} \int \psi_a(1)\psi_a(2)\psi_b(1)\psi_b(2) V(1,2) d\tau_1 d\tau_2 \tag{7.74}$$

is the exchange energy. In (7.73) and (7.74),

$$V(1,2) = \frac{e^2}{4\pi\varepsilon_0}\left(\frac{1}{r_{12}} - \frac{1}{r_{a2}} - \frac{1}{r_{b1}}\right). \tag{7.75}$$

The corresponding normalized eigenvectors are

$$\psi^\pm(1,2) = \frac{1}{\sqrt{2(1 \pm S)}}[\psi_1(1,2) \pm \psi_2(1,2)], \tag{7.76}$$

7.2 Origin and Consequences of Magnetic Order

where

$$\psi_1(1,2) = \psi_a(1)\psi_b(2), \tag{7.77}$$

$$\psi_2(1,2) = \psi_a(2)\psi_b(1). \tag{7.78}$$

So far there has been no need to discuss spin, as the Hamiltonian did not explicitly involve it. However, it is easy to see how spin enters. ψ^+ is a symmetric function in the interchange of coordinates 1 and 2, and ψ^- is an antisymmetric function in the interchange of coordinates 1 and 2. The total wave function that includes both space and spin coordinates must be antisymmetric in the interchange of all coordinates. Thus in the total wave function, an antisymmetric function of spin must multiply ψ^+, and a symmetric function of spin must multiply ψ^-. If we denote $\alpha(i)$ as the "spin-up" wave function of electron i and $\beta(j)$ as the "spin-down" wave function of electron j, then the total wave functions can be written as

$$\psi_T^+ = \frac{1}{\sqrt{2(1+S)}}(\psi_1 + \psi_2)\frac{1}{\sqrt{2}}[\alpha(1)\beta(2) - \alpha(2)\beta(1)], \tag{7.79}$$

$$\psi_T^- = \frac{1}{\sqrt{2(1-S)}}(\psi_1 - \psi_2) \begin{cases} \alpha(1)\alpha(2), \\ \frac{1}{\sqrt{2}}[\alpha(1)\beta(2) + \alpha(2)\beta(1)], \\ \beta(1)\beta(2). \end{cases} \tag{7.80}$$

Equation (7.79) has total spin equal to zero, and is said to be a *singlet* state. It corresponds to antiparallel spins. Equation (7.80) has total spin equal to one (with three projections of +1, 0, −1) and is said to describe a *triplet* state. This corresponds to parallel spins. For hydrogen atoms, J in (7.74) is called the *exchange integral* and is negative. Thus E_+ (corresponding to ψ_T^+) is lower in energy than E_- (corresponding to ψ_T^-), and hence the singlet state is lowest in energy. A calculation of $E_\pm - E_0$ for E_0 labeling the ground state of hydrogen is sketched in Fig. 7.7. Let us now pursue this two-spin case in order to write an effective spin Hamiltonian that describes the situation. Let S_1 and S_2 be the spin operators for particles 1 and 2. Then

$$(S_1 + S_2)^2 = S_1^2 + S_2^2 + 2S_1 \cdot S_2. \tag{7.81}$$

Since the eigenvalues of S_2^1 and S_2^2 are $3\hbar^2/4$ we can write for appropriate ϕ in the space of interest

$$S_1 \cdot S_2 \phi = \tfrac{1}{2}[(S_1 + S_2)^2 - \tfrac{3}{2}\hbar^2]\phi. \tag{7.82}$$

In the triplet (or parallel spin) state, the eigenvalue of $(S_1 + S_2)^2$ is $2\hbar^2$, so

$$S_1 \cdot S_2 \phi_{\text{triplet}} = \tfrac{1}{4}\hbar^2 \phi_{\text{triplet}}. \tag{7.83}$$

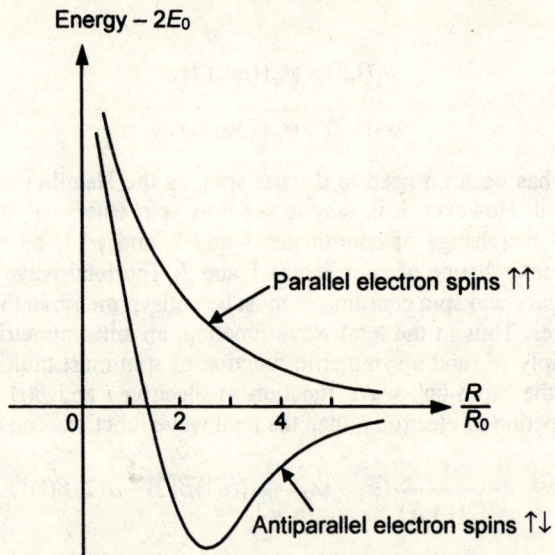

Fig. 7.7. Sketch of results of the Heitler–London theory applied to two hydrogen atoms (R/R_0 is the distance between the two atoms in Bohr radii). See also, e.g., Heitler [7.26].

In the singlet (or antiparallel spin) state, the eigenvalue of $(S_1 + S_2)^2$ is 0, so

$$S_1 \cdot S_2 \phi_{\text{singlet}} = -\tfrac{3}{4}\hbar^2 \phi_{\text{singlet}}. \tag{7.84}$$

Comparing these results to Fig. 7.7, we see we can formally write an effective spin Hamiltonian for the two electrons on the two different atoms:

$$\mathcal{H} = -2J S_1 \cdot S_2, \tag{7.85}$$

where J is often simply called the *exchange constant* and $J = J(R)$, i.e. it depends on the separation R between atoms. By suitable choice of $J(R)$, the eigenvalues of $\mathcal{H} - 2E_0$ can reproduce the curves of Fig. 7.7. Note that $J > 0$ gives the parallel-spin case the lowest energy (ferromagnetism) and $J < 0$ (the two-hydrogen-atom case – this does not *always* happen, especially in a solid) gives the antiparallel-spin case the lowest energy (antiferromagnetism). If we have many atoms on a lattice, and if there is an exchange coupling between the spins of the atoms, we assume that we can write a Hamiltonian:

$$\mathcal{H} = -\underset{\text{(electrons)}}{{\sum}'_{\alpha,\beta}} J_{\alpha,\beta} S_\alpha \cdot S_\beta. \tag{7.86}$$

If there are several electrons on the same atom and if J is constant for all electrons on the same atom, then we assume we can write

$$\sum J_{\alpha,\beta} S_\alpha \cdot S_\beta \cong \underbrace{\sum_{k,l} J_{k,l}}_{\text{(atoms)}} \underbrace{\sum_{i,j} S_{ki} \cdot S_{lj}}_{\substack{\text{(electrons} \\ \text{on } k,l \text{ atoms)}}}$$

$$= \sum_{k,l} J_{k,l} (\sum_i S_{ki})(\sum_j S_{lj}) \qquad (7.87)$$

$$= \sum_{k,l} J_{k,l} S_k^T \cdot S_l^T ,$$

where S_k^T and S_l^T refer to the spin operators associated with atoms k and l. Since $\sum' S_\alpha \cdot S_\beta J_{\alpha\beta}$ differs from $\sum S_\alpha \cdot S_\beta J_{\alpha\beta}$ by only a constant and $\sum'_{k,l} J_{k,l} S_k^T S_l^T$ differs from $\sum_{k,l} J_{k,l} S_k^T S_l^T$ by only a constant, we can write the effective spin Hamiltonian as

$$\mathcal{H} = -{\sum_{k,l}}' J_{k,l} S_k^T \cdot S_l^T , \qquad (7.88)$$

here unimportant constants have not been retained. This last expression is called the *Heisenberg Hamiltonian* for a system of interacting spins in the absence of an external field.

This form of the Heisenberg Hamiltonian already tells us two important things:

1. It is applicable to atoms with arbitrary spin.
2. Closed shells contribute nothing to the Heisenberg Hamiltonian because the spin is zero for a closed shell.

Our development of the Heisenberg Hamiltonian has glossed over the approximations that were made. Let us now return to them. The first obvious approximation was made in going from the two-spin case to the N-spin case. The presence of a third atom can and does affect the interaction between the original pair. In addition, we assumed that the exchange interaction between all electrons on the same atom was a constant.

Another difficulty with the extension of the Heitler–London method to the n-electron problem is the so-called "overlap catastrophe." This will not be discussed here as we apparently do not have to worry about it when using the simple Heisenberg theory for insulators.[7] There are also no provisions in the Heisenberg Hamiltonian for crystalline anisotropy, which must be present in any real crystal. We will discuss this concept in Sects. 7.2.2 and 7.3.1. However, so far as energy goes, the Heisenberg model does seem to contain the main contributions.

But there are also several approximations made in the Heitler–London theory itself. The first of these assumptions is that the wave functions associated with the electrons of interest are well-localized wave functions. Thus we expect the Heisenberg Hamiltonian to be more nearly valid in insulators than in metals. The assumption is necessary in order that the perturbation approach used in the Heitler–London method will be valid. It is also assumed that the electrons are in nondegenerate orbital states and that the excited states can be neglected. This makes it

---

[7] For a discussion of this point see the article by Keffer, [7.37].

harder to see what to do in states that are not "spin only" states, i.e. in states in which the total orbital angular momentum L is not zero or is not quenched. Quenching of angular momentum means that the expectation value of L (but not L^2) for electrons of interest is zero when the atom is in the solid. For the nonspin only case, we have orbital degeneracy (plus the effects of crystal fields) and thus the basic assumptions of the *simple* Heitler–London method are not met.

The Heitler–London theory does, however, indicate one useful approximation: that $J\hbar^2$ is of the same order of magnitude as the electrostatic interaction energy between two atoms and that this interaction depends on the overlap of the wave functions of the atoms. Since the overlap seems to die out exponentially, we expect the *direct* exchange interaction between any two atoms to be of rather short range. (Certain indirect exchange effects due to the presence of a third atom may extend the range somewhat and in practice these indirect exchange effects may be very important. Indirect exchange can also occur by means of the conduction electrons in metals, as discussed later.)

Before discussing further the question of the applicability of the Heisenberg model, it is useful to get a physical picture of why we expect the spin-dependent energy that it predicts. In considering the case of two interacting hydrogen atoms, we found that we had a parallel spin case and an antiparallel spin case. By the Pauli principle, the parallel spin case requires an antisymmetric spatial wave function, whereas the antiparallel case requires a symmetric spatial wave function. The antisymmetric case concentrates less charge in the region between atoms and hence the electrostatic potential energy of the electrons ($e^2/4\pi\varepsilon_0 r$) is smaller. However, the antisymmetric case causes the electronic wave function to "wiggle" more and hence raises the kinetic energy T ($T_{op} \propto \nabla^2$). In the usual situation (in the two-hydrogen-atom case and in the much more complicated case of many insulating solids) the kinetic energy increase dominates the potential energy decrease; hence the antiparallel spin case has the lowest energy and we have antiferromagnetism ($J < 0$). In exceptional cases, the potential energy decrease can dominate the kinetic energy increases, and hence the parallel spin case has the least energy and we have ferromagnetism ($J > 0$). In fact, most insulators that have an ordered magnetic state become antiferromagnets at low enough temperature.

Few rigorous results exist that would tend either to prove or disprove the validity of the Heisenberg Hamiltonian for an actual physical situation. This is one reason for doing calculations based on the Heisenberg model that are of sufficient accuracy to yield results that can usefully be compared to experiment. Dirac[8] has given an explicit proof of the Heisenberg model in a situation that is oversimplified to the point of not being physical. Dirac assumes that each of the electrons is confined to a different specified orthogonal orbital. He also assumes that these orbitals can be thought of as being localizable. It is clear that this is never the situation in a real solid. Despite the lack of rigor, the Heisenberg Hamiltonian appears to be a good starting place for any theory that is to be used to explain experimental magnetic phenomena in insulators. The situation in metals is more complex.

---

[8] See, for example, Anderson [7.1].

Another side issue is whether the exchange "constants" that work well above the Curie temperature also work well below the Curie temperature. Since the development of the Heisenberg Hamiltonian was only phenomenological, this is a sensible question to ask. It is particularly sensible since J depends on R and R increases as the temperature is increased (by thermal expansion). Charap and Boyd[9] and Wojtowicz[10] have shown for EuS (which is one of the few "ideal" Heisenberg ferromagnets) that the same set of J will fit both the low-temperature specific heat and magnetization and the high-temperature specific heat.

We have made many approximations in developing the Heisenberg Hamiltonian. The use of the Heitler–London method is itself an approximation. But there are other ways of understanding the binding of the hydrogen atoms and hence of developing the Heisenberg Hamiltonian. The Hund–Mulliken[11] method is one of these techniques. The Hund–Mulliken method should work for smaller R, whereas the Heitler–London works for larger R. However, they both qualitatively lead to a Heisenberg Hamiltonian.

We should also mention the Ising model, where $\mathcal{H} = -\sum J_{ij}\sigma_{iz}\sigma_{jz}$, and the σ are the Pauli spin matrices. Only nearest-neighbor coupling is commonly used. This model has been solved exactly in two dimensions (see Huang [7.32 p341ff]). The Ising model has spawned a huge number of calculations.

The Heisenberg Hamiltonian and its Relationship to the Weiss Mean Field Theory (B)

We now show how the mean molecular field arises from the Heisenberg Hamiltonian. If we assume a mean field γM then the interaction energy of moment μ_k with this field is

$$E_k = -\mu_0 \gamma M \cdot \mu_k. \tag{7.89}$$

Also from the Heisenberg Hamiltonian

$$E_k = -{\sum_i}' J_{ik} S_i \cdot S_k - {\sum_j}' J_{kj} S_k \cdot S_j,$$

and since $J_{ij} = J_{ji}$, and noting that j is a dummy summation variable

$$E_k = -2{\sum_i}' J_{ik} S_i \cdot S_k. \tag{7.90}$$

---

[9] See [7.10].
[10] See Wojtowicz [7.70].
[11] See Patterson [7.53 p176ff].

In the spirit of the mean-field approximation we replace S_i by its average $\bar{S}_i = S$ since the average of each site is the same. Further, we assume only nearest-neighbor interactions so $J_{ik} = J$ for each of the Z nearest neighbors. So

$$E_k \cong -2ZJS \cdot S_k. \tag{7.91}$$

But

$$\mu_k \cong -\frac{g\mu_B S_k}{\hbar} \tag{7.92}$$

(with $\mu_B = |e|\hbar/2m$), and the magnetization M is

$$M \cong -\frac{Ng\mu_B S}{\hbar}, \tag{7.93}$$

where N is the number of atomic moments per unit volume ($\equiv 1/\Omega$, where Ω is the atomic volume). Thus we can also write

$$E_k \cong -2ZJ\frac{\Omega M \cdot \mu_k}{(g\mu_B)^2}\hbar^2. \tag{7.94}$$

Comparing (7.89) and (7.94)

$$J = \frac{\mu_0 \gamma (g\mu_B)^2}{2Z\Omega\hbar^2}. \tag{7.95}$$

This not only shows how Heisenberg's theory "explains" the Weiss mean molecular field, but also gives an approximate way of evaluating the parameter J. Slight modifications in (7.95) result for other than nearest-neighbor interactions.

RKKY Interaction[12] (A)

The Ruderman, Kittel, Kasuya, Yosida, (RKKY) interaction is important for rare earths. It is an interaction between the conduction electrons with the localized moments associated with the 4f electrons. Since the spins cause the localized moments, the conduction electrons can mediate an indirect exchange interaction between the spins. This interaction is called RKKY interaction.

We assume, following previous work, that the total exchange interaction is of the form

$$\mathcal{H}_{ex}^{Total} = -\sum_{i,\alpha} J_x(r_i - R_\alpha) S_\alpha \cdot S_i, \tag{7.96}$$

---

[12] Kittel [60, pp 360-366] and White [7.68 pp 197-200].

where S_α is an ion spin and S_i is the conduction spin. For convenience we assume the S are dimensionless with $\hbar$ absorbed in the J. We assume $J_x(r_i - R_\alpha)$ is short range (the size of 4f orbitals) and define

$$J = \int J_x(r - R_\alpha) dr \,. \tag{7.97}$$

Consistent with (7.97), we assume

$$J_x(r_i - R_\alpha) = J\delta(r) \,, \tag{7.98}$$

where $r = r_i - R_\alpha$ and write

$$\mathcal{H}_{ex} = -J S_\alpha \cdot S_i \delta(r)$$

for the exchange interaction between the ion α and the conduction electron. This is the same form as the Fermi contact term, but the physical basis is different. We can regard $S_i \delta(r) = S_i(r)$ as the electronic conduction spin density. Now, the interaction between the ion spin S_α and the conduction spin S_i can be written (gaussian units, $\mu_0 = 1$)

$$-J S_\alpha \cdot S_i \delta(r) = -(-g\mu_B S_i) \cdot H_{\text{eff}}(r) \,,$$

so this defines an effective field

$$H_{\text{eff}} = -\frac{J S_\alpha}{g\mu_B} \delta(r) \,. \tag{7.99}$$

The Fourier component of the effective field can be written

$$H_{\text{eff}}(q) = \int H_{\text{eff}}(r) e^{-i q \cdot r} dr = -\frac{J}{g\mu_B} S_\alpha \,. \tag{7.100}$$

We can now determine the magnetization induced by the effective field by use of the magnetic susceptibility. In Fourier space

$$\chi(q) = \frac{M(q)}{H(q)} \,. \tag{7.101}$$

This gives us the response in magnetization of a free-electron gas to a magnetic field. It turns out that this response (at $T = 0$) is functionally just like the response to an electric field (see Sect. 9.5.3 where Friedel oscillation in the screening of a point charge is discussed).

We find

$$\chi(q) = \frac{3g^2 \mu_B^2}{8 E_F} \frac{N}{V} A(q/2k_F) \,, \tag{7.102}$$

where N/V is the number of electrons per unit volume and

$$A(q/2k_F) = \frac{1}{2} + \frac{k_F}{2q}\left(1 - \frac{q^2}{4k_F^2}\right)\ln\left|\frac{2k_F + q}{2k_F - q}\right|. \tag{7.103}$$

The magnetization $M(r)$ of the conduction electrons can now be calculated from (7.101), (7.102), and (7.103).

$$\begin{aligned} M(r) &= \frac{1}{V}\sum_q M(q)e^{i q \cdot r} \\ &= \frac{1}{V}\sum_q \chi(q)H_{\text{eff}}(q)e^{i q \cdot r} \\ &= -\frac{J}{g\mu_B V}S_\alpha \sum_q \chi(q)e^{i q \cdot r}. \end{aligned} \tag{7.104}$$

With the aid of (7.102) and (7.103), we can evaluate (7.104) to find

$$M(r) = -\frac{J}{g\mu_B}KG(r)S_\alpha, \tag{7.105}$$

where

$$K = \frac{3g^2\mu_B^2}{8E_F}\frac{N}{V}\frac{k_F^3}{16\pi}, \tag{7.106}$$

and

$$G(r) = \frac{\sin(2k_F r) - 2k_F r \cos(2k_F r)}{(k_F r)^4}. \tag{7.107}$$

The localized moment S_a causes conduction spins to develop an oscillating polarization in the vicinity of it. The spin-density oscillations have the same form as the charge-density oscillations that result when an electron gas screens a charged impurity.[13]

Let us define

$$F(x) = \frac{\sin x - x \cos x}{x^4},$$

so

$$G(r) = 2^4 F(2k_F r).$$

$F(x)$ is the basic function that describes spatial oscillating polarization induced by a localized moment in its vicinity. It is sketched in Fig. 7.8. Note as $x \to \infty$, $F(x) \to -\cos(x)/x^3$ and as $x \to 0$, $F(x) \to 1/(3x)$.

---

[13] See Langer and Vosko [7.42].

7.2 Origin and Consequences of Magnetic Order

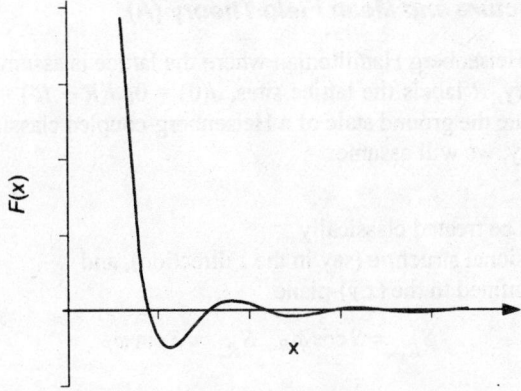

Fig. 7.8. Sketch of $F(x) = [\sin(x) - x\cos(x)]/x^4$, which describes the RKKY exchange interaction

Using (7.105), if $S(r)$ is the spin density,

$$S(r) = \frac{M(r)}{(-g\mu_B)} = \frac{J}{(g\mu_B)^2} KGS_\alpha . \tag{7.108}$$

Another localized ionic spin at S_β interacts with $S(r)$

$$\mathcal{H}^{\text{indirect}}_{\alpha \text{ and } \beta} = -JS_\beta \cdot S(r_\alpha - r_\beta) .$$

Now, summing over all α, β interactions and being careful to avoid double counting spins, we have

$$\mathcal{H}_{RKKY} = -\frac{1}{2}\sum_{\alpha,\beta} J_{\alpha\beta} S_\alpha \cdot S_\beta , \tag{7.109}$$

where

$$J_{\alpha\beta} = \frac{J^2}{(g\mu_B)^2} KG(r = r_{\alpha\beta}) . \tag{7.110}$$

For strong spin-orbit coupling, it would be more natural to express the Hamiltonian in terms of J (the total angular momentum) rather than S. $J = L + S$ and within the set of states of constant J, g_J is defined so

$$g_J \mu_B J = \mu_B(L + 2S) = \mu_B(J + S),$$

where remember the g factor for L is 1, while for spin S it is 2. Thus, we write

$$(g_J - 1)J = S .$$

If J_α is the total angular momentum associated with site α, by substitution

$$\mathcal{H}_{RKKY} = -\frac{1}{2}(g_J - 1)^2 \sum_{\alpha,\beta} J_{\alpha\beta} J_\alpha \cdot J_\beta , \tag{7.111}$$

where $(g_J - 1)^2$ is called the deGennes factor.

Magnetic Structure and Mean Field Theory (A)

We assume the Heisenberg Hamiltonian where the lattice is assumed to have transitional symmetry, R labels the lattice sites, $J(0) = 0$, $J(R - R') = J(R' - R)$. We wish to investigate the ground state of a Heisenberg-coupled classical spin system, and for simplicity, we will assume:

a. $T = 0$ K
b. The spins can be treated classically
c. A one-dimensional structure (say in the z direction), and
d. The S_R are confined to the (x,y)-plane

$$S_{R_x} = S\cos\varphi_R, \quad S_{R_y} = S\sin\varphi_R.$$

Thus, the Heisenberg Hamiltonian can be written:

$$\mathcal{H} = -\frac{1}{2}\sum_{R,R'} S^2 J(R - R')\cos(\varphi_R - \varphi_{R'}).$$

e. We are going to further consider the possibility that the spins will have a constant turn angle of qa (between each spin), so $\varphi_R = qR$, and for adjacent spins $\Delta\varphi_R = q\Delta R = qa$.

Substituting (in the Hamiltonian above), we find

$$\mathcal{H} = -\frac{NS^2}{2}J(q), \tag{7.112}$$

where

$$J(q) = \sum_R J(R)e^{iqR} \tag{7.113}$$

and $J(q) = J(-q)$. Thus, the problem of finding $\mathcal{H}_{min}$ reduces to the problem of finding $J(q)_{max}$.

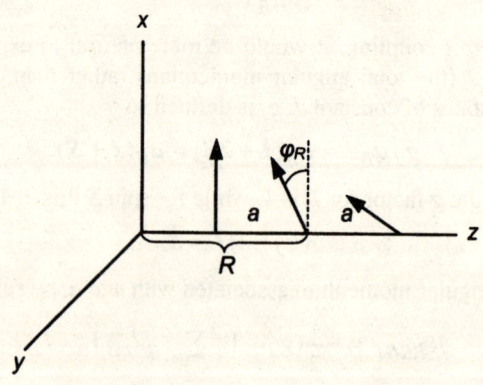

Fig. 7.9. Graphical depiction of the classical spin system assumptions

Note if $J(q) \to$ max for
$$\begin{cases} q = 0, & \text{get ferromagnetism,} \\ q = \pi/a, & \text{get antiferromagnetism,} \\ qa \neq 0 \text{ or } \pi, & \text{get heliomagnetism with } qa \\ & \text{defining the turn angles.} \end{cases}$$

It may be best to give an example. We suppose that $J(a) = J_1$, $J(2a) = J_2$ and the rest are zero. Using (7.113) we find:

$$J(q) = 2J_1 \cos(qa) + 2J_2 \cos(2qa). \tag{7.114}$$

For a minimum of energy [maximum $J(q)$] we require

$$\frac{\partial J}{\partial q} = 0 \to J_1 = -4J_2 \cos(qa) \quad \text{or} \quad q = 0 \text{ or } \frac{\pi}{a},$$

and

$$\frac{\partial^2 J}{\partial q^2} < 0 \quad \text{or} \quad J_1 \cos(qa) > -4J_2 \cos(2qa).$$

The three cases give:

$q = 0$	$q = \pi/a$	$q \neq 0, \pi/a$
$J_1 > -4J_2$	$J_1 < 4J_2$	Turn angle qa defined by
Ferromagnetism	Antiferromagnetism	$\cos(qa) = -J_1/4J_2$ and
e.g. $J_1 > 0, J_2 = 0$	e.g. $J_1 < 0, J_2 = 0$	$J_1\cos(qa) > -4J_2\cos(2qa)$

7.2.2 Magnetic Anisotropy and Magnetostatic Interactions (A)

Anisotropy

Exchange interactions drive the spins to lock together at low temperature into an ordered state, but often the exchange interaction is isotropic. So, the question arises as to why the solid magnetizes in a particular direction. The answer is that other interactions are active that lock in the magnetization direction. These interactions cause magnetic anisotropy.

Anisotropy can be caused by different mechanisms. In rare earths, because of the strong-spin orbit coupling, magnetic moments arise from both spin and orbital motion of electrons. Anisotropy, then, can be caused by direct coupling between the orbit and lattice.

There is a different situation in the iron group magnetic materials. Here we think of the spins of the 3d electrons as causing ferromagnetism. However, the spins are not directly coupled to the lattice. Anisotropy arises because the orbit "feels" the lattice, and the spins are coupled to the orbit by the spin-orbit coupling.

Let us first discuss the rare earths, which are perhaps the easier of the two to understand. As mentioned, the anisotropy comes from a direct coupling between the crystalline field and the electrons. In this connection, it is useful to consider the classical multipole expansion for the energy of a charge distribution in a potential Φ. The first three terms are given below:

$$u = q\Phi(0) - \mathbf{p} \cdot \mathbf{E}(0) - \frac{1}{6}\sum_{i,j} Q_{ij}\left(\frac{\partial E_j}{\partial x_i}\right)_0 + \text{higher-order terms.} \quad (7.115)$$

Here, q is the total charge, p is the dipole moment, Q_{ij} is the quadrupole moment, and the electric field is $E = -\nabla\Phi$. For charge distributions arising from states with definite parity, $p = 0$. (We assume this, or equivalently we assume the parity operator commutes with the Hamiltonian.) Since the term $q\Phi(0)$ is an additive constant, and since $p = 0$, the first term that merits consideration is the quadrupole term. The quadrupole term describes the interaction of the quadrupole moment with the gradient of the electric field. Generally, the quadrupole moments will vary with $|J, M\rangle$ (J = total angular momentum quantum number and M refers to the z component), which will enable us to construct an effective Hamiltonian. This Hamiltonian will include the anisotropy in which different states within a manifold of constant J will have different energies, hence anisotropy. We now develop this idea in quantum mechanics below.

We suppose the crystal field is caused by an array of charges described by $\rho(R)$. Then, the potential energy of $-e$ at the point r_i is given by

$$V(r_i) = -\int \frac{e\rho(R)\,dR}{4\pi\varepsilon_0 |r_i - R|}. \quad (7.116)$$

If we further suppose $\rho(R)$ is outside the ion in question, then in the region of the ion, $V(r)$ is a solution of the Laplace equation, and we can expand it as a solution of this equation:

$$V(r_i) = \sum_{l,m} B_l^m r^l Y_l^m(\theta,\phi), \quad (7.117)$$

where the constants B_l^m can be computed from $\rho(R)$. For rare earths, the effects of the crystal field, typically, can be adequately calculated in first-order perturbation theory. Let $|v\rangle$ be all states $|J, M\rangle$, which are formed of fixed J manifolds from $|l, m\rangle$, and $|s, m_s\rangle$ where $l = 3$ for 4f electrons. The type of matrix element that we need to evaluate can be written:

$$\langle v | \sum_i V(r_i) | v'\rangle, \quad (7.118)$$

summing over the 4f electrons. By (7.117), this eventually means we will have to evaluate matrix elements of the form

$$\langle l m_i | Y_{l'}^{m'} | l m_i'\rangle, \quad (7.119)$$

and since $l = 3$ for 4f electrons, this must vanish if $l' > 6$. Also, the parity of the functions in (7.119) is $(-)^{2l+l'}$ the matrix element must vanish if l' is odd since

7.2 Origin and Consequences of Magnetic Order

$2l = 6$, and the integral over all space is of an odd parity function is zero. For 4f electrons, we can write

$$V(r_i) = \sum_{\substack{l'=0 \\ (\text{even})}}^{6} \sum_{m'} B_{l'}^{m'} r^{l'} Y_{l'}^{m'}(\theta, \phi). \tag{7.120}$$

We define the effective Hamiltonian as

$$\mathcal{H}_A = \sum_i \langle V(r_i) \rangle_{\text{doing radial integrals only}}.$$

If we then apply the Wigner–Eckhart theorem [7.68 p33], in which one replaces (x'/r), etc. by their operator equivalents J_x, etc., we find for hexagonal symmetry

$$\mathcal{H}_A = K_1 J_z^2 + K_2 J_z^4 + K_3 J_z^6 + K_4(J_+^6 + J_-^6), \quad (J_\pm = J_x \pm iJ_y). \tag{7.121}$$

We now discuss the anisotropy that is appropriate to the iron group [7.68 p57]. This is called single-ion anisotropy. Under the action of a crystalline field we will assume the relevant atomic states include a ground state (G) of energy ε_0 and appropriate excited (E) states of energy $\varepsilon_0 + \Delta$. We will consider only one excited state, although in reality there would be several. We assume $|G\rangle$ and $|E\rangle$ are separated by energy Δ.

The states $|G\rangle$ and $|E\rangle$ are assumed to be spatial functions only and not spin functions. In our argument, we will carry the spin S along as a classical vector. The argument we will give is equivalent to perturbation theory.

We assume a spin-orbit interaction of the form $V = \lambda L \cdot S$, which mixes some of the excited state into the ground state to produce a new ground state.

$$|G\rangle \rightarrow |G_T\rangle = |G\rangle + a|E\rangle, \tag{7.122}$$

where a is in general complex. We further assume $\langle G|G\rangle = \langle E|E\rangle = 1$ and $\langle E|G\rangle = 0$ so $\langle G_T|G_T\rangle = 1$ to $O(a)$. Also note the probability that $|E\rangle$ is contained in $|G_T\rangle$ is $|a|^2$. The increase in energy due to the mixture of the excited state is (after some algebra)

$$\varepsilon_1 = \frac{\langle G_T|H|G_T\rangle}{\langle G_T|G_T\rangle} - \varepsilon_0 = \frac{\langle aE + G|H|aE + G\rangle}{1 + |a|^2} - \varepsilon_0,$$

or

$$\varepsilon_1 = |a|^2 \Delta. \tag{7.123}$$

In addition, due to first-order perturbation theory, the spin-orbit interaction will cause a change in energy given by

$$\varepsilon_2 = \lambda \langle G_T|L|G_T\rangle \cdot S. \tag{7.124}$$

We assume the angular momentum L is quenched in the original ground state so by definition $\langle G|L|G\rangle = 0$. (See also White, [7.68 p43]. White explains that if a crystal field removes the orbital degeneracy, then the matrix element of L must be zero. This does not mean the matrix element of L^2 in the same state is zero.) Thus to first order in a,

$$\varepsilon_2 = \lambda a^* \langle E|L|G\rangle \cdot S + \lambda a \langle G|L|E\rangle \cdot S. \qquad (7.125)$$

The total change in energy given by (7.123) and (7.125) is $\varepsilon = \varepsilon_1 + \varepsilon_2$. Since a and a^* are complex with two components we can treat them as linearly independent, so $\partial\varepsilon/\partial a^* = 0$, which gives

$$a = \frac{-\langle E|\lambda L|G\rangle \cdot S}{\Delta}.$$

Therefore, after some algebra $\varepsilon = \varepsilon_1 + \varepsilon_2$ becomes

$$\varepsilon = -|a|^2 \Delta = \frac{-|\langle E|\lambda L|G\rangle \cdot S|^2}{\Delta} < 0,$$

a decrease in energy. If we let

$$A = \frac{\langle E|\lambda L|G\rangle}{\sqrt{\Delta}},$$

then

$$\varepsilon = -A \cdot SA^*S = -\sum_{\mu,\nu} S_\mu B_{\mu\nu} S_\nu,$$

where $B_{\mu\nu} = A_\mu A_\nu^*$. If we let S become a spin operator, we get the following Hamiltonian for single-ion anisotropy:

$$\mathcal{H}_{spin} = -\sum_{\mu,\nu} S_\mu B_{\mu\nu} S_\nu. \qquad (7.126)$$

When we have axial symmetry, this simplifies to

$$\mathcal{H}_{spin} = -DS_z^2.$$

For cubic crystal fields, the quadratic (in S) terms go to a constant and can be neglected. In that case, we have to go to a higher order. Things are also more complicated if the ground state has orbital degeneracy. Finally, it is also possible to have anisotropic exchange. Also, as we show below, the shape of the sample can generate anisotropy.

Magnetostatics (B)

The magnetostatic energy can be regarded as the quantity whose reduction causes domains to form. The other interactions then, in a sense, control the details of how the domains form. Domain formation will be considered in Sect. 7.3. Here we will show how the domain magnetostatic interaction can cause shape anisotropy.

Consider a magnetized material in which there is no real or displacement current. The two relevant Maxwell equations can be written in the absence of external currents and in the static situation

$$\nabla \times H = 0, \tag{7.127}$$

$$\nabla \cdot B = 0. \tag{7.128}$$

Equation (7.127) implies there is a potential Φ from which the magnetic field H can be derived:

$$H = -\nabla \Phi. \tag{7.129}$$

We assume a constitutive equation linking the magnetic induction B, the magnetization M and H;

$$B = \mu_0(H + M), \tag{7.130}$$

where μ_0 is called the permeability of free space. Equations (7.128) and (7.130) become

$$\nabla \cdot H = -\nabla \cdot M. \tag{7.131}$$

In terms of the magnetic potential Φ,

$$\nabla^2 \Phi = \nabla \cdot M. \tag{7.132}$$

This is analogous to Poisson's equation of electrostatics with $\rho_M = -\nabla \cdot M$ playing the role of a magnetic source density.

By analogy to electrostatics, and in terms of equivalent surface and volume pole densities, we have

$$\Phi = \frac{1}{4\pi} \left[\int_S \frac{M \cdot dS}{r} - \int_V \frac{\nabla \cdot M}{r} dV \right], \tag{7.133}$$

where S and V refer to the surface and volume of the magnetized body. By analogy to electrostatics the magnetostatic self-energy is

$$U_M = \frac{\mu_0}{2} \int \rho_M \Phi dV = -\frac{\mu_0}{2} \int \nabla \cdot M \Phi dV = -\frac{\mu_0}{2} \int M \cdot H dV \tag{7.134}$$

$$(\text{since } \int_{\text{all space}} \nabla \cdot (M\Phi) dV = 0),$$

which also would follow directly from the energy of a dipole μ in a magnetic field $(-\mu \cdot B)$, with a 1/2 inserted to eliminate double counting. Using $\nabla \cdot M = -\nabla \cdot H$ and $\int_{\text{all space}} \nabla \cdot (H\Phi) dV = 0$, we get

$$U_M = \frac{\mu_0}{2} \int H^2 dV. \tag{7.135}$$

For ellipsoidal specimens the magnetization is uniform and

$$H_D = -DM, \qquad (7.136)$$

where H_D is the demagnetization field, D is the demagnetization factor that depends on the shape of the sample and the direction of magnetization and hence one has shape isotropy, since (7.135) would have different values for M in different directions. For ellipsoidal magnets, the demagnetization energy per unit volume is then

$$u_M = \frac{\mu_0}{2} D^2 M^2. \qquad (7.137)$$

7.2.3 Spin Waves and Magnons (B)

If there is an external magnetic field $B = \mu_0 H \hat{z}$, and if the magnetic moment of each atom is $m = 2\mu S$ ($2\mu\hbar \equiv -g\mu_B$ [14] in previous notation), then the above considerations tell us that the Hamiltonian describing an (nn) exchange coupled spin system is

$$\mathcal{H} = -J\sum_{j\Delta} S_j \cdot S_{j+\Delta} - 2\mu_0 \mu H \sum_j S_{jz}. \qquad (7.138)$$

j runs over all atoms, and δ runs over the nearest neighbors of j, and also we may redefine J so as to write (7.138) as $\mathcal{H} = (J/2)\sum\ldots$. (We do this sometimes to emphasize that (7.138) double counts each interaction.) From now on it will be assumed that there exist real solids for which (7.138) is applicable. The first term in this equation is the Heisenberg Hamiltonian and the second term is the Zeeman energy.

Let

$$S^2 = (\sum_j S_j)^2, \qquad (7.139)$$

and

$$S_z = \sum_j S_{jz}. \qquad (7.140)$$

Then it is possible to show that the total spin and the total z component of spin are constants of the motion. In other words,

$$[\mathcal{H}, S^2] = 0, \qquad (7.141)$$

and

$$[\mathcal{H}, S_z] = 0. \qquad (7.142)$$

---

[14] The minus sign comes from the negative charge on the electron.

Spin Waves in a Classical Heisenberg Ferromagnet (B)

We want to calculate the internal energy u (per spin) and the magnetization M. Assuming the magnetization is in the z direction and letting $\langle A \rangle$ stand for the quantum-statistical average of A, we have (if $H = 0$)

$$u = \frac{1}{N}\langle \mathcal{H} \rangle = -\frac{1}{2N}\sum_{i,j} J_{ij}\langle \mathbf{S}_i \cdot \mathbf{S}_j \rangle, \tag{7.143}$$

and

$$M = -\frac{g\mu_B}{V}\sum_{iz}\langle S_{iz} \rangle, \tag{7.144}$$

(with the S written in units of $\hbar$ and V is the volume of the crystal and J_{ij} absorbs an $\hbar^2$) where the Heisenberg Hamiltonian is written in the form

$$\mathcal{H} = -\frac{1}{2}\sum_{i,j} J_{ij} \mathbf{S}_i \cdot \mathbf{S}_j.$$

Using the fact that

$$S^2 = S_x^2 + S_y^2 + S_z^2,$$

assuming a ferromagnetic ground state, and very low temperatures (where spin wave theory is valid) so that S_x and S_y are very small,

$$S_z = -\sqrt{S^2 - S_x^2 - S_y^2},$$

(negative so $M > 0$) and thus

$$S_z \cong -S\left(1 - \frac{S_x^2 + S_y^2}{2S^2}\right), \tag{7.145}$$

which can be substituted in (7.144). Then by (7.143)

$$u \cong -\frac{1}{2N}\sum_{i,j} S^2 J_{ij}\left\langle\left(1 - \frac{S_{ix}^2 + S_{iy}^2}{2S^2}\right)\left(1 - \frac{S_{jx}^2 + S_{jy}^2}{2S^2}\right)\right\rangle$$

$$-\frac{1}{2N}\sum_{i,j} J_{ij}\langle S_{ix}S_{jx} + S_{iy}S_{jy}\rangle.$$

We obtain

$$M = \frac{N}{V}g\mu_B S - \frac{g\mu_B}{2SV}\sum_i \langle S_{ix}^2 + S_{iy}^2 \rangle, \tag{7.146}$$

$$u = -\frac{S^2 Jz}{2} + \frac{1}{2N}\sum_{i,j} J_{ij}\langle S_{ix}^2 + S_{iy}^2 - S_{ix}S_{jx} - S_{iy}S_{jy}\rangle, \tag{7.147}$$

where z is the number of nearest neighbors. It is now convenient to Fourier transform the spins and the exchange integral

$$S_i = \sum_k S_k e^{i k \cdot R} \qquad (7.148)$$

$$J(k) = \sum_R J(R) e^{i q \cdot R}. \qquad (7.149)$$

Using the standard crystal lattice mathematics and $S_{-kx} = S_{kx}^*$, we find:

$$M = \frac{N}{V} g\mu_B S \left\{ 1 - \frac{1}{2S} \sum_k \left\langle S_{kx} S_{kx}^* + S_{ky} S_{ky}^* \right\rangle \right\} \qquad (7.150)$$

$$u = -\frac{S^2 J z}{2} + \frac{1}{2} \sum_k (J(0) - J(k)) \left\langle S_{kx} S_{kx}^* + S_{ky} S_{ky}^* \right\rangle. \qquad (7.151)$$

We still have to evaluate the thermal averages. To do this, it is convenient to exploit the analogy of the spin waves to a set of uncoupled harmonic oscillators whose energy is proportional to the amplitude squared. We do this by deriving the equations of motion and showing in our low-temperature "spin-wave" approximation that they are harmonic oscillators. We can write the Heisenberg Hamiltonian equation as

$$\mathcal{H} = -\frac{1}{2} \sum_j \left\{ \sum_i J_{ij} \frac{S_i}{-g\mu_B} \right\} (-g\mu_B S_j), \qquad (7.152)$$

where $-g\mu_B S_j$ is the magnetic moment. The 1/2 takes into account the double counting and we therefore identify the effective field acting on S_j as

$$B_{M_j} = -\frac{1}{g\mu_B} \sum_i J_{ij} S_i. \qquad (7.153)$$

Treating the S_i as dimensionless so $\hbar S_i$ is the angular momentum, and using the fact that torque is the rate of change of angular momentum and is the moment crossed into field, we have for the equations of motion

$$\hbar \frac{dS_j}{dt} = \sum_i J_{ij} S_j \times S_i. \qquad (7.154)$$

We leave as a problem to show that after Fourier transformation the equations of motion can be written:

$$\hbar \frac{dS_k}{dt} = \sum_{k'} J(k') S_{k-k'} \times S_{k'}. \qquad (7.155)$$

For the ferromagnetic ground state at low temperature, we assume that

$$|S_{k=0}| \gg |S_{k \neq 0}|,$$

since
$$S_{k=0} = \frac{1}{N}\sum_R S_R,$$
and at absolute zero,
$$S_{k=0} = S\hat{k}, \quad S_{k\neq 0} = 0.$$
Even with small excitations, we assume $S_{0z} = S$, $S_{0x} = S_{0y} = 0$ and S_{kx}, S_{ky} are of first order. Retaining only quantities of first order, we have

$$\hbar\frac{dS_{kx}}{dt} = S[J(0) - J(k)]S_{ky} \tag{7.156a}$$

$$\hbar\frac{dS_{ky}}{dt} = -S[J(0) - J(k)]S_{kx} \tag{7.156b}$$

$$\hbar\frac{dS_{kz}}{dt} = 0. \tag{7.156c}$$

Combining (7.156a) and (7.156b), we obtain harmonic-oscillator-type equations with frequencies $\omega(k)$ and energies $\varepsilon(k)$ given by

$$\varepsilon(k) = \hbar\omega(k) = S[J(0) - J(k)]. \tag{7.157}$$

Combining this result with (7.151), we have for the average energy per oscillator,

$$u = -\frac{S^2 Jz}{2} + \frac{1}{2}\sum_k \frac{\varepsilon(k)}{S}\langle |S_{kx}|^2 + |S_{ky}|^2 \rangle$$

for z nearest neighbors. For quantized harmonic oscillators, up to an additive term, the average energy per oscillator would be

$$\frac{1}{N}\sum_k \varepsilon(k)\langle n_k \rangle.$$

Thus, we identify $\langle n_k \rangle$ as

$$\left\langle \frac{|S_{kx}|^2 + |S_{ky}|^2}{2S} \right\rangle N,$$

and we write (7.150) and (7.151) as

$$M = \frac{N}{V}g\mu_B S\left\{1 - \frac{1}{NS}\sum_k \langle n_k \rangle\right\} \tag{7.158}$$

$$u = -\frac{S^2 Jz}{2} + \frac{1}{N}\sum_k \varepsilon(k)\langle n_k \rangle. \tag{7.159}$$

Now $\langle n_k \rangle$ is the average number of excitations in mode k (magnons) at temperature T.

By analogy with phonons (which represent quanta of harmonic oscillators) we say

$$\langle n_k \rangle = \frac{1}{e^{\varepsilon(k)/kT} - 1}. \tag{7.160}$$

As an example, we work out the consequences of this for simple cubic lattices with $Z = 6$ and nearest-neighbor coupling.

$$J(k) = \sum J(R) e^{i k \cdot R} = 2J(\cos k_x a + \cos k_y a + \cos k_z a).$$

At low temperatures where only small k are important, we find

$$\varepsilon(k) = S[J(0) - J(k)] \cong SJk^2 a^2. \tag{7.161}$$

We will evaluate (7.158) and (7.159) using (7.160) and (7.161) later after treating spin waves quantum mechanically from the beginning.

The name "spin-waves" comes from the following picture. In Fig. 7.10, suppose

$$S_{kx} = S\sin(\theta)\exp[i\omega(k)t],$$

then

$$\hbar \dot{S}_{kx} = i\omega(q)\hbar S_{kx} = \omega(k)\hbar S_{ky}$$

by the equation of motion. So,

$$iS_{kx} = S_{ky}.$$

Therefore, if we had one spin-wave mode q in the x direction, e.g., then

$$S_{Rx} = \exp(i k \cdot R) S_{kx} = S\sin(\theta)\exp[i(kR_x + \omega t)],$$
$$S_{Ry} = S\sin(\theta)\exp[i(kR_x + \omega t - \pi/2)].$$

Fig. 7.10. Classical representation of a spin wave in one dimension (a) viewed from side and (b) viewed from top (along $-z$). The phase angle from spin to spin changes by ka. Adapted from Kittel C, *Introduction to Solid State Physics*, 7th edn, Copyright © 1996 John Wiley and Sons, Inc. This material is used by permission of John Wiley and Sons, Inc

7.2 Origin and Consequences of Magnetic Order

Thus, if we take the real part, we find

$$S_{Rx} = S\sin(\theta)\cos(kR_x + \omega t),$$
$$S_{Ry} = S\sin(\theta)\sin(kR_x + \omega t),$$

and the spins all spin with the same frequency but with the phase changing by ka, which is the change in kR_x, as we move from spin to spin along the x-axis.

As we have seen, spin waves are collective excitations in ordered spin systems. The collective excitations consist in the propagation of a spin deviation, θ. A localized spin at a site is said to undergo a deviation when its direction deviates from the direction of magnetization of the solid below the critical temperature. Classically, we can think of spin waves as vibrations in the magnetic moment density. As mentioned, quanta of the spin waves are called *magnons*. The concept of spin waves was originally introduced by F. Bloch, who used it to explain the temperature dependence of the magnetization of a ferromagnet at low temperatures. The existence of spin waves has now been definitely proved by experiment. Thus the concept has more validity than its derivation from the Heisenberg Hamiltonian might suggest. We will only discuss spin waves in ferromagnets but it is possible to make similar comments about them in any ordered magnetic structure. The differences between the ferromagnetic case and the antiferromagnetic case, for example, are not entirely trivial [60, p 61].

Spin Waves in a Quantum Heisenberg Ferromagnet (A)

The aim of this section is rather simple. We want to show that the quantum Heisenberg Hamiltonian can be recast, in a suitable approximation, so that its energy excitations are harmonic-oscillator-like, just as we found classically (7.161).

Here we make two transformations and a long-wavelength, low-temperature approximation. One transformation takes the Hamiltonian to a localized excitation description and the other to an unlocalized (magnon) description. However, the algebra can get a little complex.

Equation (7.138) (with $\hbar = 1$ or $2\mu = -g\mu_B$) is our starting point for the three-dimensional case, but it is convenient to transform this equation to another form for calculation. From our previous discussion, we believe that magnons are similar to phonons (insofar as their mathematical description goes), and so we might guess that some sort of second quantization notation would be appropriate. We have already indicated that the squared total spin and the z component of total spin give good quantum numbers. We can also show that S_j^2 commutes with the Heisenberg Hamiltonian so that its eigenvalues $S(S + 1)$ are good quantum numbers. This makes sense because it just says that the total spin of each atom remains constant. We assume that the spin S of every ion is the same. Although each atom has three components of each spin vector, only two of the components are independent.

The Holstein and Primakoff Transformation (A) Holstein and Primakoff[15] have developed a transformation that not only has two independent variables, but also utilizes the very convenient second quantization notation. The Holstein–Primakoff transformation is also very useful for obtaining terms that describe magnon–magnon interactions.[16] This transformation is (with $\hbar = 1$ or S representing $S/\hbar$):

$$S_j^+ \equiv S_{jx} + iS_{jy} = \sqrt{2S}\left[1 - \frac{a_j^\dagger a_j}{2S}\right]^{1/2} a_j, \qquad (7.162)$$

$$S_j^- \equiv S_{jx} - iS_{jy} = \sqrt{2S}\, a_j^\dagger \left[1 - \frac{a_j^\dagger a_j}{2S}\right]^{1/2}, \qquad (7.163)$$

$$S_{jz} \equiv S - a_j^\dagger a_j. \qquad (7.164)$$

We could use these transformation equations to attempt to determine what properties a_j and $a_j^\dagger$ must have. However, it is much simpler to define the properties of the a_j and $a_j^\dagger$ and show that with these definitions the known properties of the S_j operators are obtained. We will assume that the $a^\dagger$ and a are *boson* creation and annihilation operators (see Appendix G) and hence they satisfy the commutation relations

$$[a_j, a_l^\dagger] = \delta_j^l. \qquad (7.165)$$

We first show that (7.164) is consistent with (7.162) and (7.163). This amounts to showing that the Holstein–Primakoff transformation automatically puts in the constraint that there are only two independent components of spin for each atom. We start by dropping the subscript j for a particular atom and by using the fact that S_j^2 has a good quantum number so we can substitute $S(S+1)$ for S_j^2 (with $\hbar = 1$). We can then write

$$S(S+1) = S_x^2 + S_y^2 + S_z^2 = S_z^2 + \tfrac{1}{2}(S^+S^- + S^-S^+). \qquad (7.166)$$

By use of (7.162) and (7.163) we can use (7.166) to calculate S_z^2. That is,

$$S_z^2 = S(S+1) - S\left[\left(1 - \frac{a^\dagger a}{2S}\right)^{1/2}(1 + a^\dagger a)\left(1 - \frac{a^\dagger a}{2S}\right)^{1/2} + a^\dagger\left(1 - \frac{a^\dagger a}{2S}\right)a\right]. \qquad (7.167)$$

---

[15] See, for example, [7.38].
[16] At least for high magnetic fields; see Dyson [7.18].

Remember that we define a function of operators in terms of a power series for the function, and therefore it is clear that $a^\dagger a$ will commute with any function of $a^\dagger a$. Also note that $[a^\dagger a, a] = a^\dagger a a - a a^\dagger a = a^\dagger a a - (1 + a^\dagger a)a = -a$, and so we can transform (7.167) to give after several algebraic steps:

$$S_z^2 = (S - a^\dagger a)^2. \tag{7.168}$$

Equation (7.168) is consistent with (7.164), which was to be shown.

We still need to show that S_j^+ and S_j^- defined in terms of the annihilation and creation operators act as ladder operators should act. Let us define an eigenket of S_j^2 and S_{jz}, by (still with $\hbar = 1$)

$$S_j^2 |S, m_s\rangle = S(S+1)|S, m_s\rangle, \tag{7.169}$$

and

$$S_{jz}|S, m_s\rangle = m_s|S, m_s\rangle. \tag{7.170}$$

Let us further define a spin-deviation eigenvalue by

$$n = S - m_s, \tag{7.171}$$

and for convenience let us shorten our notation by defining

$$|n\rangle = |S, m_s\rangle. \tag{7.172}$$

By (7.162) we can write

$$S_j^+|n\rangle = \sqrt{2S}\left(1 - \frac{a_j^\dagger a_j}{2S}\right)^{1/2} a_j|n\rangle = \sqrt{2S}\left(1 - \frac{n-1}{2S}\right)^{1/2} \sqrt{n}|n-1\rangle, \tag{7.173}$$

where we have used $a_j|n\rangle = n^{1/2}|n-1\rangle$ and also the fact that

$$a_j^\dagger a_j|n\rangle = (S - S_{jz})|n\rangle = n|n\rangle. \tag{7.174}$$

By converting back to the $|S, m_s\rangle$ notation, we see that (7.173) can be written

$$S_j^+|S, m_s\rangle = \sqrt{(S - m_s)(S + m_s + 1)}\,|S, m_s + 1\rangle. \tag{7.175}$$

Therefore S_j^+ does have the characteristic property of a ladder operator, which is what we wanted to show. We can similarly show that the S_j^- has the step-down ladder properties.

Note that since (7.175) is true, we must have that

$$S^+|S, m_s = S\rangle = 0. \tag{7.176}$$

A similar calculation shows that

$$S^-|S,-m_s = S\rangle = 0 \ . \tag{7.177}$$

We needed to assure ourselves that this property still held even though we defined the S^+ and S^- in terms of the $a_j^\dagger$ and a_j. This is because we normally think of the a as operating on $|n\rangle$, where $0 \le n \le \infty$. In our situation we see that $0 \le n < 2S + 1$. We have now completed the verification of the consistency of the Holstein–Primakoff transformation. It is time to recast the Heisenberg Hamiltonian in this new notation.

Combining the results of Problem 7.10 and the Holstein–Primakoff transformation, we can write

$$\mathcal{H} =$$

$$-J\sum_{j\Delta}\left\{(S - a_j^\dagger a_j)(S - a_{j+\Delta}^\dagger a_{j+\Delta}) + S\left[a_j^\dagger\left(1 - \frac{a_j^\dagger a_j}{2S}\right)^{1/2}\left(1 - \frac{a_{j+\Delta}^\dagger a_{j+\Delta}}{2S}\right)^{1/2} a_{j+\delta}\right.\right.$$

$$\left.\left. + \left(1 - \frac{a_j^\dagger a_j}{2S}\right)^{1/2} a_j a_{j+\Delta}^\dagger \left(1 - \frac{a_{j+\Delta}^\dagger a_{j+\Delta}}{2S}\right)^{1/2}\right]\right\} + g\mu_B(\mu_0 H)\sum_j(S - a_j^\dagger a_j) .$$

$$\tag{7.178}$$

Equation (7.178) is the Heisenberg Hamiltonian (plus a term for an external magnetic field) expressed in second quantization notation. It seems as if the problem has been complicated rather than simplified by the Holstein–Primakoff transformation. Actually both (7.138) and (7.178) are equally impossible to solve exactly. Both are many-body problems. The point is that (7.178) is in a form that can be approximated fairly easily. The approximation that will be made is to expand the square roots and concentrate on low-order terms. Before this is done, it is convenient to take full advantage of translational symmetry. This will be done in the next section.

Magnons (A) The $a_j^\dagger$ create localized spin deviations at a single site (one atom per unit cell is assumed). What we need (in order to take translational symmetry into account) is creation operators that create Bloch-like nonlocalized excitations. A transformation that will do this is

$$B_k = \frac{1}{\sqrt{N}}\sum_j \exp(i\mathbf{k}\cdot\mathbf{R}_j)a_j , \tag{7.179a}$$

and

$$B_k^\dagger = \frac{1}{\sqrt{N}}\sum_j \exp(-i\mathbf{k}\cdot\mathbf{R}_j)a_j^\dagger , \tag{7.179b}$$

7.2 Origin and Consequences of Magnetic Order

where R_j is defined by (2.171) and cyclic boundary conditions are used so that the k are defined by (2.175). $N = N_1 N_2 N_3$ and so the delta function relations (2.178) to (2.184) are valid. k will be assumed to be restricted to the first Brillouin zone. Using all these results, we can derive the inverse transformation

$$a_j = \frac{1}{\sqrt{N}} \sum_k \exp(-i k \cdot R_j) B_k \,, \tag{7.180a}$$

and

$$a_j^\dagger = \frac{1}{\sqrt{N}} \sum_k \exp(i k \cdot R_j) B_k^\dagger \,. \tag{7.180b}$$

So far we have not shown that the B are boson creation and annihilation operators. To show this, we merely need to show that the B satisfy the appropriate commutation relations. The calculation is straightforward, and is left as a problem to show that the B_k obey the same commutation relations as the a_j.

We can give a very precise definition to the word *magnon*. First let us review some physical principles. Exchange coupled spin systems (e.g. ferromagnets and antiferromagnets) have low-energy states that are wave-like. These wave-like energy states are called spin waves. A spin wave is quantized into units called magnons. We may have spin waves in any structure that is magnetically ordered. Since in the low-temperature region there are only a few spin waves that are excited and thus their complicated interactions are not so important, this is the best temperature region to examine spin waves. Mathematically, precisely whatever is created by $B_k^\dagger$ and annihilated by B_k is called a *magnon*.

There is a nice theorem about the number of magnons. The total number of magnons equals the total spin deviation quantum number. This theorem is easily proved as shown below:

$$\begin{aligned}
\Delta_S &= \sum_j (S - S_{jz}) = \sum_j a^\dagger a_j \\
&= \frac{1}{N} \sum_{i,k,k'} \exp[i(k - k') \cdot R_j] B_k^\dagger B_{k'} \\
&= \sum_{k,k'} \delta_k^{k'} B_k^\dagger B_{k'} \\
&= \sum_k B_k^\dagger B_k.
\end{aligned}$$

This proves the theorem, since $B_k^\dagger B_k$ is the occupation number operator for the number of magnons in mode k.

The Hamiltonian defined by (7.178) will now be approximated. The spin-wave variables B_k will also be substituted.

At low temperatures we may expect the spin-deviation quantum number to be rather small. Thus we have approximately

$$\left\langle a_j^\dagger a_j \right\rangle \ll S \,. \tag{7.181}$$

This implies that the relation between the S and a can be approximated by

$$S_j^- \cong \sqrt{2S}\left(a_j^\dagger - \frac{a_j^\dagger a_j^\dagger a_j}{4S}\right), \quad (7.182a)$$

$$S_j^+ \cong \sqrt{2S}\left(a_j - \frac{a_j^\dagger a_j a_j}{4S}\right), \quad (7.182b)$$

and

$$S_{jz} = S - a_j^\dagger a_j. \quad (7.182c)$$

Expressing these results in terms of the B, we find

$$S_j^+ \cong \sqrt{\frac{2S}{N}}\left\{\sum_k \exp(-i\mathbf{k}\cdot\mathbf{R}_j)B_k \right.$$
$$\left. - \frac{1}{4SN}\sum_{k,k',k''}\exp[i(\mathbf{k}-\mathbf{k}'-\mathbf{k}'')\cdot\mathbf{R}_j]B_k^\dagger B_{k'} B_{k''}\right\}, \quad (7.183a)$$

$$S_j^- \cong \sqrt{\frac{2S}{N}}\left\{\sum_k \exp(i\mathbf{k}\cdot\mathbf{R}_j)B_k \right.$$
$$\left. - \frac{1}{4SN}\sum_{k,k',k''}\exp[i(\mathbf{k}+\mathbf{k}'-\mathbf{k}'')\cdot\mathbf{R}_j]B_k^\dagger B_{k'}^\dagger B_{k''}\right\}, \quad (7.183b)$$

and

$$S_{jz} = S - \frac{1}{N}\sum_{k,k'}\exp[i(\mathbf{k}-\mathbf{k}')\cdot\mathbf{R}_j]B_k^\dagger B_{k'}. \quad (7.183c)$$

The details of the calculation begin to get rather long at about this stage. The approximate Hamiltonian in terms of spin-wave variables is obtained by substituting (7.183) into (7.178). Considerable simplification results from the delta function relations. Terms of order $(\langle a_i^\dagger a_i\rangle/S)^2$ are to be neglected for consistency. The final result is

$$\mathcal{H} = \mathcal{H}_0 + \mathcal{H}_{\text{ex}}, \quad (7.184)$$

neglecting a constant term, where Z is the number of nearest neighbors, $\mathcal{H}_0$ is the term that is bilinear in the spin wave variables and is given by

$$\mathcal{H}_0 = -JSZ[\sum_k (\alpha_k(1+B_k^\dagger B_k) + \alpha_{-k}B_k^\dagger B_k - 2B_k^\dagger B_k)]$$
$$+ g\mu_B(\mu_0 H)\sum_k B_k^\dagger B_k, \quad (7.185)$$

$$\alpha_k = \frac{1}{Z}\sum_\Delta \exp(i\mathbf{k}\cdot\mathbf{\Delta}), \quad (7.186)$$

and $\mathcal{H}_{ex}$ is called the *exchange interaction* Hamiltonian and is biquadratic in the spin-wave variables. It is given by

$$\mathcal{H}_{ex} \propto Z\frac{J}{N}\sum_{k_1 k_2 k_3 k_4} \delta^{k_2+k_3}_{k_1+k_4}(B_{k_1} B^{\dagger}_{k_2} - \delta^{k_2}_{k_1})B^{\dagger}_{k_3} B_{k_4}(\alpha_{k_1} - \alpha_{k_1-k_2}). \quad (7.187)$$

Note that $\mathcal{H}_0$ describes magnons without interactions and $\mathcal{H}_{ex}$ includes terms that describe the effect of interactions. Mathematically, we do not want to consider interactions. Physically, it makes sense to believe that interactions should not be important at low temperatures. We can show that $\mathcal{H}_{ex}$ can be neglected for long-wavelength magnons, which should be the only important magnons at low temperature. We will therefore neglect $\mathcal{H}_{ex}$ in all discussions below.

$\mathcal{H}_0$ can be somewhat simplified. Incidentally, the formalism that is being used assumes only one atom per unit cell and that all atoms are equally spaced and identical. Among other things, this precludes the possibility of having "optical magnons." This is analogous to the lattice vibration problem where we do not have optical phonons in lattices with one atom per unit cell.

$\mathcal{H}_0$ can be simplified by noting that if the crystal has a center of symmetry, then $\alpha_k = \alpha_{-k}$, and also

$$\sum_k \alpha_k = \frac{1}{Z}\sum_\Delta \sum_k \exp(i k \cdot \Delta) = \frac{N}{Z}\sum_\Delta \delta^0_\Delta = 0,$$

where the last term is zero because Δ, being the vector to nearest-neighbor atoms, can never be zero. Also note that $BB^\dagger - 1 = B^\dagger B$. Using these results and defining (with $H = 0$)

$$\hbar \omega_k = 2JSZ(1 - \alpha_k), \quad (7.188)$$

we find

$$H_0 = \sum_k \hbar \omega_k n_k, \quad (7.189)$$

where n_k is the occupation number operator for the magnons in mode k.

If the wavelength of the spin waves is much greater than the lattice spacing, so that atomic details are not of much interest, then we are in a classical region. In this region, it makes sense to assume that $k \cdot \Delta \ll 1$, which is also the long-wavelength approximation made in neglecting $\mathcal{H}_{ex}$. Thus we find

$$\hbar \omega_k \cong JS \sum_\Delta (k \cdot \Delta)^2. \quad (7.190)$$

If further we have a simple cubic, bcc, or fcc lattice, then

$$\hbar \omega_k = \frac{\hbar^2 k^2}{2m^*}, \quad (7.191)$$

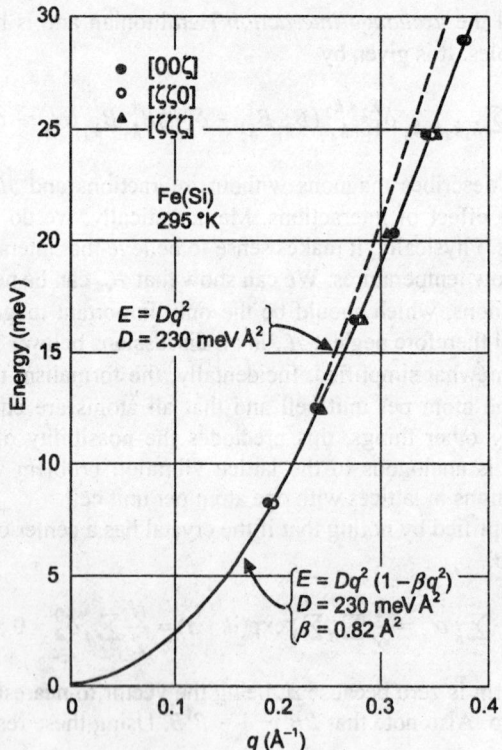

Fig. 7.11. Fe (12 at. % Si) room-temperature spin-wave dispersion relations at low energy. Reprinted with permission from Lynn JW, *Phys Rev B* **11**(7), 2624 (1975). Copyright 1975 by the American Physical Society

where

$$m^* \propto (2ZJSa^2)^{-1}, \tag{7.192}$$

and a is the lattice spacing. The reality of spin-wave dispersion has been shown by inelastic neutron scattering. See Fig. 7.11.

Specific Heat of Spin Waves (A) With

$$\frac{\langle a_i^\dagger a_i \rangle}{S} \ll 1, \quad ka \ll 1, \quad H = 0,$$

and assuming we have a monatomic lattice, the magnons were found to have the energies

$$\hbar\omega_k = Ck^2, \tag{7.193}$$

7.2 Origin and Consequences of Magnetic Order

where C is a constant. Thus apart from notation (7.161) and (7.193) are identical. We also know that the magnons behave as bosons. We can return to (7.158), (7.159), (7.160), and (7.161) to evaluate the magnetization as well as the internal energy due to spin waves.

Now in (7.158) we can replace a sum with an integral because for large N the number of states is fairly dense and in $d\mathbf{k}$ per unit volume is $d\mathbf{k}/(2\pi)^3$. So

$$\sum_k \frac{1}{\exp(JSk^2a^2/k_BT)-1} \to \frac{V}{(2\pi)^3}\int\frac{d\mathbf{k}}{\exp(JSk^2a^2/k_BT)-1}$$

$$\to \frac{V}{(2\pi)^3}\int_0^\infty \frac{k^2 dk}{\exp(JSk^2a^2/k_BT)-1}.$$

Also we have used that at low T the upper limit can be set to infinity without appreciable error. Changing the integration variable to $x = (JS/k_BT)^{1/2}ka$, we find at low temperature

$$\sum_k \frac{1}{\exp(JSk^2a^2/k_BT)-1} \to \frac{V}{(2\pi)^3}\left(\sqrt{\frac{k_BT}{JS}}\frac{1}{a}\right)^3 N_1,$$

where

$$N_1 = \int_0^\infty \frac{x^2 dx}{\exp(x^2)-1}.$$

Similarly

$$\sum_k \frac{JSk^2a^2}{\exp(JSk^2a^2/k_BT)-1} \to \frac{V}{(2\pi)^3}\left(\sqrt{\frac{k_BT}{JS}}\frac{1}{a}\right)^5 N_2,$$

where

$$N_2 = \int_0^\infty \frac{x^4 dx}{\exp(x^2)-1}.$$

N_1 and N_2 are numbers that can be evaluated in terms of gamma functions and Riemann zeta functions. We thus find

$$M = \frac{N}{V}g\mu_B S\left\{1 - \frac{V}{2\pi^2 SN}\left(\frac{k_B}{JSa^2}\right)^{3/2} N_1 T^{3/2}\right\}, \qquad (7.194)$$

and

$$u = -\frac{S^2 Jz}{2} + \frac{V}{2\pi^2 N}\left(\frac{k_B}{JSa^2}\right)^{5/2} N_2 T^{5/2}. \qquad (7.195)$$

Thus, from (7.195) by taking the temperature derivative we find the low-temperature magnon specific heat, as first shown by Bloch, is

$$C_V \propto T^{3/2}. \tag{7.196}$$

Similarly, by (7.194) the low-temperature deviation from saturation goes as $T^{3/2}$. these results only depend on low-energy excitations going as k^2.

Also at low T, we have a lattice specific heat that goes as T^3. So at low T we have

$$C_V = aT^{3/2} + bT^3,$$

where a and b are constants. Thus

$$C_V T^{-3/2} = a + bT^{3/2},$$

so theoretically, plotting $CT^{-3/2}$ vs $T^{3/2}$ will yield a straight line at low T. Experimental verification is shown in Fig. 7.12 (note this is for a ferrimagnet for which the low-energy $\hbar\omega_k$ is also proportional to k^2).

At higher temperatures there are deviations from the $3/2$ power law and it is necessary to make refinements in the above theory. One source of deviations is spin-wave interactions. We also have to be careful that we do not approximate away the kinematical part, i.e. the part that requires the spin-deviation quantum number on a given site not to exceed $(2S_j + 1)$. Then, of course, in a more careful analysis we would have to pay more attention to the geometrical shape of the Brillouin zone. Perhaps our worst error involves (7.191), which leads to an approximate density of states and hence to an approximate form for the integral in the calculation of C_V and ΔM.

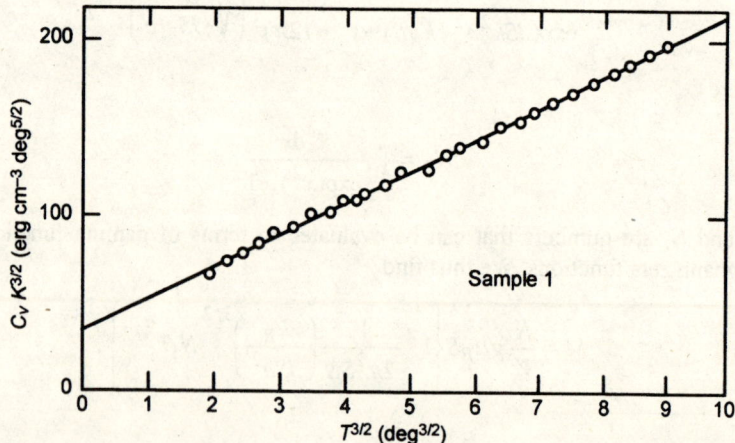

Fig. 7.12. C_V at low T for ferrimagnet YIG. After Elliott RJ and Gibson AF, *An Introduction to Solid State Physics and Applications*, Macmillan, 1974, p 461. Original data from Shinozaki SS, *Phys Rev* **122**, 388 (1961))

Table 7.1. Summary of spin-wave properties (low energy and low temperature)

	Dispersion relation	$\Delta M = M_s - M$ magnetization	C magnetic Sp. Ht.
Ferromagnet	$\omega = A_1 k^2$	$B_1 T^{3/2}$	$B_2 T^{3/2}$
Antiferromagnet	$\omega = A_2 k$	$B_2 T^2$ (sublattice)	$C_2 T^3$

A_i and B_i are constants. For discussion of spin waves in more complicated structures see, e.g., Cooper [7.13].

Equation (7.193) predicts that the density of states (up to cutoff) is proportional to the magnon energy to the $1/2$ power. A similar simple development for antiferromagnets [it turns out that the analog of (7.193) only involves the first power of $|k|$ for antiferromagnets] also leads to a relatively smooth dependence of the density of states on energy. In any case, a determination from analyzing the neutron diffraction of an actual magnetic substance will show a result that is not so smooth (see Fig. 7.13). Comparison of spin-wave calculations to experiment for the specific heat for EuS is shown in Fig. 7.14.[17] EuS is an ideal Heisenberg ferromagnet.

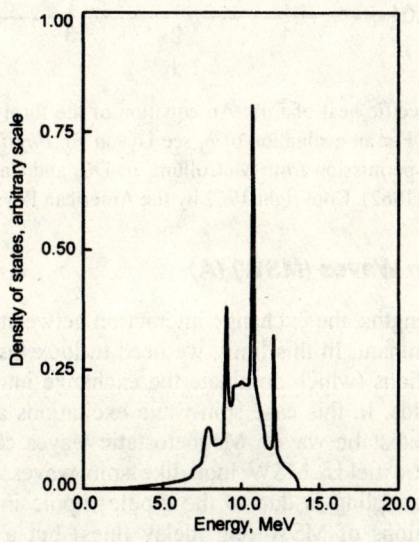

Fig. 7.13. Density of states for magnons in Tb at 90 K. The curve is a smoothed computer plot. [Reprinted with permission from Moller HB, Houmann JCG, and Mackintosh AR, *Journal of Applied Physics*, 39(2), 807 (1968). Copyright 1968, American Institute of Physics.]

---

[17] A good reference for the material in this chapter on spin waves is an article by Kittel [7.38]

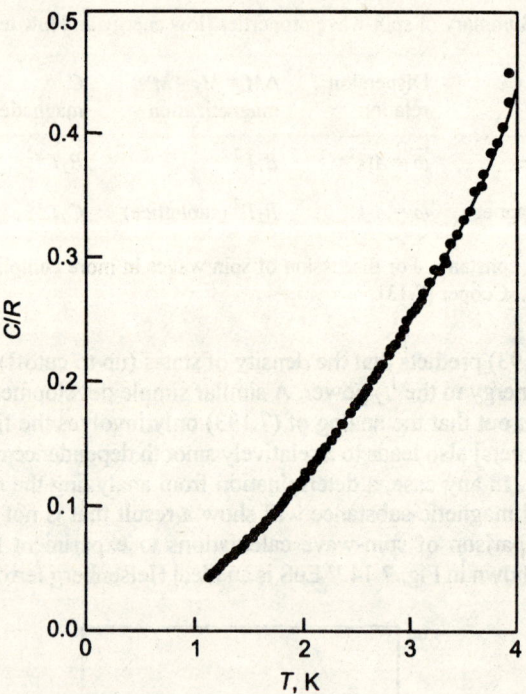

Fig. 7.14. Spin wave specific heat of EuS. An equation of the form $C/R = aT^{3/2} + bT^{5/2}$ is needed to fit this curve. For an evaluation of b, see Dyson FJ, *Physical Review*, **102**, 1230 (1956). [Reprinted with permission from McCollum, Jr. DC, and Callaway J, *Physical Review Letters*, **9** (9), 376 (1962). Copyright 1962 by the American Physical Society.]

Magnetostatic Spin Waves (MSW) (A)

For very large wavelengths, the exchange interaction between spins no longer can be assumed to be dominant. In this limit, we need to look instead at the effect of dipole–dipole interactions (which dominate the exchange interactions) as well as external magnetic fields. In this case spin-wave excitations are still possible but they are called magnetostatic waves. Magnetostatic waves can be excited by inhomogeneous magnetic fields. MSW look like spin waves of very long wavelength, but the spin coupling is due to the dipole–dipole interaction. There are many device applications of MSW (e.g. delay lines) but a discussion of them would take us too far afield. See, e.g., Auld [7.3], and Ibach and Luth [7.33]. Also see Kittel [7.38 p471ff], and Walker [7.65]. There are also surface or Damon–Eshbach wave solutions.[18]

---

[18] Damon and Eshbach [7.17].

7.2.4 Band Ferromagnetism (B)

Despite the obvious lack of rigor, we have justified qualitatively a Heisenberg Hamiltonian for insulators and rare earths. But what can we do when we have ferromagnetism in metals? It seems to be necessary to take into account the band structure. This topic is very complicated, and only limited comments will be made here. See Mattis [7.48], Morrish [68] and Yosida [7.72] for more discussion.

In a metal, one might hope that the electrons in unfilled core levels would interact by the Heisenberg mechanism and thus produce ferromagnetism. We might expect that the conduction process would be due to electrons in a much higher band and that there would be little interaction between the ferromagnetic electrons and conduction electrons. This is not always the case. The core levels may give rise to a band that is so wide that the associated electrons must participate in the conduction process. Alternatively, the core levels may be very tightly bound and have very narrow bands. The core wave functions may interact so little that they could not directly have the Heisenberg exchange between them. That such materials may still be ferromagnetic indicates that other electrons such as the conduction electrons must play some role (we have discussed an example in Sect. 7.2.1 under *RKKY Interaction*). Obviously, a localized spin model cannot be good for all types of ferromagnetism. If it were, the saturation magnetization per atom would be an integral number of Bohr magnetons. This does not happen in Ni, Fe, and Co, where the number of electrons per atom contributing to magnetic effects is not an integer.

Despite the fact that one must use a band picture in describing the magnetic properties of metals, it still appears that a Heisenberg Hamiltonian often leads to predictions that are approximately experimentally verified. It is for this reason that many believe the Heisenberg Hamiltonian description of magnetic materials is much more general than the original derivation would suggest.

As an approach to a theory of ferromagnetism in metals it is worthwhile to present one very simple band theory of ferromagnetism. We will discuss *Stoner's theory*, which is also known as the *theory of collective electron ferromagnetism*. See Mattis [7.48 Vol. I p250ff] and Herring [7.56 p256ff]. The two basic assumptions of Stoner's theory are:

1. The ferromagnetic electrons or holes are free-electron-like (at least near the Fermi energy); hence their density of states has the form of a constant times $E^{1/2}$, and the energy is

$$E = \frac{\hbar^2 k^2}{2m^*}. \tag{7.197a}$$

2. There is still assumed to be some sort of exchange interaction between the (free) electrons. This interaction is assumed to be representable by a molecular field M. If γ is the molecular field constant, then the exchange interaction energy of the electrons is (SI)

$$E = \pm \mu_0 \gamma M \mu, \tag{7.197b}$$

where μ represents the magnetic moment of the electrons, + indicates electrons with spin parallel, and − indicates electrons with spin antiparallel to M.

The magnetization equals μ (here the magnitude of the magnetic moment of the electron = μ_B) times the magnitude of the number of parallel spin electrons per unit volume minus the number of antiparallel spin electrons per unit volume. Using the ideas of Sect. 3.2.2, we can write

$$M = \left| \mu \int [f(E - \mu_0 \gamma M \mu) - f(E + \mu_0 \gamma M \mu)] \frac{K\sqrt{E}}{2V} dE \right|, \qquad (7.198)$$

where f is the Fermi function. The above is the basic equation of Stoner's theory, with the sum of the parallel and antiparallel electrons being constant. For $T = 0$ and sufficiently strong exchange coupling the magnetization has as its saturation value $M = N\mu$. For sufficiently weak exchange coupling the magnetization vanishes. For intermediate values of the exchange coupling the magnetization has intermediate values. Deriving M as a function of temperature from the above equation is a little tedious. The essential result is that the Stoner theory also allows the possibility of a phase transition. The qualitative details of the M versus T curves do not differ enormously from the Stoner theory to the Weiss theory. We develop one version of the Stoner theory below.

The Hubbard Model and the Mean-Field Approximation (A)

So far, except for Pauli paramagnetism, we have not considered the possibility of nonlocalized electrons carrying a moment, which may contribute to the magnetization. Consistent with the above, starting with the ideas of Pauli paramagnetism and adding an exchange interaction leads us to the type of band ferromagnetism called the Stoner model. Stoner's model for band ferromagnetism is the nonlocalized mean field counterpart of Weiss' model for localized ferromagnetism. However, Stoner's model has neither the simplicity, nor the wide applicability of the Weiss approach.

Just as a mean-field approximation to the Heisenberg Hamiltonian gives us the Weiss model, there exists another Hamiltonian called the Hubbard Hamiltonian, whose mean-field approximation gives rise to a Stoner model. Also, just as the Heisenberg Hamiltonian gives good insight to the origin of the Weiss molecular field. So, the Hubbard model gives some physical insight concerning the exchange field for the Stoner model.

The Hubbard Hamiltonian as originally introduced was intended to bridge the gap between a localized and a mobile electron point of view. In general, in a suitable limit, it can describe either case. If one does not go to the limit, it can (in a sense) describe all cases in between. However, we will make a mean-field approximation and this displays the band properties most effectively.

One can give a derivation, of sorts, of the Hubbard Hamiltonian. However, so many assumptions are involved that it is often clearer just to write the Hamiltonian down as an assumption. This is what we will do, but even so, one cannot solve it exactly for cases that approach realism. Here we will solve it within the mean-field approximation, and get, as we have mentioned, the Stoner model of itinerant ferromagnetism.

7.2 Origin and Consequences of Magnetic Order

In a common representation, the Hubbard Hamiltonian is

$$\mathcal{H} = \sum_{k,\sigma} \varepsilon_k a^{\dagger}_{k\sigma} a_{k\sigma} + \frac{I}{2}\sum_{\alpha,\sigma} n_{\alpha\sigma} n_{\alpha,-\sigma}, \qquad (7.199)$$

where σ labels the spin (up or down), k labels the band energies, and α labels the lattice sites (we have assumed only one band—say an s-band—with ε_k being the band energy for wave vector k). The $a^{\dagger}_{k\sigma}$ and $a_{k\sigma}$ are creation and annihilation operators and I defines the interaction between electrons on the same site.

It is important to notice that the Hubbard Hamiltonian (as written above) assumes the electron–electron interactions are only large when the electrons are on the same site. A narrow band corresponds to localization of electrons. Thus, the Hubbard Hamiltonian is often said to be a narrow s-band model. The $n_{\alpha\sigma}$ are Wannier site-occupation numbers. The relation between band and Wannier (site localized) wave functions is given by the use of Fourier relations:

$$\psi_k = \frac{1}{\sqrt{N}} \sum_{R_\alpha} \exp(-i k \cdot R_\alpha) W(r - R_\alpha), \qquad (7.200a)$$

$$W(r - R_\alpha) = \frac{1}{\sqrt{N}} \sum_k \exp(i k \cdot R_\alpha) \psi_k(r). \qquad (7.200b)$$

Since the Bloch (or band) wave functions ψ_k are orthogonal, it is straightforward to show that the Wannier functions $W(r - R_\alpha)$ are also orthogonal. The Wannier functions $W(r - R_\alpha)$ are localized about site α and, at least for narrow bands, are well approximated by atomic wave functions.

Just as $a^{\dagger}_{k\sigma}$ creates an electron in the state ψ_k [with spin σ either $+$ or $\uparrow$ (up) or $-\downarrow$ (down)], so $c^{\dagger}_{\alpha\sigma}$ (the site creation operator) creates an electron in the state $W(r - R_\alpha)$, again with the spin either up or down. Thus, occupation number operators for the localized Wannier states are $n^{\dagger}_{\alpha\sigma} = c^{\dagger}_{\alpha\sigma} c_{\alpha\sigma}$ and consistent with (7.200a) the two sets of annihilation operators are related by the Fourier transform

$$a_{k\sigma} = \frac{1}{\sqrt{N}} \sum_{R_\alpha} \exp(i k \cdot R_\alpha) c_{\alpha\sigma}. \qquad (7.201)$$

Substituting this into the Hubbard Hamiltonian and defining

$$T_{\alpha\beta} = \frac{1}{N} \sum_k \varepsilon_k \exp[i k \cdot (R_\alpha - R_\beta)], \qquad (7.202)$$

we find

$$\mathcal{H} = \sum_{\alpha,\beta,\sigma} T_{\alpha\beta} c^{\dagger}_{\beta\sigma} c_{\alpha\sigma} + \frac{I}{2}\sum_{\alpha,\sigma} n^{+}_{\alpha\sigma} n_{\alpha-\sigma}. \qquad (7.203)$$

This is the most common form for the Hubbard Hamiltonian. It is often further assumed that $T_{\alpha\beta}$ is only nonzero when α and β are nearest neighbors. The first term then represents nearest-neighbor hopping.

Since the Hamiltonian is a many-electron Hamiltonian, it is not exactly solvable for a general lattice. We solve it in the mean-field approximation and thus replace

$$\frac{I}{2}\sum_{\alpha,\sigma} n_{\alpha\sigma} n_{\alpha,-\sigma},$$

with

$$I\sum_{\alpha,\sigma} n_{\alpha\sigma} \langle n_{\alpha,-\sigma}\rangle,$$

where $\langle n_{\alpha,-\sigma}\rangle$ is the thermal average of $n_{\alpha,-\sigma}$. We also assume $\langle n_{\alpha,-\sigma}\rangle$ is independent of site and so write it down as $n_{-\sigma}$ in (7.204).

Itinerant Ferromagnetism and the Stoner Model (B)

The mean-field approximation has been criticized on the basis that it builds in the possibility of an ordered ferromagnetic ground state regardless of whether the Hubbard Hamiltonian exact solution for a given lattice would predict this. Nevertheless, we continue, as we are more interested in the model we will eventually reach (the Stoner model) than in whether the theoretical underpinnings from the Hubbard model are physical. The mean-field approximation to the Hubbard model gives

$$\mathcal{H} = \sum_{\alpha,\beta,\sigma} T_{\alpha\beta} c^\dagger_{\beta\sigma} c_{\alpha\sigma} + I\sum_{\alpha,\sigma} n_{-\sigma} n_{\alpha\sigma}. \qquad (7.204)$$

Actually, in the mean-field approximation, the band picture is more convenient to use. Since we can show

$$\sum_\alpha n_{\alpha\sigma} = \sum_k n_{k\sigma},$$

the Hubbard model in the mean field can then be written as

$$\mathcal{H} = \sum_{k,\sigma}(\varepsilon_k + I n_{-\sigma}) n_{k\sigma}. \qquad (7.205)$$

The single-particle energies are given by

$$E_{k,\sigma} = \varepsilon_k + I n_{-\sigma}. \qquad (7.206)$$

The average number of electrons per site n is less than or equal to 2 and $n = n_+ + n_-$, while the magnetization per site n is $M = (n_+ - n_-)\mu_B$, where μ_B is the Bohr magneton.

Note: In order not to introduce another "–" sign, we will say "spin up" for now. This really means "moment up" or spin down, since the electron has a negative charge.

Note $n + (M/\mu_B) = 2n_+$ and $n - (M/\mu_B) = 2n_-$. Thus, up to an additive constant

$$E_{k\pm} = \varepsilon_k + I\left(\mp \frac{M}{2\mu_B}\right). \tag{7.207}$$

Note (7.207) is consistent with (7.197b). If we then define $H_{\text{eff}} = IM/2\mu_B^2$, we write the following basic equations for the Stoner model:

$$M = \mu_B(n_\uparrow - n_\downarrow), \tag{7.208}$$

$$E_{k,\sigma} = \varepsilon_k \mp \mu_B H_{\text{eff}}, \tag{7.209}$$

$$H_{\text{eff}} = \frac{IM}{2\mu_B^2}, \tag{7.210}$$

$$n_\sigma = \frac{1}{N}\sum_k \frac{1}{\exp[(E_{k\sigma} - M\mu)/kT]+1}, \tag{7.211}$$

$$n_\uparrow + n_\downarrow = n. \tag{7.212}$$

Although these equations are easy to write down, it is not easy to obtain simple convenient solutions from them. As already noted, the Stoner model contains two basic assumptions: (1) The electronic energy band in the metal is described by a known ε_k. By standard means, one can then derive a density of states. For free electrons, $N(E) \propto (E)^{1/2}$. (2) A molecular field approximately describes the effects of the interactions and we assume Fermi–Dirac statistics can be used for the spin-up and spin-down states. Much of the detail and even standard notation has been presented by Wohlfarth [7.69]. See also references to Stoner's work in the works by Wohlfarth.

The only consistent way to determine ε_k and, hence, $N(E)$ is to derive it from the Hubbard Hamiltonian. However, following the usual Stoner model we will just use an $N(E)$ for free electrons.

The maximum saturation magnetization (moment per site) is $M_0 = \mu_B n$ and the actual magnetization is $M = \mu_B(n_\uparrow - n_\downarrow)$. For the Stoner model, a relative magnetization is defined below:

$$\xi = \frac{M}{M_0} = \frac{n_\uparrow - n_\downarrow}{n}. \tag{7.213}$$

Using (7.212) and (7.213), we have

$$n_+ = n_\uparrow = (1+\xi)\frac{n}{2}, \tag{7.214a}$$

$$n_- = n_\downarrow = (1-\xi)\frac{n}{2}. \tag{7.214b}$$

It is also convenient to define a temperature θ', which measures the strength of the exchange interaction

$$k\theta'\xi = \mu_B H_{\text{eff}}. \tag{7.215}$$

We now suppose that the exchange energy is strong enough to cause an imbalance in the number of spin-up and spin-down electrons. We can picture the situation with constant Fermi energy $\mu = E_F$ (at $T = 0$) and a rigid shifting of the up N_+ and the down N_- density states as shown in Fig. 7.15.

The ↑ represents the "spin-up" (moment up actually) band and the ↓ the "spin-down" band. The shading represents states filled with electrons. The exchange energy causes the splitting of the two bands. We have pictured the density of states by a curve that goes to zero at the top and bottom of the band unlike a free-electron density of states that goes to zero only at the bottom.

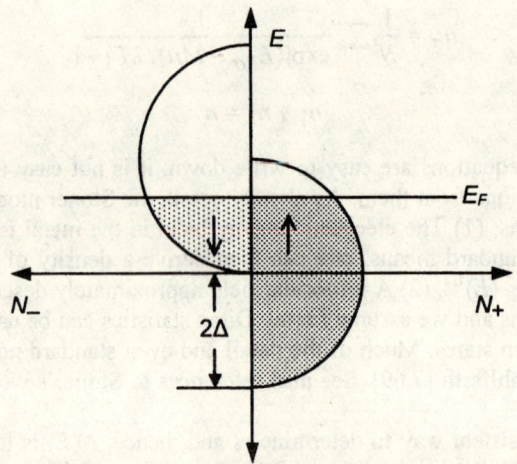

Fig. 7.15. Density states imbalanced by exchange energy

At $T = 0$, we have

$$n_+ = (1+\xi)\frac{n}{2} = \int_{\text{occ. states}} N_+(E)dE, \tag{7.216a}$$

$$n_- = (1-\xi)\frac{n}{2} = \int_{\text{occ. states}} N_-(E)dE. \tag{7.216b}$$

This can be easily worked out for free electrons if $E = 0$ at the bottom of both bands,

$$N_\pm(E) = \frac{1}{2}N_{\text{total}}(E) = \frac{1}{4\pi^2}\left(\frac{2m}{\hbar^2}\right)^{3/2}\sqrt{E} \equiv N(E). \tag{7.217}$$

We now derive conditions for which the magnetized state is stable at $T = 0$. If we just use a single-electron picture and add up the single-electron energies, we find, with the $(-)$ band shifted up by Δ and the $(+)$ band shifted down by Δ, for the energy per site

$$E = n_-\Delta + \int_0^{E_F^-} EN(E)dE - n_+\Delta + \int_0^{E_F^+} EN(E)dE \,.$$

The terms involving Δ are the exchange energy. We can rewrite it from (7.208), (7.213), and (7.215) as

$$-\frac{M}{\mu_B}\Delta = -nk\theta'\xi^2 \,.$$

However, just as in the Hartree–Fock analysis, this exchange term has double counted the interaction energies (once as a source of the field and once as interaction with the field). Putting in a factor of 1/2, we finally have for the total energy

$$E = \int_0^{E_F^+} EN(E)dE + \int_0^{E_F^-} EN(E)dE - \frac{1}{2}nk\theta'\xi^2 \,. \tag{7.218}$$

Differentiating $(d/d\xi)$ (7.216) and (7.218) and combining the results, we can show

$$\frac{1}{n}\frac{dE}{d\xi} = \frac{1}{2}(E_F^+ - E_F^-) - k\theta'\xi \,. \tag{7.219}$$

Differentiating (7.219) a second time and again using (7.216), we have

$$\frac{1}{n}\frac{d^2E}{d\xi^2} = \frac{n}{4}\left(\frac{1}{N(E_F^+)} + \frac{1}{N(E_F^-)}\right) - k\theta' \,. \tag{7.220}$$

Setting $dE/d\xi = 0$, just gives the result that we already know

$$2k\theta'\xi = (E_F^+ - E_F^-) = 2\mu_B H_{\text{eff}} = 2\Delta \,.$$

Note if $\xi = 0$ (paramagnetism) and $dE/d\xi = 0$, while $d^2E/d\xi^2 < 0$ the paramagnetism is unstable with respect to ferromagnetism. $\xi = 0$, $dE/d\xi = 0$ implies $E_F^+ = E_F^-$ and $N(E_F^-) = N(E_F^+) = N(E_F)$. So by (7.220) with $d^2E/d\xi^2 \leq 0$ we have

$$k\theta' \geq \frac{n}{2N(E_F)} \,. \tag{7.221}$$

For a parabolic band with $N(E) \propto E^{1/2}$, this implies

$$\frac{k\theta'}{E_F} \geq \frac{2}{3} \,. \tag{7.222}$$

We now calculate the relative magnetization (ξ_0) at absolute zero for a parabolic band where $N(E) = K(E)^{1/2}$ where K is a constant. From (7.216)

$$(1+\xi_0)\frac{n}{2} = \frac{2}{3}K(E_F^+)^{3/2},$$

$$(1-\xi_0)\frac{n}{2} = \frac{2}{3}K(E_F^-)^{3/2}.$$

Also

$$n = \frac{4}{3}KE_F^{3/2}.$$

Eliminating K and using $E_F^+ - E_F^- = 2k\theta'\xi_0$, we have

$$\frac{k\theta'}{E_F} = \frac{1}{2\xi_0}[(1+\xi_0)^{2/3} - (1-\xi_0)^{2/3}], \qquad (7.223)$$

which is valid for $0 \leq \xi_0 \leq 1$. The maximum ξ_0 can be is 1 for which $k\theta'/E_F = 2^{-1/3}$, and at the threshold for ferromagnetism ξ_0 is 0. So, $k\theta'/E_F = 2/3$ as already predicted by the Stoner criterion.

Summary of Results at Absolute Zero
We have three ranges:

$$\frac{k\theta'}{E_F} < \frac{2}{3} = 0.667 \quad \text{and} \quad \xi_0 = \frac{M}{n\mu_B} = 0,$$

$$\frac{2}{3} < \frac{k\theta'}{E_F} < \frac{1}{2^{1/3}} = 0.794, \quad 0 < \xi_0 = \frac{M}{n\mu_B} < 1,$$

$$\frac{k\theta'}{E_F} > \frac{1}{2^{1/3}} \quad \text{and} \quad \xi_0 = \frac{M}{n\mu_B} = 1.$$

The middle range, where $0 < \xi_0 < 1$ is special to Stoner ferromagnetism and not to be found in the Weiss theory. This middle range is called "unstructured" or "weak" ferromagnetism. It corresponds to having electrons in both ↑ and ↓ bands. For very low, but not zero, temperatures, one can show for weak ferromagnetism that

$$M = M_0 - CT^2, \qquad (7.224)$$

where C is a constant. This is particularly easy to show for very weak ferromagnetism, where $\xi_0 \ll 1$ and is left as an exercise for the reader.

7.2 Origin and Consequences of Magnetic Order

We now discuss the case of strong ferromagnetism where $k\theta'/E_F > 2^{-1/3}$. For this case, $\zeta_0 = 1$, and $n_\uparrow = n$, $n_\downarrow = 0$. There is now a gap E_g between E_F^+ and the bottom of the spin-down band. For this case, by considering thermal excitations to the $n_\downarrow$ band, one can show at low temperature that

$$M = M_0 - K''T^{3/2}\exp(-E_g/kT), \qquad (7.225)$$

where K'' is a constant. However, spin-wave theory says $M = M_0 - C'T^{3/2}$, where C' is a constant, which agrees with low-temperature experiments. So, at best, (7.225) is part of a correction to low-temperature spin-wave theory.

Within the context of the Stoner model, we also need to talk about exchange enhancement of the paramagnetic susceptibility χ_P (*gaussian units with $\mu_0 = 1$*)

$$M = \chi_P B_{\text{eff}}^{\text{Total}}, \qquad (7.226)$$

where M is the magnetization and χ_P the Pauli susceptibility, which for low temperatures, has a very small αT^2 term. It can be written

$$\chi_P = 2\mu_B^2 N(E_F)(1+\alpha T^2), \qquad (7.227)$$

where $N(E)$ is the density of states for one subband. Since

$$B_{\text{eff}}^{\text{Total}} = H_{\text{eff}} + B = \gamma B + B,$$

it is easy to show that (gaussian with $B = H$)

$$\chi = \frac{M}{B} = \frac{\chi_P}{1 - \gamma\chi_P}, \qquad (7.228)$$

where $1/(1-\gamma\chi_P)$ is the exchange enhancement factor.

We can recover the Stoner criteria from this at $T = 0$ by noting that paramagnetism is unstable if

$$\chi_P^0 \gamma \geq 1. \qquad (7.229)$$

By using $\gamma = k\theta'/n\mu_B^2$ and $\chi_P^0 = 2\mu_B^2 N(E_F)$, (7.229) just gives the Stoner criteria. At finite, but low temperatures where $(\alpha = -|a|)$

$$\chi_P = \chi_P^0(1 - |a|T^2),$$

if we define

$$\theta^2 = \frac{\gamma\chi_P^0 - 1}{\gamma\chi_P^0 |a|},$$

and suppose $|a|T^2 \ll 1$, it is easy to show

$$\chi = \frac{1}{\gamma|a|}\frac{1}{T^2 - \theta^2}.$$

Thus, as long as $T \cong \theta$, we have a Curie–Weiss-like law:

$$\chi = \frac{1}{2\theta\gamma|a|} \frac{1}{T-\theta}. \tag{7.230}$$

At very high temperatures, one can also show that an ordinary Curie–Weiss-like law is obtained:

$$\chi = \frac{n\mu_B^2}{k} \frac{1}{T-\theta}. \tag{7.231}$$

Summary Comments About the Stoner Model

1. The low-temperature results need to be augmented with spin waves. Although in this book we only derive the results of spin waves for the localized model, it turns out that spin waves can also be derived within the context of the itinerant electron model.
2. Results near the Curie temperature are never qualitatively good in a mean-field approximation because the mean-field approximation does not properly treat fluctuations.
3. The Stoner model gives a simple explanation of why one can have a fractional number of electrons contributing to the magnetization (the case of weak ferromagnetism where $\zeta_0 = M_{T=0}/n\mu_B$ is between 0 and 1).
4. To apply these results to real materials, one usually needs to consider that there are overlapping bands (e.g. both s and d bands), and not all bands necessarily split into subbands. However, the Stoner model does seem to work for $ZrZn_2$.

7.2.5 Magnetic Phase Transitions (A)

Simple ideas about spin waves break down as T_c is approached. We indicate here one way of viewing magnetic phenomena near the $T = T_c$ region. In this Section we will discuss magnetic phase transitions in which the magnetization (for ferromagnets with $H = 0$) goes continuously to zero as the critical temperature is approached from below. Thus at the critical temperature (Curie temperature for a ferromagnet) the ordered (ferromagnetic) phase goes over to the disordered (paramagnetic) phase. This "smooth" transition from one phase (or more than one phase in more general cases) to another is characteristic of the behavior of many substances near their critical temperature. In such continuous phase transitions there is no latent heat and these phase transitions are called second-order phase transitions. All second-order phase transitions show many similarities. We shall consider only phase transitions in which there is no latent heat.

No complete explanation of the equilibrium properties of ferromagnets near the magnetic critical temperature (T_c) has yet been given, although the renormalization technique, referred to later, comes close. At temperatures well below T_c we know that the method of spin waves often yields good results for describing the magnetic behavior of the system. We know that high-temperature expansions of the partition function yield good results. The Green function method provides results for interesting physical quantities at all temperatures. However, the Green function results (in a usable approximation) are not valid near T_c. Two methods (which are not as straightforward as one might like) have been used. These are the use of scaling laws[19] and the use of the Padé approximant.[20] These methods often appear to give good quantitative results without offering much in the way of qualitative insight. Therefore we will not discuss them here. The renormalization group, referenced later, in some ways is a generalization of scaling laws. It seems to offer the most in the way of understanding.

Since the region of lack of knowledge (around the phase transition) is only near $\tau = 1$ ($\tau = T/T_c$, where T_c is the critical temperature) we could forget about the region entirely (perhaps) if it were not for the fact that very unusual and surprising results happen here. These results have to do with the behavior of the various quantities as a function of temperature. For example, the Weiss theory predicts for the (zero field) magnetization that $M \propto (T_c - T)^{+1/2}$ as $T \to T_c^-$ (the minus sign means that we approach T_c from below), but experiment often seems to agree better with $M \propto (T_c - T)^{+1/3}$. Similarly, the Weiss theory predicts for $T > T_c$ that the zero-field susceptibility behaves as $\chi \propto (T - T_c)^{-1}$, whereas experiment for many materials agrees with $\chi \propto (T - T_c)^{-4/3}$ as $T \to T_c^+$. In fact, the Weiss theory fails very seriously above T_c because it leaves out the short-range ordering of the spins. Thus it predicts that the (magnetic contribution to the) specific heat should vanish above T_c, whereas the zero-field magnetic specific heat does not so vanish. Using an improved theory that puts in some short-range order above T_c modifies the specific heat somewhat, but even these improved theories [92] do not fit experiment well near T_c. Experiment appears to suggest (although this is not settled yet) that for many materials $C \cong \ln |(T - T_c)|$ as $T \to T_c^+$ (the exact solution of the specific heat of the two-dimensional Ising ferromagnet shows this type of divergence), and the concept of short-range order is just not enough to account for this logarithmic or near logarithmic divergence. Something must be missing. It appears that the missing concept that is needed to correctly predict the "critical exponents" and/or "critical divergences" is the concept of (anomalous) fluctuations. [The exponents $1/3$ and $4/3$ above are critical exponents, and it is possible to set up the formalism in such a way that the logarithmic divergence is consistent with a certain critical exponent being zero.] Fluctuations away from the thermodynamic equilibrium appear to play a very dominant role in the behavior of thermodynamic functions near the phase transition. Critical-point behavior is discussed in more detail in the next section.

Additional insight into this behavior is given by the Landau theory.[19] The Landau theory appears to be qualitatively correct but it does not predict correctly the critical exponents.

---

[19] See Kadanoff et al [7.35].
[20] See Patterson et al [7.54] and references cited therein.

Critical Exponents and Failures of Mean-Field Theory (B)

Although mean-field theory has been extraordinarily useful and in fact, is still the "workhorse" of theories of magnetism (as well as theories of the thermodynamics behavior of other types of systems that show phase transitions), it does suffer from several problems. Some of these problems have become better understood in recent years through studies of critical phenomena, particularly in magnetic materials, although the studies of "critical exponents" relates to a much broader set of materials than just magnets as referred to above. It is helpful now to define some quantities and to introduce some concepts.

A sensitive test of mean-field theory is in predicting critical exponents, which define the nature of the singularities of thermodynamic variables at critical points of second-order phase transitions. For example,

$$\phi \sim \left|\frac{T_c - T}{T_c}\right|^{\beta} \quad \text{and} \quad \xi = \left|\frac{T_c - T}{T_c}\right|^{-\nu},$$

for $T < T_c$, where β, ν are critical exponents, ϕ is the order parameter, which for ferromagnets is the average magnetization M and ξ is the correlation length. In magnetic systems, the correlation length measures the characteristic length over which the spins are ordered, and we note that it diverges as the Curie temperature T_c is approached. In general, the order parameter ϕ is just some quantity whose value changes from disordered phases (where it may be zero) to ordered phases (where it is nonzero). Note for ferromagnets that ϕ is zero in the disordered paramagnetic phase and nonzero in the ordered ferromagnetic situation.

Mean-field theory can be quite good above an upper critical (spatial) dimension where by definition it gives the correct value of the critical exponents. Below the upper critical dimension (UCD), thermodynamic fluctuations become very important, and mean-field theory has problems. In particular, it gives incorrect critical exponents. There also exists a lower critical dimension (LCD) for which these fluctuations become so important that the system does not even order (by definition of the LCD). Here, mean-field theory can give qualitatively incorrect results by predicting the existence of an ordered phase. The lower critical dimension is the largest dimension for which long-range order is not possible. In connection with these ideas, the notion of a universality class has also been recognized. Systems with the same spatial dimension d and the same dimension of the order parameter D are usually in the same universality class. Range and symmetry of the interaction potential can also play a role in determining the universality class. Quite dissimilar systems in the same universality class will, by definition, exhibit the same critical exponents. Of course, the order parameter itself as well as the critical temperature T_c, may be quite different for systems in the same universality class. In this connection, one also needs to discuss concepts like the renormalization group, but this would take us too far afield. Reference can be made to excellent statistical mechanics books like the one by Huang.[21]

---

[21] See Huang [7.32, p441ff].

Critical exponents for magnetic systems have been defined in the following way. First, we define a dimensionless temperature that is small when we are near the critical temperature.

$$t = (T - T_c)/T_c.$$

We assume $B = 0$ and define critical exponents by the behavior of physical quantities such as M:

Magnetization (order parameter): $M \sim |t|^\beta$.

Magnetic susceptibility: $\chi \sim |t|^{-\gamma}$.

Specific heat: $C \sim |t|^{-\alpha}$.

There are other critical exponents, such as the one for correlation length (as noted above), but this is all we wish to consider here. Similar critical exponents are defined for other systems, such as fluid systems. When proper analogies are made, if one stays within the same universality class, the critical exponents have the same value. Under rather general conditions, several inequalities have been derived for critical exponents. For example, the Rushbrooke inequality is

$$\alpha + 2\beta + \gamma \geq 2.$$

It has been proposed that this relation also holds as an equality. For mean-field theory $\alpha = 0$, $\beta = \frac{1}{2}$, and $\gamma = 1$. Thus, the Rushbrooke relation is satisfied as an equality. However, except for α being zero, the critical exponents are wrong. For ferromagnets belonging to the most common universality class, experiment, as well as better calculations than mean field, suggest, as we have mentioned (Sect. 7.2.5), $\beta = \frac{1}{3}$, and $\gamma = \frac{4}{3}$. Note that the Rushbrooke equality is still satisfied with $\alpha = 0$. The most basic problem mean-field theory has is that it just does not properly treat fluctuations nor does it properly treat a related aspect concerning short-range order. It must include these for agreement with experiment. As already indicated, short-range correlation gives a tail on the specific heat above T_c, while the mean-field approximation gives none.

The mean-field approximation also fails as $T \rightarrow 0$ as we have discussed. An elementary calculation from the properties of the Brillouin function shows that ($s = 1/2$)

$$M = M_0[1 - 2\exp(-2T_c/T)],$$

whereas for typical ferromagnets, experiment agrees better with

$$M = M_0(1 - aT^{3/2}).$$

As we have discussed, this dependence on temperature can be derived from spin wave theory.

Although considerable calculation progress has been made by high-temperature series expansions plus Padé Approximants, by scaling, and renormalization group arguments, most of this is beyond the scope of this book. Again,

Huang's excellent text can be consulted.[21] Tables 7.2 and 7.3 summarize some of the results.

Table 7.2. Summary of mean-field theory

Failures	Successes
Neglects spin-wave excitations near absolute zero.	Often used to predict the type of magnetic structure to be expected above the lower critical dimension (ferromagnetism, ferrimagnetism, antiferromagnetism, heliomagnetism, etc.).
Near the critical temperature, it does not give proper critical exponents if it is below the upper critical dimension.	
May predict a phase transition where there is none if below the lower critical dimension. For example, a one-dimension isotropic Heisenberg magnet would be predicted to order at a finite temperature, which it does not.	Predicts a phase transition, which certainly will occur if above the lower critical dimension.
	Gives at least a qualitative estimate of the values of thermodynamic quantities, as well as the critical exponents – when used appropriately.
	Serves as the basis for improved calculations.
Predicts no tail in the specific heat for typical magnets.	The higher the spatial dimension, the better it is.

Table 7.3. Critical exponents (calculated)

	α	β	γ
Mean field	0	0.5	1
Ising (3D)	0.11	0.32	1.24
Heisenberg (3D)	–0.12	0.36	1.39

Adapted with permission from Chaikin PM and Lubensky TC, *Principles of Condensed Matter Physics*, Cambridge University Press, 1995, p. 231.

Two-Dimensional Structures (A)

Lower-dimensional structures are no longer of purely theoretical interest. One way to realize two dimensions is with thin films. Suppose the thin film is of thickness t and suppose the correlation length of the quantity of interest is c. When the thickness is much less than the correlation length ($t \ll c$), the film will behave two dimensionally and when $t \gg c$ the film will behave as a bulk three-dimensional material. If there is a critical point, since c grows without bound as the critical point is approached, a thin film will behave two-dimensionally near the two-dimensional critical point. Another way to have two-dimensional behavior is in layered magnetic materials in which the coupling between magnetic layers, of spacing d, is weak. Then when $c \ll d$, all coupling between the layers can be neglected and one sees 2D behavior, whereas if $c \gg d$, then interlayer coupling can no longer be neglected. This means with magnetic layers, a two-dimensional critical point will be modified by 3D behavior near the critical temperature.

In this chapter we are mainly concerned with materials for which the three-dimensional isotropic systems are a fairly good or at least qualitative model. However, it is interesting that two-dimensional isotropic Heisenberg systems can be shown to have no spontaneous (sublattice – for antiferromagnets) magnetization [7.49]. On the other hand, it can be shown [7.26] that the highly anisotropic two-dimensional Ising ferromagnet (defined by the Hamiltonian $\mathcal{H} \propto \sum_{ij(\text{nn.})} \sigma_i^z \sigma_j^z$, where the σs refer to Pauli spin matrices, the i and j refer to lattice sites) *must* show spontaneous magnetization.

We have just mentioned the two-dimensional Heisenberg model in connection with the Mermin–Wagner theorem. The planar Heisenberg model is in some ways even more interesting. It serves as a model for superfluid helium films and predicts the long-range order is destroyed by formation of vortices [7.40].

Another common way to produce two-dimensional behavior is in an electronic inversion layer in a semiconductor. This is important in semiconductor devices.

Spontaneously Broken Symmetry (A)

A Heisenberg Hamiltonian is invariant under rotations, so the ensemble average of the magnetization is zero. For every M there is a $-M$ of the same energy. Physically this answer is not correct since magnets do magnetize. The symmetry is spontaneously broken when the ground state does not have the same symmetry as the Hamiltonian, The symmetry is recovered by having degenerate ground states whose totality recovers the rotational symmetry. Once the magnet magnetizes, however, it does not go to another degenerate state because all the magnets would have to rotate spontaneously by the same amount. The probability for this to happen is negligible for a realistic system. Quantum mechanically in the infinite limit, each ground state generates a separate Hilbert space and transitions between them are forbidden—a super selection rule. Because of the symmetry there are excited states that are wave-like in the sense that the local ground state changes slowly over space (as in a wave). These are the Goldstone excitations and they are orthogonal to any ground state. Actually each of the (infinite) number of ground states is orthogonal to each other: The concept of spontaneously broken symmetry is much more general than just for magnets. For ferromagnets the rotational symmetry is broken and spin waves or magnons appear. Other examples include crystals (translation symmetry is broken and phonons appear), and superconductors (local gauge symmetry is broken and a Higgs mode appears—this is related to the Meissner effect – see Chap. 8).[22]

---

[22] See Weinberg [7.67].

7.3 Magnetic Domains and Magnetic Materials (B)

7.3.1 Origin of Domains and General Comments[23] (B)

Because of their great practical importance, a short discussion of domains is merited even though we are primarily interested in what happens in a single domain.

We want to address the following questions: What are the domains? Why do they form? Why are they important? What are domain walls? How can we analyze the structure of domains, and domain walls? Is there more than one kind of domain wall?

Magnetic domains are small regions in which the atomic magnetic moments are lined up. For a given temperature, the magnetization is saturated in a single domain, but ferromagnets are normally divided into regions with different domains magnetized in different directions.

When a ferromagnet splits into domains, it does so in order to lower its free energy. However, the free energy and the internal energy differ by TS and if T is well below the Curie temperature, TS is small since also the entropy S is small because the order is high. Here we will neglect the difference between the internal energy and the free energy. There are several contributions to the internal energy that we will discuss presently.

Magnetic domains can explain why the overall magnetization can vanish even if we are well below the Curie temperature T_c. In a single domain the M vs. T curve looks somewhat like Fig. 7.16.

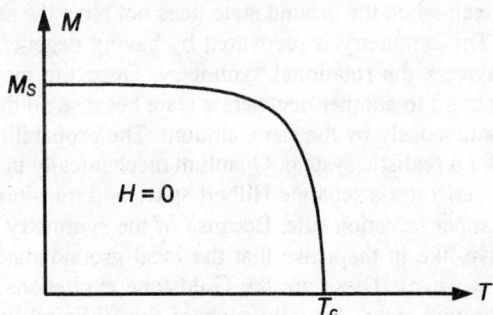

Fig. 7.16. M vs. T curve for a single magnetic domain

For reference, the Curie temperature of iron is 1043 K and its saturation magnetization M_S is 1707 G. But when there are several domains, they can point in different directions so the overall magnetization can attain any value from zero up to saturation magnetization. In a magnetic field, the domains can change in size (with those that are energetically preferred growing). Thus the phenomena of hysteresis, which we sketch in Fig. 7.17 starting from the ideal demagnetized state, can be understood (see Section *Hysteresis, Remanence, and Coercive Force*).

---

[23] More details can be found in Morrish [68] and Chikazumi [7.11].

7.3 Magnetic Domains and Magnetic Materials (B)   421

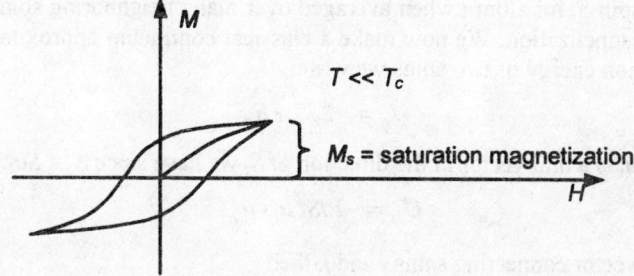

Fig. 7.17. M vs. H curve showing magnetic hysteresis

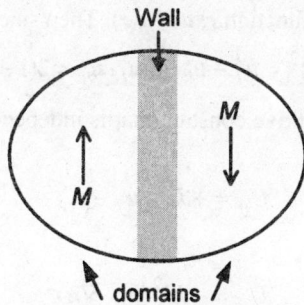

Fig. 7.18. Two magnetic regions (domains) separated by a *domain wall*, where size is exaggerated

In order for some domains to grow at the expense of others, the domain walls separating the two regions must move. Domain walls are transition regions that separate adjacent regions magnetized in different directions. The idea is shown in Fig. 7.18.

We now want to analyze the four types of energy involved in domain formation. We consider (1) exchange energy, (2) magnetostatic energy, (3) anisotropy energy, and (4) magnetostrictive energy. Domain structures with the lower sum of these energies are the most stable.

Exchange Energy (B)

We have seen (see Section *The Heisenberg Hamiltonian and its Relationship to the Weiss Mean-Field Theory*) that quantum mechanics indicates that there may be an interaction energy between atomic spins S_i that is proportional to the scalar product of the spins. From this, one obtains the Heisenberg Hamiltonian describing the interaction energy. Assuming J is the proportionality constant (called the exchange integral) and that only nearest-neighbor (nn) interactions need be considered, the Heisenberg Hamiltonian becomes

$$\mathcal{H} = -J \sum_{\substack{i,j \\ (nn)}} \mathbf{S}_i \cdot \mathbf{S}_j , \qquad (7.232)$$

where the spin S_i for atom i when averaged over many neighboring spins gives us the local magnetization. We now make a classical continuum approximation. For the interaction energy of two spins we write:

$$U_{ij} = -2JS_i \cdot S_j. \qquad (7.233)$$

Assuming u_i is a unit vector in the direction of S_i we have since $S_i = Su_i$:

$$U_{ij} = -2JS^2 u_i \cdot u_j. \qquad (7.234)$$

If r_{ji} is the vector connecting spins i and j, then

$$u_j = u_i + r_{ji} \cdot (\nabla u)_i, \qquad (7.235)$$

treating u as a continuous function r, $u = u(r)$. Then since

$$(u_j - u_i)^2 = u_j^2 + u_i^2 - 2u_i \cdot u_j = 2(1 - u_i \cdot u_j), \qquad (7.236)$$

we have, neglecting an additive constant that is independent of the directions of u_i and u_j,

$$U_{ij} = +JS^2 (u_j - u_i)^2.$$

So

$$U_{ij} = +JS^2 (r_{ji} \cdot \nabla u)^2. \qquad (7.237)$$

Thus the total interaction energy is

$$U = \tfrac{1}{2} \sum U_{ij} = \frac{JS^2}{2} \sum_{i,j} (r_{ji} \cdot \nabla u)^2, \qquad (7.238)$$

where we have inserted a 1/2 so as not to count bonds twice. If

$$u = \alpha_1 i + \alpha_2 j + \alpha_3 k,$$

where the α_i are the direction cosines, for $r_{ji} = ai$, for example:

$$\sum_{\pm ai} (r_{ji} \cdot \nabla u)^2 = 2a^2 \left(\frac{\partial \alpha_1}{\partial x} i + \frac{\partial \alpha_2}{\partial x} j + \frac{\partial \alpha_3}{\partial x} k \right)^2$$
$$= 2a^2 \left[\left(\frac{\partial \alpha_1}{\partial x} \right)^2 + \left(\frac{\partial \alpha_2}{\partial x} \right)^2 + \left(\frac{\partial \alpha_3}{\partial x} \right)^2 \right]. \qquad (7.239)$$

For a simple cubic lattice where we must also include neighbors at $r_{ji} = \pm aj$ and $\pm ak$, we have:[24]

$$U = \frac{JS^2}{a} \sum_{i \text{ (all spins)}} [(\nabla \alpha_1)^2 + (\nabla \alpha_2)^2 + (\nabla \alpha_3)^2]_i a^3, \qquad (7.240)$$

---

[24] An alternative derivation is based on writing $U \propto \sum \mu_i B_i$, where μ_i is the magnetic moment $\propto S_i$ and B_i is the effective exchange field $\propto \sum_{j(nn)} J_{ij} S_j$, treating the S_j in a continuum spatial approximation and expanding S_j in a Taylor series ($S_j = S_i + a \partial S_i / \partial x +$ etc. to 2nd order). See (7.275) and following.

or in the continuum approximation:

$$U = \frac{JS^2}{a} \int [(\nabla \alpha_1)^2 + (\nabla \alpha_2)^2 + (\nabla \alpha_3)^2] dV . \qquad (7.241)$$

For variation of M only in the y direction, and using spherical coordinates r, θ, φ, a little algebra shows that ($M = M(r, \theta, \varphi)$)

$$\frac{\text{Energy}}{\text{Volume}} = A \left\{ \left(\frac{\partial \theta}{\partial y} \right)^2 + \sin^2 \theta \left(\frac{\partial \varphi}{\partial y} \right)^2 \right\}, \qquad (7.242)$$

where $A = JS^2/a$ and has the following values for other cubic structures ($A_{fcc} = 4A$, and $A_{bcc} = 2A$). We have treated the exchange energy first because it is this interaction that causes the material to magnetize.

Magnetostatic Energy (B)

We have already discussed magnetostatics in Sect. 7.2.2. Here we want to mention that along with the exchange interaction it is one of the two primary interactions of interest in magnetism. It is the driving mechanism for the formation of domains. Also, at very long wavelengths, as we have mentioned, it can be the causative factor in spin-wave motion (magnetostatic spin waves). A review of magnetostatic fields of relevance for applications is given by Bertram [7.6].

Anisotropy (B)

Because of various energy-coupling mechanisms, certain magnetic directions are favored over others. As discussed in Sect. 7.2.2, the physical origin of crystalline anisotropy is a rather complicated subject. As discussed there, a partial understanding, in some materials, relates it to spin-orbit coupling in which the orbital motion is coupled to the lattice. Anisotropy can also be caused by the shape of the sample or the stress it is subjected to, but these two types are not called crystalline anisotropy. Regardless of the physical origin, a ferromagnetic material will have preferred (least energy) directions of magnetization. For uniaxial symmetry, we can write

$$H_{anis} = -D_a \sum_i (\mathbf{k} \cdot \mathbf{S}_i)^2 , \qquad (7.243)$$

where $\mathbf{k}$ is the unit vector along the axis of symmetry. If we let $K_1 = D_a S^2/a^3$, where a is the atom–atom spacing, then since $\sin^2\theta = 1 - \cos^2\theta$ and neglecting unimportant additive terms, the anisotropy energy per unit volume is

$$u_{anis} = K_1 \sin^2 \theta . \qquad (7.244)$$

Also, for proper choice of K_1, this may describe hexagonal crystals, e.g. cobalt (hcp) where θ is the angle between M and the hexagonal axis. Figure 7.19 shows some data related to anisotropy. Note Fe with a bcc structure has easy directions in $\langle 100 \rangle$ and Ni with fcc has easy directions in $\langle 111 \rangle$.

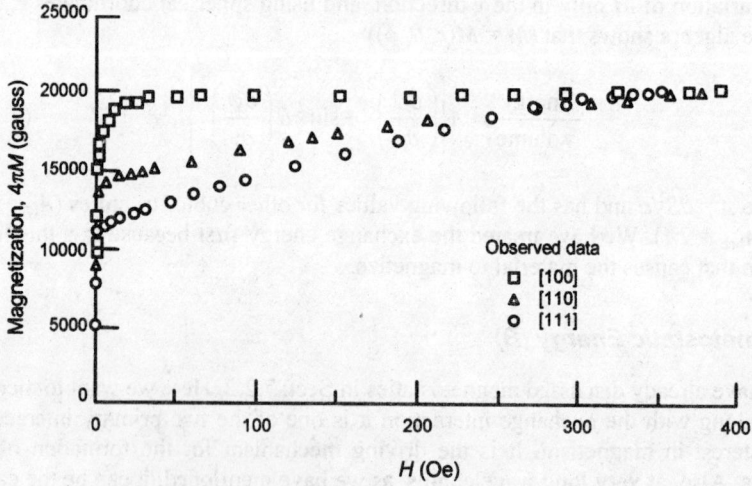

Fig. 7.19. Magnetization curves showing anisotropy for single crystals of iron with 3.85% silicon [Reprinted with permission from Williams HJ, *Phys Rev* **52**, 1 (1937). Copyright 1937 by the American Physical Society.]

Wall Energy (B)

The wall energy is an additive combination of exchange and anisotropy energy, which are independent. Exchange favors parallel moments and a wide wall. Anisotropy prefers moments along an easy direction and a narrow wall. Minimizing the sum of the two determines the width of the wall. Consider a uniaxial ferromagnet with the magnetization varying only in the y direction. If the energy per unit volume is (using spherical coordinates, see, e.g., (7.242) and Fig. 7.25)

$$w = A\left[\left(\frac{\partial \theta}{\partial y}\right)^2 + \left(\sin\theta \frac{\partial \varphi}{\partial y}\right)^2\right] + K_1 \sin^2\theta, \tag{7.245}$$

where

$$A = \alpha_1 \frac{JS^2}{a} \quad \text{and} \quad K_1 = \kappa_1 \frac{D_a S^2}{a^3}, \tag{7.246}$$

and α_1, κ_1 differ for different crystal structures, but both are approximately unity. For simplicity in what follows we will set α_1 and κ_1 equal to one.

Using $\delta \int w\, dy = 0$ we get two Euler–Lagrange equations. Inserting (7.245) in the Euler–Lagrange equations, we get the results indicated by the arrows.

$$\frac{\partial w}{\partial \theta} - \frac{d}{dy}\frac{\partial w}{\partial \frac{\partial \theta}{\partial y}} = 0 \quad \rightarrow \quad \frac{d}{d\theta} K_1 \sin^2\theta = 2A \frac{d}{dy}\frac{\partial \theta}{\partial y}, \qquad (7.247)$$

$$\frac{\partial w}{\partial \varphi} - \frac{d}{dy}\frac{\partial w}{\partial \frac{\partial \varphi}{\partial y}} = 0 \quad \rightarrow \quad 2\frac{d}{dy}\left(\sin^2\theta \frac{\partial \varphi}{\partial y}\right) = 0. \qquad (7.248)$$

For Bloch walls by definition, $\varphi = 0$, which is a possible solution. The first equation (7.247) has a first integral of

$$\sqrt{\frac{A}{K_1}}\frac{d\theta}{dy} = \sin\theta, \qquad (7.249)$$

which integrates in turn to

$$\theta = 2\arctan(e^{y/\Delta_0}), \quad \Delta_0 = \sqrt{\frac{A}{K_1}}. \qquad (7.250)$$

The effective wall width is obtained by approximating $d\theta/dy$ by its value at the midpoint of the wall, where $\theta = \pi/2$.

$$\frac{d\theta}{dy} = \sqrt{\frac{K_1}{A}} \cong \frac{1}{a}\sqrt{\frac{D}{J}}, \qquad (7.251)$$

so the wall width/a is

$$\frac{\text{wall width}}{a} = \pi\sqrt{\frac{D}{J}}.$$

One can also show the wall width per unit area (perpendicular to the y-axis in Fig. 7.25) is $4(AK_1)^{1/2}$. For Iron, the wall energy per unit area is of order 1 erg/cm^2, and the wall width is of order 500 Å.

Magnetostrictive Energy (B)

Magnetostriction is the variation of size of a magnetic material when its magnetization varies. Magnetostriction implies a coupling between elastic and magnetic effects caused by the interaction of atomic magnetic moments and the lattice. The magnetostrictive coefficient λ is $\delta l/l$, where δl is the change in length associated with the magnetization change. In general λ can be either sign and is typically of the order of 10^{-5} or so. There may also be a change in volume due to changing magnetization. In any case the deformation is caused by a lowering of the energy.

Magnetostriction is a very complex matter and a detailed description is really outside the scope of this book. We needed to mention it because it has a bearing on domains. See, e.g., Gibbs [7.24].

Formation of Magnetic Domains (B)

We now give a qualitative account of the formation of domains. Consider a cubic material, originally magnetized along an easy direction as shown in Fig. 7.20. Because the magnetization M and demagnetizing fields have opposite directions (7.136), this configuration has large magnetostatic energy. The magnetostatic energy can be reduced if the material splits into domains as shown in Fig. 7.21

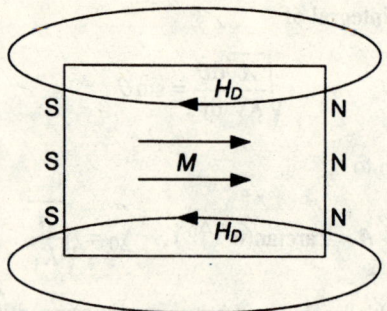

Fig. 7.20. Magnetic domain formation within a material

Since the density of surface poles is $+M \cdot n$ where n_M is the *outward* normal, at an interface the net *magnetic charge* per unit area is

$$(M_2 - M_1) \cdot n_{M_2},$$

where n_{M_2} is a unit vector pointing from region 1 to region 2. Thus when $M \cdot n$ is continuous, there are no demagnetizing fields (assuming also M is uniform in the interior). Thus (for typical magnetic materials with cubic symmetry) the magnetostatic energy can be further reduced by forming domains of closure, as shown in Fig. 7.22. The overall magnetostrictive and strain energy can be reduced by the formation of more domains of closure (see Fig. 7.23). That is, this splitting into smaller domains reduces the extra energy caused by the internal strain brought about by the spontaneous strain in the direction of magnetization. This process will not continue forever because of the increase in the wall energy (due to exchange and anisotropy). An actual material will of course have many imperfections as well as other complications that will cause irregularities in the domain structure.

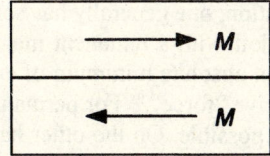

Fig. 7.21. Magnetic-domain splitting within a material

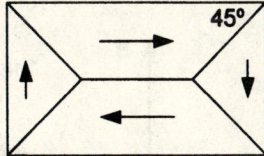

Fig. 7.22. Formation of magnetic domains of closure

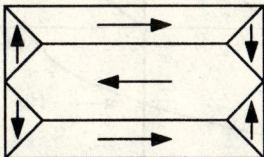

Fig. 7.23. Formation of more magnetic domains of closure

Hysteresis, Remanence, and Coercive Force (B)

Consider an unmagnetized ferromagnet well below its Curie temperature. We can understand the material being unmagnetized if it consists of a large number of domains, each of which is spontaneously magnetized, but that have different directions of magnetization so the net magnetization averages to zero.

The magnetization changes from one domain to another through thin but finite-width domain walls. Typically, domain walls are of thickness of about 10^{-7} meters or some hundreds of atomic spacings, while the sides of the domains are a few micrometers and larger.

The hysteresis loop can be visualized by plotting M vs. H or $B = \mu_0(H + M)$ (in SI) = $H + 4\pi M$ (in Gaussian units) (see Fig. 7.24). The *virgin curve* is obtained by starting in an ideal demagnetized state in which one is at the absolute minimum of energy.

When an external field is turned on, "favorable" domains have lower energy than "unfavorable" ones, and thus the favorable ones grow at the expense of the unfavorable ones.

Imperfections determine the properties of the hysteresis loop. Moving a domain wall generally increases the energy of a ferromagnetic material due to a complex combination of interactions of the domain wall with dislocations, grain boundaries, or other kinds of defects. Generally the first part of the virgin curve is reversible, but as the walls sweep past defects one enters an irreversible region, then in the final approach to saturation, one generally has some rotation of domains. As H is reduced to zero, one is left with a remanent magnetization (in a metastable state with a "local" rather than absolute minimum of energy) at $H = 0$ and B only goes to zero at $-H_c$, the coercive "force".[25] For permanent magnetic materials, M_R and H_c should be as large as possible. On the other hand, soft magnets will have very low coercivity. The hysteresis and domain properties of magnetic materials are of vast technological importance, but a detailed discussion would take us too far afield. See Cullity [7.16].

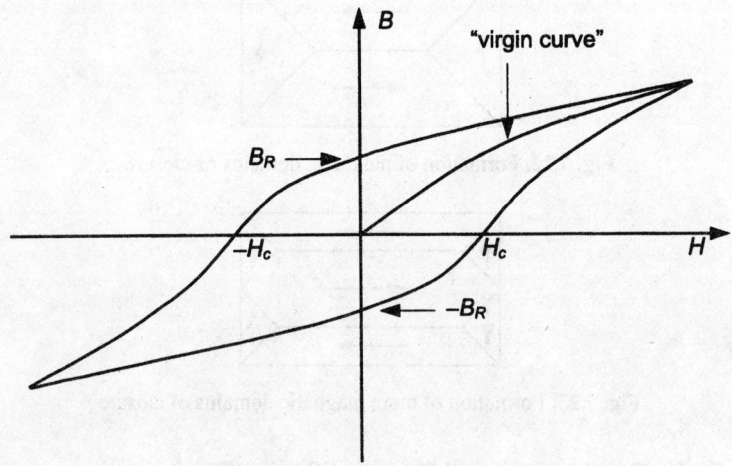

Fig. 7.24. Magnetic hysteresis loop identifying the virgin curve
$H_c \equiv$ coercive "force".
$B_R =$ remanence.
$M_s = [(B - H)/4\pi]_{H \to \infty} =$ saturation magnetization.
$M_R = B_R/4\pi =$ remanent magnetization.

Néel and Bloch Walls (B)

Figure 7.25 provides a convenient way to distinguish Bloch and Néel walls. Bloch walls have $\varphi = 0$, while Néel walls have $\varphi = \pi/2$. Néel walls occur in thin films of materials such as permalloy in order to reduce surface magnetostatic energy as suggested by Fig. 7.26. There are many other complexities involved in domain-wall structures. See, e.g., Malozemoff and Slonczewski [7.44].

---

[25] Some authors define H_c as the field that reduces M to zero.

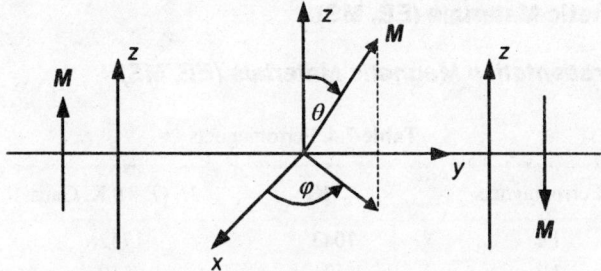

Fig. 7.25. Bloch wall: $\varphi = 0$; Néel wall: $\varphi = \pi/2$

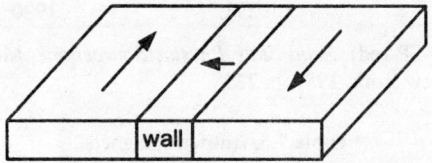

Fig. 7.26. Néel wall in thin film

Methods of Observing Domains (EE, MS)

We briefly summarize five methods.

1. Bitter patterns–a colloidal suspension of particles of magnetite is placed on a polished surface of the magnetic material to be examined. The particles are attracted to regions of nonuniform magnetization (the walls) and hence the walls are readily seen by a microscope.

2. Faraday and Kerr effects–these involve rotation of the plane of polarization on transmission and reflection (respectively) from magnetic substances.

3. Neutrons–since neutrons have magnetic moments they experience interaction with the internal magnetization and its direction, see Bacon GE, "Neutron Diffraction," Oxford 1962 (p355ff).

4. Transmission electron microscopy (TEM)–Moving electrons are influenced by forces due to internal magnetic fields.

5. Scanning electron microscopy (SEM)–Moving secondary electrons sample internal magnetic fields.

7.3.2 Magnetic Materials (EE, MS)

Some Representative Magnetic Materials (EE, MS)

Table 7.4. Ferromagnets

Ferromagnets	T_c (K)	M_s ($T = 0$ K, Gauss)
Fe	1043	1752
Ni	631	510
Co	1394	1446
EuO	77	1910
Gd	293	1980

From Parker SP (ed), *Solid State Physics Sourcebook*, McGraw-Hill Book Co., New York, 1987, p. 225.

Table 7.5. Antiferromagnets

Antiferromagnets	T_N (K)
MnO	122
NiO	523
CoO	293

From Cullity BD, *Introduction to Magnetic Materials*, Addison-Wesley Publ Co, Reading, Mass, 1972, p. 157.

Table 7.6. Ferrimagnets

Ferrimagnets	T_c (K)	M_s ($T = 0$ K, Gauss)	
YIG ($Y_3Fe_5O_{12}$)	560	195	a garnet
Magnetite (Fe_3O_4)	858	510	a spinel

(From *Solid State Physics Sourcebook*, op cit p. 225)

We should emphasize that these classes do not exhaust the types of magnetic order that one can find. At suitably low temperatures the heavy rare earths, may show helical or conical order. and there are other types of order, as for example, spin glass order. Amorphous ferromagnets show many kinds of order such as speromagnetic and asperomagnetic. (See, e.g., *Solid State Physics Source Book*, op cit p 89.)

Ferrites are perhaps the most common type of ferrimagnets. Magnetite, the oldest magnetic material that is known, is a ferrite also called lodestone. In general, ferrites are double oxides of iron and another metal such as Ni or Ba (e.g. nickel ferrite: $NiOFe_2O_3$ and barium ferrite: $BaO \cdot 6Fe_2O_3$). Many ferrites find application in high-frequency devices because they have high resistivity and hence do not have appreciable eddy currents. They are used in microwave devices, radar, etc. Barium ferrite, a hard magnet, is one of the materials used for magnetic recording that is a very large area of application of magnets (see, e.g., Craik [7.15 p. 379]).

Hard and Soft Magnetic Materials (EE, MS) The clearest way to distinguish between hard and soft magnetic materials is by a hysteresis loop (see Fig. 7.27). Hard permanent magnets are hard to magnetize and demagnetize, and their coercive forces can be of the order of 10^6 A/m or larger. For a soft magnetic material, the coercive force can be of order 1 A/m or smaller. For conversions: 1 A/m is $4\pi \times 10^{-3}$ Oersted, 1 kJ/m³ converts to MGOe (mega Gauss Oersted) if we multiply by 0.04π, 1 Tesla = 10^4 G.

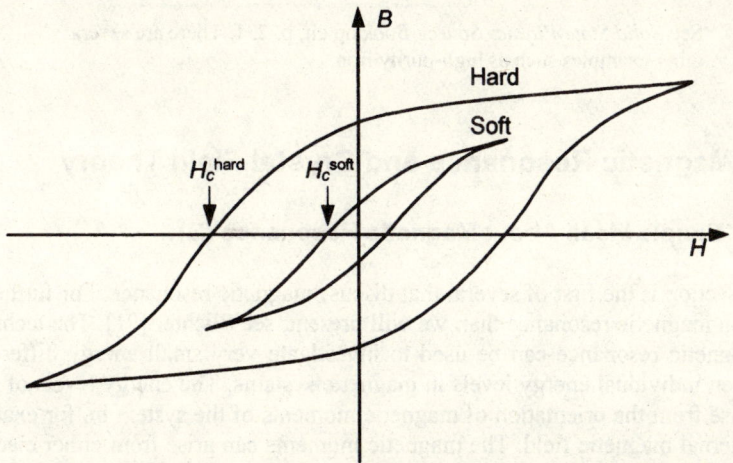

Fig. 7.27. Hard and soft magnetic material hysteresis loops (schematic)

Permanent Magnets (EE, MS) There are many examples of permanent magnetic materials. The largest class of magnets used for applications are permanent magnets. They are used in electric motors, in speakers for audio systems, as wiggler magnets in synchrotrons, etc. We tabulate here only two examples that have among the highest energy products $(BH)_{max}$.

Table 7.7. Permanent Magnets

	T_c (K)	M_s (kA m^{-1})	H_c (kA m^{-1})	$(BH)_{max}$ (kJ m^{-3})
(1) SmCo$_5$	997	768	700–800	183
(2) Nd$_2$Fe$_{14}$B	~583	—	~880	~290

(1) Craik [7.15 pp. 385, 387]. Sm$_2$Co$_{17}$ is in some respects better, see [7.15 p. 388].
(2) *Solid State Physics Source Book* op cit p 232. Many other hard magnetic materials are mentioned here such as the AlNiCos, barium ferrite, etc. See also Herbst [7.29].

Soft Magnetic Materials (EE, MS) There are also many kinds of soft magnetic materials. They find application in communication materials, motors, generators, transformers, etc. Permalloys form a very common class of soft magnets. These are Ni-Fe alloys with sometimes small additions of other elements. 78 Permalloy means, e.g., 78% Ni and 22% Fe.

Table 7.8. Soft Magnet

	T_c (K)	H_c (A m^{-1})	B_s (T)
78 Permalloy	873	4	1.08

See *Solid State Physics Source Book* op cit, p. 231. There are several other examples such as high-purity iron.

7.4 Magnetic Resonance and Crystal Field Theory

7.4.1 Simple Ideas About Magnetic Resonance (B)

This Section is the first of several that discuss magnetic resonance. For further details on magnetic resonance than we will present, see Slichter [91]. The technique of magnetic resonance can be used to investigate very small energy differences between individual energy levels in magnetic systems. The energy levels of interest arise from the orientation of magnetic moments of the system in, for example, an external magnetic field. The magnetic moments can arise from either electrons or nuclei.

Consider a particle with magnetic moment μ and total angular momentum J and assume that the two are proportional so that we can write

$$\mu = \gamma J, \tag{7.252}$$

where the proportionality constant γ is called the *gyromagnetic ratio* and equals $-g\mu_B/\hbar$ (for electrons, it would be + for protons) in previous notation. We will then suppose that we apply a magnetic induction B in the z direction so that the Hamiltonian of the particle with magnetic moment becomes

$$\mathcal{H}_0 = -\gamma\mu_0 H J_z, \tag{7.253}$$

where we have used (7.252), and $B = \mu_0 H$, where H is the magnetic field. If we define j (which are either integers or half-integers) so that the eigenvalues of J^2 are $j(j+1)\hbar^2$, then we know that the eigenvalues of $\mathcal{H}_0$ are

$$E_m = -\gamma\hbar\mu_0 H m, \tag{7.254}$$

where $-j \leq m \leq j$.

From (7.254) we see that the difference between adjacent energy levels is determined by the magnetic field and the gyromagnetic ratio. We can induce transitions between these energy levels by applying an alternating magnetic field (perpendicular to the z direction) of frequency ω, where

$$\hbar\omega = |\gamma|\hbar\mu_0 H \quad \text{or} \quad \omega = |\gamma|\mu_0 H. \tag{7.255}$$

These results follow directly from energy conservation and they will be discussed further in the next section. It is worthwhile to estimate typical frequencies that are involved in resonance experiments for a convenient size magnetic field. For an electron with charge e and mass m, if the gyromagnetic ratio γ is defined as the ratio of magnetic moment to *orbital angular* momentum, it is given by

$$\gamma = e/2m, \quad \text{for } e < 0. \tag{7.256}$$

For an electron with spin but no orbital angular momentum, the ratio of magnetic moment to spin angular momentum is $2\gamma = e/m$. For an electron with both orbital and spin angular momentum, the contributions to the magnetic moment are as described and are additive. If we use (7.255) and (7.256) with magnetic fields of order 8000 G, we find that the resonance frequency for electrons is in the *microwave* part of the spectrum. Since nuclei have much greater mass, the resonance frequency for nuclei lies in the *radio frequency* part of the spectrum. This change in frequency results in a considerable change in the type of equipment that is used in observing electron or nuclear resonance.

Abbreviations that are often used are NMR for nuclear magnetic resonance and EPR or ESR for electron paramagnetic resonance or electron spin resonance.

7.4.2 A Classical Picture of Resonance (B)

Except for the concepts of spin-lattice and spin-spin relaxation times (to be discussed in the Section on the Bloch equations) we have already introduced many of the most basic ideas connected with magnetic resonance. It is useful to present a classical description of magnetic resonance [7.39]. This description is more pictorial than the quantum description. Further, it is true (with a suitable definition of the time derivative of the magnetic moment operator) that the classical magnetic moment in an external magnetic field obeys the same equations of motion as the magnetic moment operator. We shall not prove this theorem here, but it is because of it that the classical picture of resonance has considerable use. The simplest way of presenting the classical picture of resonance is by use of the concept of the rotating coordinate system. It also should be pointed out that we will leave out of our discussion any relaxation phenomena until we get to the Section on the Bloch equations.

As before, let a magnetic system have angular momentum J and magnetic moment μ, where $\mu = \gamma J$. By classical mechanics, we know that the time rate of change of angular momentum equals the external torque. Therefore we can write for a magnetic moment in an external field H,

$$\frac{dJ}{dt} = \mu \times \mu_0 H. \tag{7.257}$$

Since $\mu = \gamma J$ ($\gamma < 0$ for electrons), we can write

$$\frac{d\mu}{dt} = \mu \times (\gamma\mu_0)H . \qquad (7.258)$$

This is the general equation for the motion of the magnetic moment in an external magnetic field.

To obtain the solution to (7.258) and especially in order to picture this solution, it is convenient to use the concept of the rotating coordinate system. Let

$$A = \hat{i}A_x + \hat{j}A_y + \hat{k}A_z$$

be any vector, and let $\hat{i}, \hat{j}, \hat{k}$ be unit vectors in a rotating coordinate system. If Ω is the angular velocity of the rotating coordinate system relative to a fixed coordinate system, then relative to a fixed coordinate system we can show that

$$\frac{d\hat{i}}{dt} = \Omega \times \hat{i} . \qquad (7.259)$$

This implies that

$$\frac{dA}{dt} = \frac{\delta A}{\delta t} + \Omega \times A , \qquad (7.260)$$

where $\delta A/\delta t$ is the rate of change of A relative to the rotating coordinate system and dA/dt is the rate of change of A relative to the fixed coordinate system.

By using (7.260), we can write (7.258) in a rotating coordinate system. The result is

$$\frac{\delta \mu}{\delta t} = \mu \times (\Omega + \gamma\mu_0 H) . \qquad (7.261)$$

Equation (7.261) is the same as (7.258). The only difference is that in the rotating coordinate system the effective magnetic field is

$$H_{\text{eff}} = H + \frac{\Omega}{\gamma\mu_0} . \qquad (7.262)$$

If H is constant and Ω is chosen to have the constant value $\Omega = -\gamma\mu_0 H$, then $\delta\mu/\delta t = 0$. This means that the spin precesses about H with angular velocity $\gamma\mu_0 H$. Note that this is the same as the frequency for magnetic resonance absorption. We will return to this point below.

It is convenient to get a little closer to the magnetic resonance experiment by supposing that we have a static magnetic field H_0 along the z direction and an alternating magnetic field $H_x(t) = 2H'\cos(\omega t)$ along the x-axis. We can resolve the alternating field into two rotating magnetic fields (one clockwise, one counterclockwise) as shown in Fig. 7.28. Simple vector addition shows that the two fields add up to $H_x(t)$ along the x-axis.

With the static magnetic field along the z direction, the magnetic moment will precess about the z-axis. The moment will precess in the same sense as one of the rotating magnetic fields. Now that we have both constant and alternating magnetic

7.4 Magnetic Resonance and Crystal Field Theory

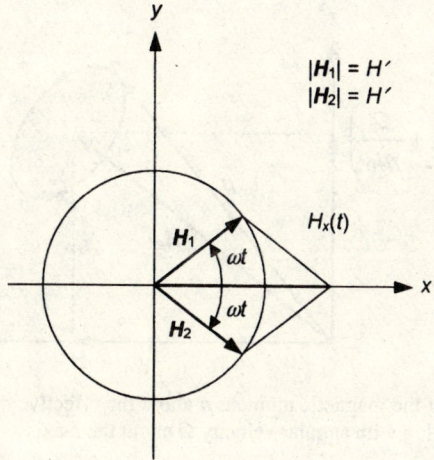

Fig. 7.28. Decomposition of an alternating magnetic field into two rotating magnetic fields

fields, something interesting begins to happen. The component of the alternating magnetic field that rotates in the same direction as the magnetic moment is the important component [91]. Near resonance, the magnetic moment and one of the circularly polarized components of the alternating magnetic field rotate with almost the same angular velocity. In this situation the rotating magnetic field exerts an almost constant torque on the magnetic moment and tends to tip it over. Physically, this is what happens in resonance absorption.

Let us be a little more quantitative about this problem. If we include only one component of the rotating magnetic field and if we assume that Ω is the cyclic frequency of the alternating magnetic field, then we can write

$$\frac{\delta\mu}{\delta t} = \mu \times [\Omega + \gamma\mu_0(\hat{i}H' + \hat{k}H_0)] \,. \tag{7.263}$$

This can be further written as

$$\frac{\delta\mu}{\delta t} = \mu \times H_{\text{eff}} \,, \tag{7.264}$$

where now

$$H_{\text{eff}} \equiv \hat{k}\left(H_0 + \frac{\Omega}{\gamma\mu_0}\right) + \hat{i}H' \,.$$

Since in the rotating coordinate system μ precesses about H_{eff}, we have the picture shown in Fig. 7.29. If we adjust the static magnetic field so that

$$H_0 = -\frac{\Omega}{\gamma\mu_0} \,,$$

then we have satisfied the conditions of resonance. In this situation H_{eff} is along the x-axis (in the rotating coordinate system) and the magnetic moment flops up and down with frequency $\gamma\mu_0 H'$.

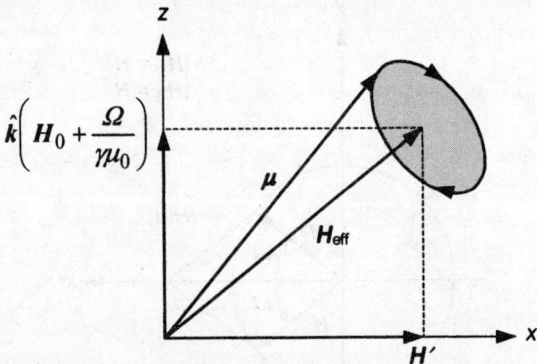

Fig. 7.29. Precession of the magnetic moment μ about the effective magnetic field H_{eff} in a coordinate system rotating with angular velocity Ω about the z-axis

Similar quantum-mechanical calculations can be done in a rotating coordinate system, but we shall not do them as they do not add much that is new. What we have done so far is useful in forming a pictorial image of magnetic resonance, but it is not easy to see how to put in spin-lattice interactions, or other important interactions. In order to make progress in interpreting experiments, it is necessary to generalize our formalism somewhat.

7.4.3 The Bloch Equations and Magnetic Resonance (B)

These equations are used for a qualitative and phenomenological discussion of NMR and EPR. In general, however, it is easier to describe NMR than EPR. This is because the nuclei do not interact nearly so strongly with their surroundings as do the electrons. We shall later devote a Section to discussing how the electrons interact with their surroundings.

The Bloch equations are equations that describe precessing magnetic moments, and various relaxation mechanisms. They are almost purely phenomenological, but they do provide us a means of calculating the power absorbed versus the frequency. Without the interactions responsible for the relaxation times, this plot would be a delta function. Such a situation would not be very interesting. It is the relaxation times that give us information about what is going on in the solid.[26]

Definition of Bloch Equations and Relaxation Times (B)

The theory of the resonance of free spins in a magnetic field is simple but it holds little inherent interest. To relate to more physically interesting phenomena it is necessary to include the interactions of the spins with their environment. The Bloch equations include these interactions in a phenomenological way.

---

[26] See Manenkov and Orbach (eds) [7.45].

When we include a relaxation time (or an interaction process), we find that the time rate of change of the magnetization (along the field) is proportional to the deviation of the magnetization from its equilibrium value. This guarantees a relaxation of magnetization along the field. If we add an alternating magnetic field along the x- or y-axes, it is also necessary to add a term $(M \times H)_z$ that is proportional to the torque. Thus for the component of magnetization along the constant external magnetic field, it is reasonable to write

$$\frac{dM_z}{dt} = \frac{M_0 - M_z}{T_1} + (\gamma\mu_0)(M \times H)_z. \tag{7.265}$$

As noted, (7.265) has a built-in relaxation process of M_z to M_0, the spin-lattice relaxation time T_1. However, as we approach equilibrium in a static magnetic field $H_0\hat{k}$, we will want both M_x and M_y to tend to zero. For this purpose, a new term with a relaxation time T_2 is often introduced. We write

$$\frac{dM_x}{dt} = \gamma\mu_0(M \times H)_x - \frac{M_x}{T_2}, \tag{7.266}$$

and

$$\frac{dM_y}{dt} = \gamma\mu_0(M \times H)_y - \frac{M_y}{T_2}. \tag{7.267}$$

Equations (7.265), (7.266), and (7.267) are called the *Bloch equations*. T_2 is often called the *spin-spin relaxation time*. The idea is that the term involving T_1 is caused by the interaction of the spin system with the lattice or phonons, while the term involving T_2 is caused by something else. The physical origin of T_2 is somewhat complicated. Consider, for example, two nuclei precessing in an external static magnetic field. The precession of one nucleus produces a varying magnetic field at the second nucleus and hence tends to "flip" the spin of the second nucleus (and vice versa). Waller[27] first pointed out that there are two different types of spin relaxation processes.

Ferromagnetic Resonance (B)

Using a simple quantum picture, for an atomic system, we have already argued (see (7.258))

$$\frac{d\mu}{dt} = \gamma\mu \times B_a, \tag{7.268}$$

where $B_a = \mu_0 H$. This implies a precession of μ and M about the constant magnetic field B_a with frequency $\omega = \gamma B_a$ the Larmor frequency, as already noted. For

---

[27] See Waller [7.66]. Discussion of ways to calculate T_1 and T_2 is contained in White [7.68 p124ff and 135ff].

ferromagnetic resonance (FMR) all spins precess together and $M = N\mu$, where N is the number of spins per unit volume. Thus by (7.268)

$$\frac{dM}{dt} = \gamma M \times B_a. \qquad (7.269)$$

Several comments can be made. The above equation is valid also for $M = M(r)$ varying slowly in space. We will also use this equation for spin-wave resonance when the wavelengths of the waves are long compared to the atom to atom spacing that allows the classical approach to be valid. One generalizes the above equation by replacing B_a by B where

$B = B_a$ (applied)
$\quad + B_{rf}$ (due to a radio-frequency applied field)
$\quad + B_{demag}$ (from demagnetizing fields that depend on geometry)
$\quad + B_{exchange}$ (as derived from the Heisenberg Hamiltonian)
$\quad + B_{anisotropy}$ (an effective field arising from interactions producing anisotropy).

We should also include dissipative or damping and relaxation effects.

We start with all fields zero or negligible except for the applied field (note here $B_{exchange} \propto M$, which is assumed to be uniform, so $M \times B_{exchange} = 0$). This gives resonance at the natural precessional frequency of the uniform precessional mode. With $B = B_0 \hat{k}$ we have

$$\frac{dM_x}{dt} = \gamma M_y B_0, \quad \frac{dM_y}{dt} = -\gamma M_x B_0, \quad \frac{dM_z}{dt} = 0. \qquad (7.270)$$

We look for solutions with

$$M_x = A_1 e^{-i\omega t},$$
$$M_y = A_2 e^{-i\omega t}, \qquad (7.271)$$
$$M_z = \text{constant},$$

and so we have a solution provided

$$\begin{vmatrix} -i\omega & -\gamma B_0 \\ \gamma B_0 & -i\omega \end{vmatrix} = 0, \qquad (7.272)$$

or

$$|\omega| = |\gamma B_0|, \qquad (7.273)$$

7.4 Magnetic Resonance and Crystal Field Theory

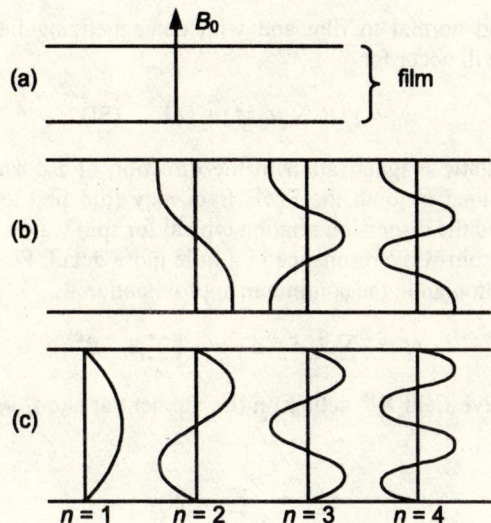

Fig. 7.30. (a) Thin film with magnetic field, (b) "Unpinned" spin waves, (c) "Pinned" spin waves

which as expected is just the Larmor precessional frequency. In actual situations we also need to include demagnetization fields and hence shape effects, which will alter the resonant frequencies. FMR typically occurs at microwave frequencies. Antiferromagnetic resonance (AFMR) has also been studied as a way to determine anisotropy fields.

Spin-Wave Resonance (A)

Spin-wave resonance is a direct way to experimentally prove the existence of spin waves (as is inelastic neutron scattering – see Kittel [7.39 pp456-458]). Consider a thin film with a magnetic field B_0 perpendicular to the film (Fig. 7.30a). In the simplest picture, we view the spin waves as "vibrations" in the spin between the surfaces of the film. Plotting the amplitude versus position, Fig. 7.30b is obtained for unpinned spins. Except for the uniform mode, these have no net interaction (absorption) with the electromagnetic field. The pinned case is a little different (Fig. 7.30c). Here only waves with an even number of half-wavelengths will show no net interaction energy with the field while the ones with an odd number of half-wavelengths ($n = 1, 3$, etc.) will absorb energy. (Otherwise the induced spin flippings will absorb and emit equal amounts of energy).

We get absorption when

$$n\frac{\lambda}{2} = T \quad \{n \text{ odd}, T \text{ thickness of film}\},$$

$$k = \frac{2\pi}{\lambda} = \frac{n\pi}{T} \quad \text{or} \quad k = (2n+1)\frac{\pi}{T} \quad \{n = 0, 1, 2...\}.$$

With applied field normal to film and with demagnetizing field and exchange $D'k^2$, absorption will occur for

$$\omega_0 = \gamma(B_0 - \mu_0 M) + D'k^2 \quad \text{(SI)},$$

where M is the static magnetization in the direction of B_0. The spin-wave frequency is determined by both the FMR frequency (the first term including demagnetization) and the dispersion relation typical for spin waves.

We now analyze spin-wave resonance in a little more detail. First we develop the Heisenberg Hamiltonian in the continuum approximation,

$$\mathcal{H} = -\sum J_{ij} \mathbf{S}_i \cdot \mathbf{S}_j = -\tfrac{1}{2}\sum \boldsymbol{\mu}_i \cdot \mathbf{B}_i^{\text{ex}} \qquad (7.274)$$

defines the effective field $\mathbf{B}_i^{\text{ex}}$ acting on the moment at site i, $\boldsymbol{\mu}_i = \gamma \mathbf{S}_i$. ($\gamma < 0$ for electrons)

$$\mathbf{B}_i^{\text{ex}} = \frac{2}{\gamma} \sum_j J_{ij} \mathbf{S}_j. \qquad (7.275)$$

Assuming nearest neighbors (nn) at distance a and nn interactions only. We find for a simple cubic (SC) structure after expansion, and using cancellation resulting from symmetry

$$\gamma \mathbf{B}_i^{\text{ex}} = 12 J \mathbf{S}_i + 2 J a^2 \nabla^2 \mathbf{S}_i.$$

Consistent with the classical continuum approximation

$$\frac{\mathbf{M}}{M} = \frac{\mathbf{S}_i}{S}, \qquad (7.276)$$

$$\mathbf{B}^{\text{ex}} = \lambda \mathbf{M} + K' \nabla^2 (\mathbf{M}/M), \qquad (7.277)$$

where

$$\lambda = \frac{12JS}{\gamma M}, \quad K' = \frac{2Ja^2 S}{\gamma}. \qquad (7.278)$$

As an aside we note B^{ex} is consistent with results obtained before (Sect. 7.3.1). Since

$$U = -\tfrac{1}{2}\sum \boldsymbol{\mu}_i \cdot \mathbf{B}_i^{\text{ex}}, \qquad (7.279)$$

neglecting constant terms (resulting from the magnitude of the magnetization being constant) we have

$$U = -\frac{JS^2}{a}\int \mathbf{m}\nabla^2 \mathbf{m}\, dV \quad \{\mathbf{m} = \mathbf{M}/M\}. \qquad (7.280)$$

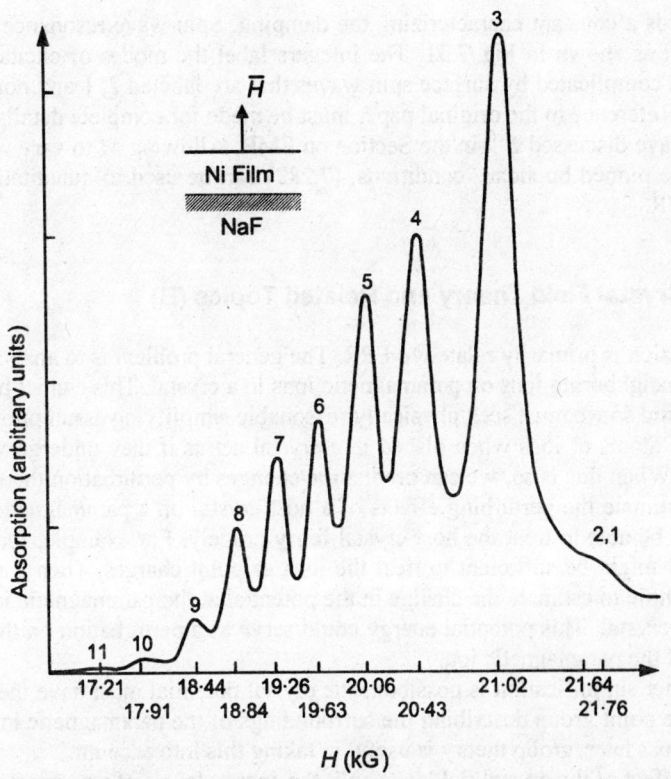

Fig. 7.31. Spin wave resonance spectrum for Ni film, room temperature, 17 GHz. After Puszharski H, "Spin Wave Resonance", *Magnetism in Solids Some Current Topics*, Scottish Universities Summer School in Physics, 1981, p. 287, by permission of SUSSP. Original data in Mitra DP and Whiting JSS, *J Phys F: Metal Physics*, **8**, 2401 (1978)

Assuming $\int m_x \nabla m_x \cdot dA$ etc = 0 for a large surface we can also recast the above as

$$U = \frac{JS^2}{a} \int [(\nabla \alpha_1)^2 + (\nabla \alpha_2)^2 + (\nabla \alpha_3)^2] dV \qquad (7.281)$$

which is the same as we obtained before, with a slightly different analysis. The α_i are of course the direction cosines.

The anisotropy energy and effective field can be written in the same way as before, and no further comments need be made about it.

When one generalizes the equation for the time development of M, one has the Landau–Lifshitz equations. Damping causes broadening of the absorption lines. Then

$$\dot{M} = \gamma M \times B^{\text{eff}} + \frac{\alpha}{M} M \times (M \times B^{\text{eff}}), \qquad (7.282)$$

where α is a constant characterizing the damping. Spin-wave resonance has been observed as shown in Fig. 7.31. The integers label the modes of excitation. The figure is complicated by surface spin waves that are labeled 2, 1 and not fully resolved. Reference to the original paper must be made for complete details.

We have discussed B^{eff} in the Section on FMR. Allowing M to vary with r and using the pinned boundary conditions, (7.282) can be used to quantitatively discuss SWR.

7.4.4 Crystal Field Theory and Related Topics (B)

This Section is primarily related to EPR. The general problem is to analyze the effects of neighboring ions on paramagnetic ions in a crystal. This cannot be exactly solved, and so we must seek physically reasonable simplifying assumptions.

Some atoms or ions when placed in a crystal act as if they undergo very little change. When this is so, we can predict the changes by perturbation theory. In order to estimate the perturbing effects of a host crystal on a paramagnetic ion, we ought to be able to treat the host crystal fairly crudely. For example, for an ionic crystal it might be sufficient to treat the ions as point charges. Then it would be fairly simple to estimate the change in the potential at the paramagnetic ion due to the host crystal. This potential energy could serve as a perturbation on the Hamiltonian of the paramagnetic ion.

Another simplification is possible. The crystal potential must have the symmetry of the point group describing the surroundings of the paramagnetic ion. As we will discuss later, group theory is useful in taking this into account.

The effect of the crystal field is to split the energy levels of a paramagnetic ion. In order to show how this comes about, it is useful to know what we mean by the energy levels. The best way to do this is to write down the Hamiltonian (whose eigenvalues are the energy levels) for the electrons. With no external field, the Hamiltonian has a form similar to

$$\mathcal{H} = \sum_i \left[\frac{P_i^2}{2m} - \frac{Ze^2}{4\pi\varepsilon_0 r_i} + a_i J_i \cdot I - e\phi_c(r_i) \right]$$
$$+ \sum_{i,j}' \frac{1}{2} \frac{e^2}{4\pi\varepsilon_0 r_{ij}} + \sum_{i,j} \lambda_{ij} L_i \cdot S_j .$$
(7.283)

The origin of the coordinate system for (7.283) is the nucleus of the paramagnetic ion. The sum over i and j is a sum over electronic coordinates. The first term is the kinetic energy. The second term is the potential energy of the electrons in the field of the nucleus. The third term is the hyperfine interaction of the electron (with total angular momentum J_i) with the nucleus that has angular momentum I. The fourth term is the crystal field energy. The fifth term is the potential energy of the electrons interacting with themselves. The last term is the spin (S_j)-orbit (angular momentum L_i) interaction (see Appendix F) of the electrons. By the unperturbed energy levels of the paramagnetic ion, one often means the energy eigenstates of the first, second, and fifth terms obtained perhaps by Hartree–Fock calculations. The rest of the terms are usually thought of as perturbations. In the discussion that follows, the hyperfine interaction will be neglected.

To avoid complicated many-body effects, we will assume that the sources of the crystal field ($E_c \equiv -\nabla \phi_c$) are external to the paramagnetic ion. Thus in the vicinity of the paramagnetic ion, it can be assumed that $\nabla^2 \phi_c = 0$.

Weak, Medium, and Strong Crystal Fields (B)

In discussing the effect of the crystal field on the energy levels, which is important to EPR, three cases can be distinguished [47].

Weak crystal fields are by definition those for which the spin-orbit interaction is stronger than the crystal field interaction. This is often realized when the electrons of the paramagnetic shell of the ion lie "fairly deep" within the ion, and hence are shielded from the crystalline field by the outer electrons. This may happen in ionic compounds of the rare earths. Rare earths have atomic numbers (Z) from 58 to 71. Examples are Ce, Pr, and Ne, which have incomplete 4f shells.

By a *medium crystal field* we mean that the crystal field is stronger than the spin-orbit interaction. This happens when the paramagnetic electrons of the ion are mainly distributed over the outer portions of the ion and hence are not well shielded. In this situation something else may occur. The potential that the paramagnetic ions move in is no longer even approximately spherically symmetric, and hence the orbital angular momentum is not conserved. We say that the orbital angular momentum is (at least partially) "quenched" (this means $\langle \psi | L | \psi \rangle = 0$, $\langle \psi | L^2 | \psi \rangle \neq 0$). Paramagnetic crystals that have iron group elements ($Z = 21$ to 29, e.g., Cr, Mn, and Fe that have an incomplete 3d shell) are typical examples of the medium-field case.

Strong crystal field by definition means covalent bonding. In this situation, the wave functions for the paramagnetic ion electrons overlap considerably with the wave functions of the other electrons of the crystal. Crystal field theory does not work here. This type of situation will not be discussed in this chapter.

As we will see, group theory can be an aid in understanding how energy levels are split by perturbations.

Miscellaneous Theorems and Facts (In Relation to Crystal Field Theory) (B)

The theorems below will not be proved. They are stated because they are useful in carrying out actual crystal field calculations.

The Equivalent Operator Theorem. This theorem is used in calculating needed matrix elements in crystal field calculations. The theorem states that within a manifold of states for which l is constant, there are simple relations between the matrix elements of the crystal-field potential and appropriate angular momentum operators. For constant l, the rule says to replace the x by L_x (operator, in this case L_x is the x operator equivalent) and so forth for other coordinates. If the result is a product in which the order of the factors is important, then we must use all possible different permutations. There is a similar rule for manifolds of constant J

(where we include both the orbital angular momentum and the spin angular momentum).

There is a straightforward way of generating operator equivalents (OpEq) by using

$$[L^+, Op\ Eq\ Y_l^M] \propto Op\ Eq\ Y_l^{M+1},$$

and

$$[L^-, Op\ Eq\ Y_l^M] \propto Op\ Eq\ Y_l^{M-1}. \quad (7.284)$$

The constants of proportionality can be computed from a knowledge of the Clebsch–Gordon coefficients.

Table 7.4. Effective magneton number for some representative trivalent lanthanide ions

Ion	Configuration	Ground state	$g\sqrt{J(J+1)}$ *
Pr (3+)	...$4f^2\ 5s^2\ 5p^6$	$^3H_4$	3.58
Nd (3+)	...$4f^3\ 5s^2\ 5p^6$	$^4I_{9/2}$	3.62
Gd (3+)	...$4f^7\ 5s^2\ 5p^6$	$^8S_{7/2}$	7.94
Dy (3+)	...$4f^9\ 5s^2\ 5p^6$	$^6H_{15/2}$	10.63

* $g = g(\text{Lande}) = 1 + \dfrac{J(J+1) + S(S+1) - L(L+1)}{2J(J+1)}$

Table 7.5. Effective magneton number for some representative iron group ions*

Ion	Configuration	Ground state	$2\sqrt{S(S+1)}$
Fe (3+)	...$3d^5$	$^6S_{5/2}$	5.92
Fe (2+)	...$3d^6$	$^5D_4$	4.90
Co (2+)	...$3d^7$	$^4F_{9/2}$	3.87
Ni (2+)	...$3d^8$	$^3F_4$	2.83

* Quenching with $J = S$, $L = 0$ (so $g = 2$) is assumed for better agreement with experiment

Kramers' Theorem. This theorem tells us about systems that must have a degeneracy. The theorem says that the systems with an odd number of electrons on which a purely electrostatic field is acting can have no energy levels that are less than two-fold degenerate. If a magnetic field is imposed, this two-fold degeneracy can be lifted.

Jahn–Teller effect. This effect tells us that high degeneracy may be unlikely. The theorem states that a nonlinear molecule that has a (orbitally) degenerate ground state is unstable, and tends to distort itself so as to lift the degeneracy. Because of the Jahn–Teller effect, the symmetry of a given atomic environment in a solid is frequently slightly different from what one might expect. Of course, the Jahn–Teller effect does not remove the fundamental Kramers' degeneracy.

Hund's rules. Assuming Russel–Sanders coupling, these rules tell us what the ground state of an atomic system is. Hund's rules were originally obtained from spectroscopic evidence, but they have been confirmed by atomic calculations that include the Coulomb interactions between electrons. The rules state that in figuring out how electrons fill a shell in the ground state we should (1) assign a maximum S allowed by the Pauli principle, (2) assign maximum L allowed by S, (3) assign $J = L - S$ when the shell is not half-full, and $J = L + S$ when the shell is over half-full. See Problems 7.17 and 7.18. Results from the use of Hund's rules are shown in Tables 7.9 and 7.10.

Energy-Level Splitting in Crystal Fields by Group Theory (A)

In this Section we introduce enough group theory to be able to discuss the relation between degeneracies (in the energies of atoms) and symmetries (of the environment of the atoms). The fundamental work in the field was done by H. A. Bethe (see, e.g., Von der Lage and Bethe [7.64]). For additional material see Knox and Gold [61, in particular see Table 1-2 pp. 5-8 for definitions].

We have already discussed some of the more elementary ideas related to groups in Chap. 1 (see Sect. 1.2.1). The most important new concept that we will introduce here is the concept of group representations. A group representation starts with a set of nonsingular square matrices. For each group element g_i there is a matrix R_i such that $g_i g_j = g_k$ implies that $R_i R_j = R_k$. Briefly stated, a representation of a group is a set of matrices with the same multiplication table as the original group.

Two representations (R', R) of g that are related by

$$R'(g) = S^{-1} R(g) S \tag{7.285}$$

are said to be *equivalent*. In (7.285), S is any nonsingular matrix.

We define

$$R(g) \equiv R^{(1)}(g) \oplus R^{(2)}(g) \equiv \begin{pmatrix} R^{(1)}(g) & 0 \\ 0 & R^{(2)}(g) \end{pmatrix}. \tag{7.286}$$

In (7.286) we say that the representation $R(g)$ is reducible because it can be reduced to the direct sum of at least two representations. If $R(g)$ is of the form (7.286), it is said to be in *block diagonal form*. If a matrix representation can be brought into block diagonal form by a similarity transformation, then the representation is *reducible*. If no matrix representation reduces the representation to block diagonal form, then the matrix representation is *irreducible*. In considering any representation that is reducible, the most interesting information is to find out what irreducible representations are contained in the given reducible representation. We should emphasize that when we say a given representation $R(g)$ is reducible, we mean that a single S in (7.285) will put $R'(g)$ in block diagonal form for all g in the group.

In a typical problem in crystal field theory, a reducible representation (with respect to some group) of interest might be the irreducible representation $R^{(l)}$ of the three-dimensional rotation group. That is, we would like to know what irreducible representations of a group of interest is contained in a given irreducible representation of $R^{(l)}$ for some l. As we will see later, this can tell us a good deal about what happens to the electronic energy levels of a spherical atom in a crystal field.

It is worthwhile to give an explicit example of the irreducible representations of a group. Let us consider the group D_3 already defined in Chap. 1 (see Table 1.2).

In Table 7.6 note that $R^{(1)}$ and $R^{(2)}$ are *unfaithful* (many elements of the group correspond to the same matrix) representations while $R^{(3)}$ is a *faithful* (there is a one-to-one correspondence between group elements and matrices) representation. $R^{(1)}$ is, of course, the trivial representation.

Table 7.6. The irreducible representations of D_3

D_3	g_1	g_2	g_3	g_4	g_5	g_6
$R^{(1)}$	1	1	1	1	1	1
$R^{(2)}$	1	1	1	-1	-1	-1
$R^{(3)}$	$\begin{pmatrix} 1, & 0 \\ 0, & 1 \end{pmatrix}$	$\frac{1}{2}\begin{pmatrix} -1, & +\sqrt{3} \\ -\sqrt{3}, & -1 \end{pmatrix}$	$\frac{1}{2}\begin{pmatrix} -1, & -\sqrt{3} \\ \sqrt{3}, & -1 \end{pmatrix}$	$\frac{1}{2}\begin{pmatrix} 1, & -\sqrt{3} \\ -\sqrt{3}, & -1 \end{pmatrix}$	$\frac{1}{2}\begin{pmatrix} 1, & \sqrt{3} \\ \sqrt{3}, & -1 \end{pmatrix}$	$\begin{pmatrix} -1, & 0 \\ 0, & 1 \end{pmatrix}$

Since a similarity transformation will induce so many equivalent irreducible representations, a quantity that is invariant to similarity transformation might be (and in fact is) of considerable interest. Such a quantity is the *character*. The character of a group element is the trace of the matrix representing that group element.

It is elementary to show that the trace is invariant to similarity transformation. A similar argument shows that all group elements in the same class[28] have the same character. The argument goes as indicated below:

$$\mathrm{Tr}(R(g)) = \mathrm{Tr}(R(g)SS^{-1}) = \mathrm{Tr}(S^{-1}R(g)S) = \mathrm{Tr}(R'(g)),$$

if $R'(g)$ is defined by (7.285).

In summary the characters are defined by

$$\chi^{(i)}(g) = \sum_\alpha R^{(i)}_{\alpha\alpha}(g). \qquad (7.287)$$

Equation (7.287) defines the character of the group element g in the ith representation. The characters still serve to distinguish various representations. As an example, the character table for the irreducible representation of D_3 is shown in Table 7.7. In Table 7.7, the top row labels the classes.

---

[28] Elements in the same class are conjugate to each other that means if g_1 and g_2 are in the same class there exists a $g \in G \ni g_1 = g^{-1}g_2g$.

7.4 Magnetic Resonance and Crystal Field Theory

Table 7.7. The character table of D_3

C_1		C_2		C_3		
g_1		g_2	g_3	g_4	g_5	g_6
$\chi^{(1)}$	1	1	1	1	1	1
$\chi^{(2)}$	1	1	1	-1	-1	-1
$\chi^{(3)}$	2	-1	-1	0	0	0

Below we summarize some important rules for constructing the character table for the irreducible representations. These results will not be proved, since they are readily available.[29] These rules are:

1. The number of classes s in the group is equal to the number of irreducible representations of the group.

2. If n_i is the dimension of the ith irreducible representation, then $n_i = \chi_i(E)$, where E is the identity of the group and $\sum_i^s n_i^2 = h$, where h is the order of the group G. For small finite groups, this rule obviously greatly restricts what the n_i can be.

3. If B_k is the number of group elements in the class C_k, then the characters for each class obey the relationship

$$\sum_{k=1}^{s} B_k \chi^{(l)*}(C_k) \chi^{(j)}(C_k) = h \delta_l^j ,$$

where δ_l^j is the Kronecker delta. This relation is often called the *orthogonality relation for characters*.

4. Suppose the order of a group element g is m (i.e. suppose $g^m = E$). Further suppose that the dimension of a representation (which need not be irreducible) is n. It then follows that $\chi(g)$ equals the sum of n, mth roots of unity.

5. The one-dimensional representation is always present.

Finally it is worth giving the criterion for determining the irreducible representations in a given reducible representation. The rule is if

$$R = \sum_i C_i' R^i , \qquad (7.288)$$

then C_i' (which is the number of times that irreducible representation i appears in the reducible representation R) is given by

$$C_i' = (1/h) \sum_k B_k \chi^{(i)}(C_k)^* \chi(C_k) , \qquad (7.289)$$

where χ denotes character relative to R and the sum over k is a sum over classes. When a reducible representation is expressible in the form (7.288) it is said to be in, *reduced form*. Putting it into such a form as (7.288) is called *reduction*.

---

[29] See Mathews and Walker [7.47].

A frequent use of these results occurs when the representation R is formed by taking *direct products* (see Sect. 1.2.1 for a definition) of the representations $R^{(i)}$. We can then evaluate (7.289) by remembering that the trace of a direct product is the product of the traces.

There are many ways that group theory has been used as an aid in actual calculations. No doubt there remain other ways that have not yet been discovered. The basic ideas that we will use in our physical calculations involve:

1. The physical system determines a symmetry group with irreducible representations that can be found by group theory.

2. Except for what is called by definition "accidental degeneracy" we have a distinct eigenvalue for each (occurrence of an) irreducible representation. (It is possible for the same irreducible representation to occur many times. The meaning of the word "occur" will be given later.)

3. The dimension of the irreducible representation is the degeneracy of each corresponding eigenvalue.

For a brief insight into the above, let the eigenfunctions of $\mathcal{H}$ corresponding to the eigenvalue E_n be labeled ψ_{ni} ($i \to 1$ to d). E_n is thus d-fold degenerate. Thus

$$\mathcal{H}\psi_{ni} = E_n\psi_{ni}. \tag{7.290}$$

If g is an element of the symmetry group G, it follows that

$$[g, \mathcal{H}] = 0. \tag{7.291}$$

From this,

$$\mathcal{H}(g\psi_{ni}) = E_n(g\psi_{ni}). \tag{7.292}$$

Comparing (7.290) and (7.291), we see that

$$g\psi_{ni} = \sum_{i=1}^{d} C_i^n \psi_{in}. \tag{7.293}$$

It can be shown that C_i^n matrices are a representation of the group G. We thus have the desired connection between energy levels, degeneracy, and representations.

Let us consider the physically interesting problem of an atom with one 4f electron. Let us place this atom in a potential with trigonal symmetry. The group appropriate to trigonal symmetry is our old friend D_3. We want to neglect spin and discover what happens (or may happen) to the 4f energy levels when the atom is placed in a trigonal field. This is a problem that could be directly attacked by perturbation theory, but it is interesting to see what type of statements can be made by the group theory.

If you think a little about the ideas we have introduced and about our problem, you should come to the conclusion that what we have to find is the irreducible representations of D_3 generated by (in previous notation) $\psi_{(4f)i}$. Here i runs from 1 to 7. This problem can be solved by using (7.289).

7.4 Magnetic Resonance and Crystal Field Theory

The first thing we need to know is the character of our rotation group. This is given by [61] for the lth irreducible representation

$$\chi^{(l)}(\phi) = \frac{\sin[(l+\tfrac{1}{2})\phi]}{\sin(\phi/2)}. \tag{7.294}$$

In (7.294), ϕ is an appropriate rotation angle. Since we are dealing with a 4f level we are interested only in the case $l = 3$:

$$\chi^{(3)}(\phi) = \frac{\sin(7\phi/2)}{\sin(\phi/2)}. \tag{7.295}$$

By (7.289), we need to evaluate (7.294) only for ϕ in each of the three classes of D_3. Since the classes of D_3 correspond to the identity, three-fold rotations and two-fold rotations, we have

$$\chi^{(3)}(0) = 7, \tag{7.296a}$$

$$\chi^{(3)}(2\pi/3) = +1, \tag{7.296b}$$

$$\chi^{(3)}(\pi) = -1. \tag{7.296c}$$

Table 7.13. Character table for calculating C'_i

		C_1	C_2	C_3
	B_k	1	2	3
	$\chi^{(1)}$	1	1	1
D_3	$\chi^{(2)}$	1	1	-1
	$\chi^{(3)}$	2	-1	0
Rotation Group	$\chi^{(3)}$	7	+1	-1

We can now construct **Fehler! Verweisquelle konnte nicht gefunden werden..** Applying (7.289) we have

$$C'_1 = \tfrac{1}{6}[(7)(1)(1) + 2(+1)(1) + 3(-1)(1)] = 1,$$
$$C'_2 = \tfrac{1}{6}[7(1)(1) + 2(+1)(1) + 3(-1)(-1)] = 2,$$
$$C'_3 = \tfrac{1}{6}[7(1)(2) + 2(+1)(-1) + 3(-1)(0)] = 2.$$

Thus

$$R = 2R^{(3)} + R^{(1)} + 2R^{(2)}. \tag{7.297}$$

By (7.297) we expect the 4f level to split into two doubly degenerate levels plus three nondegenerate levels. The two levels corresponding to $R^{(3)}$ that occur twice and the two $R^{(2)}$ levels will probably not have the same energy.

7.5 Brief Mention of Other Topics

7.5.1 Spintronics or Magnetoelectronics (EE)[30]

We are concerned here with spin-polarized transport for which the name spintronics is sometimes used. We need to think back to the ideas of band ferromagnetism as contained for example in the Stoner model. Here one assumes that an exchange interaction can cause the spin-up and spin-down density of states to split apart as shown in the schematic diagram (for simplicity we consider that the majority spin-up band is completely filled). Thus, the number of electrons at the Fermi level with spin up (N_{up}) can differ considerably from the number with spin down (N_{down}). See (7.242) and Fig. 7.32 (spins and moments have opposite directions due to the negative charge of the electron – the spins are drawn in the bands). This results in two phenomena: (a) a net magnetic moment, and (b) a net spin polarization in transport defined by

$$P = \frac{N_{up} - N_{down}}{N_{up} + N_{down}}. \tag{7.298}$$

Fe, Ni, and Co typically have P of order 50%.

In the figures, the $D(E)$ describe the density of states of up and down spins. As shown also in Fig. 7.33 one can use this idea to produce a "spin valve," which preferentially transmits electrons with one spin direction. Spin valves have many possible device applications (see, e.g., Prinz [7.55]). In Fig. 7.33 we show transport from a magnetized metal to a magnetized metal through a nonmagnetic metal.

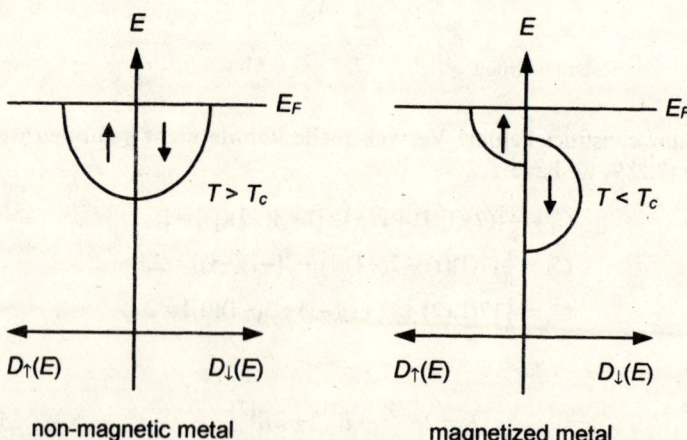

Fig. 7.32. Exchange coupling causes band ferromagnetism. The D are the density of states of the spin-up and spin-down bands. EF is the Fermi energy. Adapted from Prinz [7.55]

---

[30] A comprehensive review has recently appeared, Zutic et al [7.73].

The two ferromagnets are still exchange coupled through the metal separating them. For the case of the second metal being antialigned with the first, the current is reduced and the resistance is high. The electrons with moment up can go from (a) to (b) but are blocked from (b) to (c). The moment-down electrons are inhibited from movement by the gap from (a) to (b). If the second magnetized metal were aligned, the resistance would be low. Since the second ferromagnet's magnetization direction can be controlled by an external magnetic field, this is the principle used in GMR (giant magnetoresistance, discovered by Baibich et al in 1988). See Baibich et al [7.4].

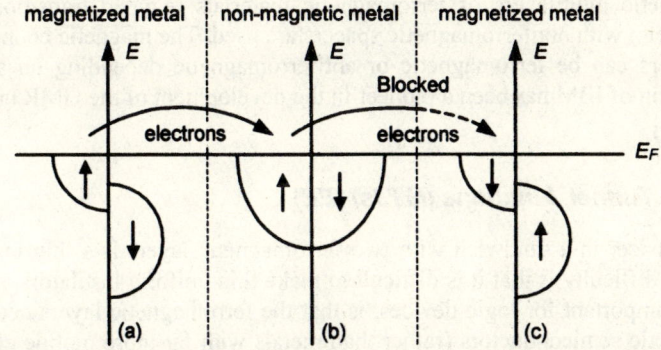

Fig. 7.33. Due to preferential transmission of spin orientation, the resistance is high if the second ferromagnet is antialigned. Adapted from Prinz [7.55]

One should note that spintronic devices are possible because the spin diffusion length that is the square root of the diffusion constant times the spin relaxation time can be fairly large, e.g. 0.1 mm in Al at 40 K. This means that the spin polarization of the transport will typically last over these distances when the polarized current is injected into a nonmagnetized metal or semiconductor. Only in 1988 was it realized that electronic current flowing into an ordinary metal from a ferromagnet could preserve spin, so that spin could be transported just as charge is.

We should also mention that control of spin is important in efforts to achieve quantum computing. Quantum computers perform a series of sequences of unitary transformation on sets of "qubits" – see Bennett [7.5] for a definition. In essence, this holds out the possibility of something like massive parallel computation. Quantum computing is a huge subject; see, e.g., Bennett [7.5].

Hard Drives (EE)

In 1997 IBM introduced another innovation – the giant magnetoresistance (GMR) read head for use in magnetic hard drives – in which magnetic and nonmagnetic materials are layered one in the read head, roughly doubling or tripling its sensitivity. By layering one can design the device with the desired GMR properties. The device works on the quantum-mechanical principle, already mentioned, that when the layers are magnetized in the same direction, the spin-dependent scattering is small, and when the layers are alternatively magnetized in opposite directions, the electrons experience a maximum of spin-dependent scattering (and hence much higher resistivity). Thus, magnetoresistance can be used to read the state of a magnetic bit in a magnetic disk drive. The direction of soft layer in the read head can be switched by the direction of the magnetization in the storing media. The magnetoresistance is thus changed and the direction of storage is then read by the size of the current in the read head. Sandwiches of Co and Cu can be used with the widths of the layers typically of the order of nanometers (a few atoms say) as this is the order of the wavelength of electrons in solids. More generally, magnetic multilayers of ferromagnetic materials (e.g. 3d transition metal ferromagnets) with nonferromagnetic spacers are used. The magnetic coupling between layers can be ferromagnetic or antiferromagnetic depending on spacing. Stuart Parkin of IBM has been a pioneer in the development of the GMR hard disk drive [7.52]

Magnetic Tunnel Junctions (MTJs) (EE)

Here the spacer in a sandwich with two ferromagnetic layers is a thin insulating layer. One difficulty is that it is difficult to make thin uniform insulators. Another difficulty, important for logic devices, is that the ferromagnetic layers need to be ferromagnetic semiconductors (rather than metals with far more mobile electrons than in semiconductors) so that a large fraction of the spin-aligned electrons can get into the rest of the device (made of semiconductors). GaMnAs and $TiCoO_2$ are being considered for use as ferromagnetic semiconductors for these devices.

The tunneling current depends on the relative magnetization directions of the ferromagnetic layers. It should be mentioned here that in the usual GMR structures the current typically flows parallel to the layers (but electrons undergo a random walk, and sample more than one layer so GMR can still operate), while in a MTJ sandwich the current typically flows perpendicular to the layers.

For the typical case, the resistance of the MTJ is lower when the moments of the ferromagnetic layers are aligned parallel and higher when the moments are antiparallel. This produces tunneling magnetic resistance TMR that may be 40% or so larger than GMR. MTJ holds out the possibility of making nonvolatile memories.

Spin-dependent tunneling through the FM-I-FM (ferromagnetic-insulator-ferromagnetic) sandwich had been predicted by Julliere [7.34] and Slonczewski [7.61].

Colossal Magnetoresistance (EE)

Magnetoresistance (MR) can be defined as

$$MR = \frac{\rho(H) - \rho(0)}{\rho(0)}, \tag{7.299}$$

where ρ is the resistivity and H is the magnetic field.

Typically, MR is a few per cent, while GMR may be a few tens of per cent. Recently, materials with so-called colossal magnetoresistance (CMR) of 100 per cent or more have been discovered. CMR occurs in certain oxides of manganese – manganese perovskites (e.g. $La_{0.75}Ca_{0.25}MnO_3$). Space does not permit further discussion here. See Fontcuberta [7.23]. See also Salamon and Jaime [7.58].

7.5.2 The Kondo Effect (A)

Scattering of conduction electrons by localized moments due to s-d exchange can produce surprising effects as shown by J. Kondo in 1964. Although, this would appear to be a very simple basic phenomena that could be easily understood, at low temperature Kondo carried the calculation beyond the first Born approximation and showed that as the temperature is lowered the scattering is enhanced. This led to an explanation of the old problem of the resistance minimum as it occurred in, e.g., dilute solutions of Mn in Cu.

The Kondo temperature is defined as the temperature at which the Kondo effect clearly appears and for which Kondo's result is valid (see (7.301)). It is given approximately by

$$T_k = E_F \sqrt{J} \exp\left(-\frac{1}{nJ}\right), \tag{7.300}$$

where T_k is the Kondo temperature, E_F is the Fermi energy, J characterizes the strength of the exchange interaction, and n is the density of states. Generally T_k is below the resistance minimum that can be estimated from the approximate expression giving the resistivity ρ,

$$\rho = a - b\ln(T) + cT^5. \tag{7.301}$$

The $\ln(T)$ term contains the spin-dependent Kondo scattering and cT^5 characterizes the resistivity due to phonon scattering at low temperature (the low temperature is also required for a sharp Fermi surface), and a, b and c are constants with b being proportional to the exchange interaction. This leads to a resistivity minimum at approximately

$$T_M = \left(\frac{b}{5c}\right)^{1/5}. \tag{7.302}$$

In actual practice the Kondo resistivity does not diverge at extremely low temperatures, but rather at temperatures well below the Kondo temperature, the resistivity approaches a constant value as the conduction electrons and impurity spins form a singlet. Wilson has used renormalized group theory to explain this. There are actually three regimes that need to be distinguished. The logarithmic regime is above the Kondo temperature, the crossover region is near the Kondo temperature, and the plateau of the resistivity occurs at the lowest temperatures. To discuss this in detail would take us well beyond the scope of this book. See, e.g., Kirk WP,

"Kondo Effect," pp. 162-165 in [24] and references contained therein. Using quantum dots as artificial atoms and studying them with scanning tunneling microscopes has revived interest in the Kondo effect. See Kouwenhoven and Glazman [7.41].

7.5.3 Spin Glass (A)

Another class of order that may occur in magnetic materials at low temperatures is spin glass. The name is meant to suggest frozen in (long-range) disorder. Experimentally the onset of a spin glass is signaled by a cusp in the magnetic susceptibility at T_f (the freezing temperature) in zero magnetic field. Below T_f there is no long-range order. The classic examples of spin glasses are dilute alloys of iron in gold (Au:Fe, also Cu:Mn, Ag:Mn, Au:Mn and several other examples). The critical ingredients of a spin glass seem to be (a) a competition among interactions as to the preferred direction of a spin (frustration), and (b) a randomness in the interaction between sites (disorder). There are still many questions surrounding spin glasses such as do they have a unique ground state and if the spin glass transition is a true phase transition to a new state (see Bitko [7.7]).

For spin glasses, it is common to define an order parameter by summing over the average spin's squared:

$$q = \frac{1}{N}\sum_{i=1}^{N}\langle S_i \rangle^2, \tag{7.303}$$

and for $T > T_f$, $q = 0$, while $q \neq 0$ for $T < T_f$. Much further detail can be found in Fischer and Hertz [7.22]. See also the article by Young [7.8 pp. 331-346].

Randomness and frustration (where two paths linking the same pair of spins do not have the same net effective sign of exchange coupling) are shared by many other systems besides spin glasses. Or another way of saying this is that the study of spin glasses fall in the broad category of the study of disordered systems, including random field systems (like diluted antiferromagnets), glasses, neural networks, optimization and decision problems. Other related problems include combinatorial optimization problems, such as the traveling salesman problem, and other problems involving complexity. For the neural network problem see for example, Muller and Reinhardt [7.50]. The book by Fischer and Hertz, already mentioned has a chapter on the physics of complexity with references. Another reference to get started in this general area is Chowdhury [7.12]. Mean-field theories of spin glasses have been promising, but there is no general consensus as to how to model spin glasses.

It is worth looking at a few experimental results to show real spin glass properties. Figure 7.34 shows the cusp in the susceptibility for CuMn. The true T_f occurs as the ac frequency goes to zero. Figure 7.35 shows the temperature dependence of the magnetization and order parameter for CuMn and AuFe. Figure 7.36 shows the magnetic specific heat for two CuMn samples.

7.5 Brief Mention of Other Topics

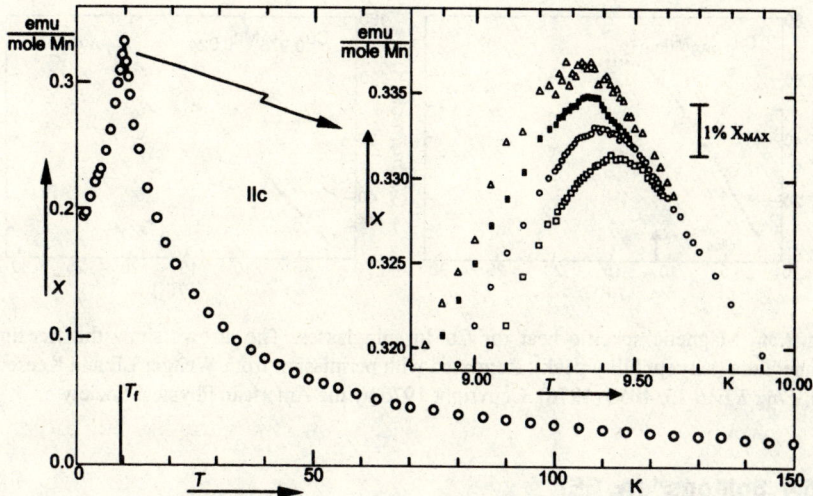

Fig. 7.34. The ac susceptibility as a function of T for CuMn (1 at %). Measuring frequencies: □, 1.33 kHz; ○, 234 Hz; ■, 10.4 Hz; and △, 2.6 Hz. From Mydosh JA, "Spin-Glasses – The Experimental Situation" *Magnetism in Solids Some Current Topics*, Scottish Universities Summer School in Physics, 1981, p. 95, by permission of SUSSP. Data from Mulder CA, van Duyneveldt AJ, and Mydosh JA, *Phys Rev*, **B23**, 1384 (1981)

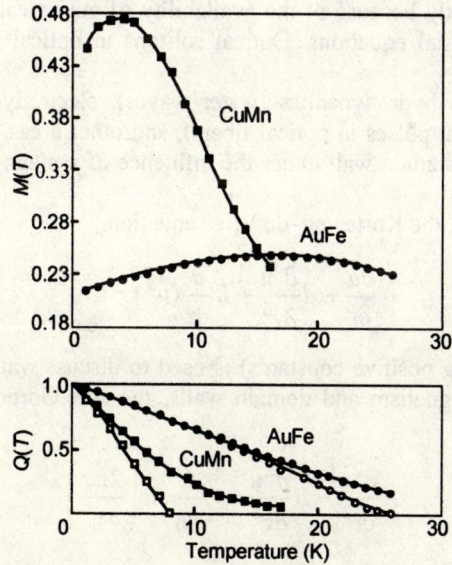

Fig. 7.35. The temperature variation of the magnetization $M(T,H)$ and order parameter $Q(T,H)$ with vanishing field (open symbols) and with 16 kG applied external magnetic field (full symbols) for Cu–0.7 at% Mn and Au–6.6 at% Fe; $M(T,H=0)$ is zero. After Mookerjee A and Chowdhury D, *J Physics F, Metal Physics* **13**, 365 (1983), by permission of the Institute of Physics

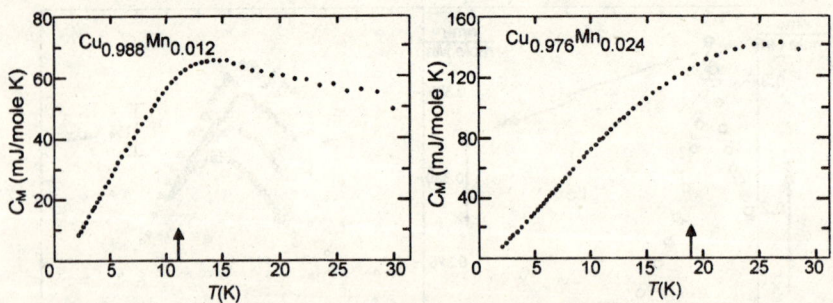

Fig. 7.36. Magnetic specific heat for CuMn spinglasses. The arrows show the freezing temperature (susceptibility peak). Reprinted with permission from Wenger LE and Keesom PH, *Phys Rev B* **13**, 4053 (1976). Copyright 1976 by the American Physical Society

7.5.4 Solitons[31] (A, EE)

Solitary waves are large-amplitude, localized, stable propagating disturbances. If in addition they preserve their identity upon interaction they are called solitons. They are particle-like solutions of nonlinear partial differential equations. They were first written about by John Scott Russell, in 1834, who observed a peculiar stable shallow water wave in a canal. They have been the subject of much interest since the 1960s, partly because of the availability of numerical solutions to relevant partial differential equations. Optical solitons in optical fibers are used to transmit bits of data.

Solitons occur in hydrodynamics (water waves), electrodynamics (plasmas), communication (light pulses in optical fibers), and other areas. In magnetism the steady motion of a domain wall under the influence of a magnetic field is an example of a soliton.[32]

In one dimension, the Korteweg–de Vries equation

$$\frac{\partial u}{\partial t} + A\frac{\partial^3 u}{\partial x^3} + B\frac{\partial}{\partial x}(u^2) = 0$$

(with A and B being positive constants) is used to discuss water waves. In other areas, including magnetism and domain walls, the sine-Gordon equation is encountered

$$A\frac{\partial^2 u}{\partial t^2} - B\frac{\partial^2 u}{\partial x^2} = -\frac{C}{u_0}\sin\left(\frac{2\pi u}{u_0}\right)$$

---

[31] See Fetter and Walecka [7.20] and Steiner, "Linear and non linear modes in 1d magnets," in [7.14, p199ff].

[32] See the article by Krumhansl in [7.8, pp. 3–21] who notes that static solutions are also solitons.

(with A, B, C, and u_0 being positive constants). Generalization to higher dimension have been made. The solitary wave owes its stability to the competition of dispersion and nonlinear effects (such as a tendency to steepen waves). The solitary wave propagates with a velocity that depends on amplitude.

Problems

7.1 Calculate the demagnetization factor of a sphere.

7.2 In the mean-field approximation in dimensionless units for spin 1/2 ferromagnets the magnetization (m) is given by

$$m = \tanh\left(\frac{m}{t}\right),$$

where $t = T/T_c$ and T_c is the Curie temperature. Show that just below the Curie temperature $t < 1$,

$$m = \sqrt{3}\sqrt{1-t}.$$

7.3 Evaluate the angular momentum L and magnetic moment μ for a sphere of mass M (mass uniformly distributed through the volume) and charge Q (uniformly distributed over the surface), assuming a radius r and an angular velocity ω. Thereby, obtain the ratio of magnetic moment to angular momentum.

7.4 Derive Curie's law directly from a high-temperature expansion of the partition function. For paramagnets, Curie's law is

$$\chi = \frac{C}{T} \text{ (The magnetic susceptibility),}$$

where Curie's constant is

$$C = \frac{\mu_0 N g^2 \mu_B^2 j(j+1)}{3k}.$$

N is the number of moments per unit volume, g is Lande's g factor, μ_B is the Bohr magneton, and j is the angular momentum quantum number.

7.5 Prove (7.155).

7.6 Prove (7.156).

7.7 In one spatial dimension suppose one assumes the Heisenberg Hamiltonian

$$\mathcal{H} = -\frac{1}{2}\sum_{R,R'} J(R-R') S_R \cdot S_{R'}, \quad J(0) = 0,$$

where $R - R' = \pm a$ for nearest neighbor and $J_1 \equiv J(\pm a) > 0$, $J_2 \equiv J(\pm 2a) = -J_1/2$ with the rest of the couplings being zero. Show that the stable ground state is helical and find the turn angle. Assume classical spins. For simplicity, assume the spins are confined to the (x,y)-plane.

7.8 Show in an antiferromagnetic spin wave that the neighboring spins precess in the same direction and with the same angular velocity but have different amplitudes and phases. Assume a one-dimensional array of spins with nearest-neighbor antiferromagnetic coupling and treat the spins classically.

7.9 Show that (7.163) is a consistent transformation in the sense that it obeys a relation like (7.175), but for S_j^-.

7.10 Show that (7.138) can be written as

$$\mathcal{H} = -J\sum_{j\delta}[S_{jz}S_{j+\delta,z} + \tfrac{1}{2}(S_j^- S_{j+\delta}^+ + S_j^+ S_{j+\delta}^+)] - 2\mu_0 \mu H \sum_j S_{jz}.$$

7.11 Using the definitions (7.179), show that

$$[b_k, b_{k'}^\dagger] = \delta_k^{k'},$$
$$[b_k, b_{k'}] = 0,$$
$$[b_k^\dagger, b_{k'}^\dagger] = 0.$$

7.12 (a) Apply Hund's rules to find the ground state of Nd^{3+} ($4f^3 5s^2 p^6$).
(b) Calculate the Lande g-factor for this case.

7.13 By use of Hund's rules, show that the ground state of Ce^{3+} is $^2F_{5/2}$, of Pm^{3+} is $^5I_4$, and of Eu^{3+} is $^7F_0$.

7.14 Explain what the phrases "$3d^1$ configuration" and "^2D term" mean.

7.15 Give a rough order of magnitude estimate of the magnetic coupling energy of two magnetic ions in EuO ($T_c \cong 69$ K). How large an external magnetic field would have to be applied so that the magnetic coupling energy of a single ion to the external field would be comparable to the exchange coupling energy (the effective magnetic moment of the magnetic Eu^{2+} ions is 7.94 Bohr magnetons)?

7.16 Estimate the Curie temperature of EuO if the molecular field were caused by magnetic dipole interactions rather than by exchange interactions.

7.17 Prove the Bohr–van Leeuwen theorem that shows the absence of magnetism with purely classical statistics. Hint – look at Chap. 4 of Van Vleck [7.63].

8 Superconductivity

8.1 Introduction and Some Experiments (B)

In 1911 H. Kamerlingh Onnes measured the electrical resistivity of mercury and found that it dropped to zero below 4.15 K. He could do this experiment because he was the first to liquefy helium and thus he could work with the low temperatures required for superconductivity. It took 46 years before Bardeen, Cooper, and Schrieffer (BCS) presented a theory that correctly accounted for a large number of experiments on superconductors. Even today, the theory of superconductivity is rather intricate and so perhaps it is best to start with a qualitative discussion of the experimental properties of superconductors.

Superconductors can be either of type I or type II, whose different properties we will discuss later, but simply put the two types respond differently to external magnetic fields. Type II materials are more resistant, in a sense, to a magnetic field that can cause destruction of the superconducting state. Type II superconductors are more important for applications in permanent magnets. We will introduce the Ginzburg–Landau theory to discuss the differences between type I and type II.

The superconductive state is a macroscopic state. This has led to the development of superconductive quantum interference devices that can be used to measure very weak magnetic fields. We will briefly discuss this after we have laid the foundation by a discussion of tunneling involving superconductors.

We will then discuss the BCS theory and show how the electron–phonon interaction can give rise to an energy gap and a coherent motion of electrons without resistance at sufficiently low temperatures.

Until 1986 the highest temperature that any material stayed superconducting was about 23 K. In 1986, the so-called high-temperature ceramic superconductors were found and by now, materials have been discovered with a transition temperature of about 140 K (and even higher under pressure). Even though these materials are not fully understood, they merit serious discussion. In 2001 MgB_2, an intermetallic material was discovered to superconduct at about 40 K and it was found to have several unusual properties. We will also discuss briefly so-called heavy-electron superconductors.

Besides the existence of superconductivity, Onnes further discovered that a superconducting state could be destroyed by placing the superconductor in a large enough magnetic field. He also noted that sending a large enough current through the superconductor would destroy the superconducting state. Silsbee later suggested that these two phenomena were related. The disruption of the superconductive state is caused by the magnetic field produced by the current at the surface of the wire. However, the critical current that destroys superconductivity is very structure sensitive (see below) so that it can be regarded for some purposes as an independent parameter. The critical magnetic field (that destroys superconductivity) and the critical temperature (at which the material becomes superconducting) are related in the sense that the highest transition temperature occurs when there is no external magnetic field with the transition temperature decreasing as the field increases. We will discuss this a little later when we talk about type I and type II superconductors. Figure 8.1 shows at low temperature the difference in behavior of a normal metal versus a superconductor.

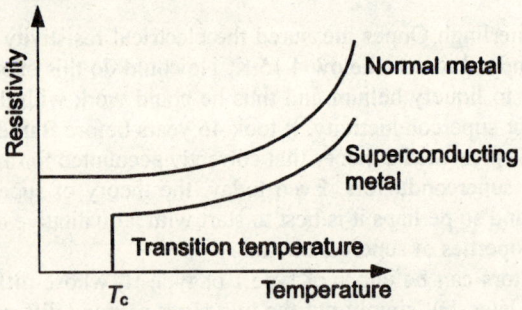

Fig. 8.1. Electrical resistivity in normal and superconducting metals (schematic)

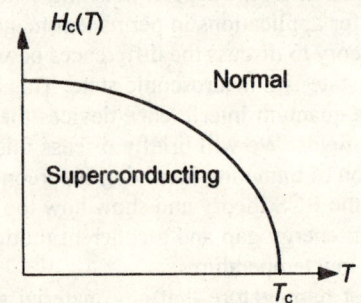

Fig. 8.2. Schematic of critical field vs. temperature for Type I superconductors

In 1933, Meissner and Ochsenfeld made another fundamental discovery. They found that superconductors expelled magnetic flux when they were cooled below the transition temperature. This established that there was more to the superconducting state than perfect conductivity (which would require $E = 0$); it is also a state of perfect diamagnetism or $B = 0$. For a long, thin superconducting specimen, $B = H + 4\pi M$ (cgs). Inside $B = 0$, so $H + 4\pi M = 0$ and $H_{in} = B_a$ (the externally applied B field) by the boundary conditions of H along the length being continuous. Thus, $B_a + 4\pi M = 0$ or $\chi = M/B_a = -1/(4\pi)$, which is the case for a perfect diamagnet. Exclusion of the flux is due to perfect diamagnetism caused by surface currents, which are always induced so as to shield the interior from external magnetic fields. A simple application of Faraday's law for a perfect conductor would lead to a constant flux rather than excluded flux. A plot of critical field *versus* temperature typically (for type I as we will discuss) looks like Fig. 8.2. The equation describing the critical fields dependence on temperature is often empirically found to obey

$$H_c(T) = H_c(0)\left[1 - \left(\frac{T}{T_c}\right)^2\right]. \tag{8.1}$$

In 1950, H. Fröhlich discussed the electron–phonon interaction and considered the possibility that this interaction might be responsible for the formation of the superconducting state. At about the same time, Maxwell and Reynolds, Serin, Wright, and Nesbitt found that the superconducting transition temperature depended on the isotopic mass of the atoms of the superconductor. They found $M^a T_c \cong$ constant. This experimental result gave strong support to the idea that the electron–phonon interaction was involved in the superconducting transition. In the simplest model, $\alpha = 1/2$.

In 1957, Bardeen, Cooper, and Schrieffer (BCS) finally developed a formalism that contained the correct explanation of the superconducting state in common superconductors. Their ideas had some similarity to Fröhlich's. A key idea of the BCS theory was developed by Cooper in 1956. Cooper analyzed the electron–phonon interaction in a different way from Fröhlich. Fröhlich had discussed the effect of the lattice vibrations on the self-energy of the electrons. Cooper analyzed the effect of lattice vibrations on the effective interaction between electrons and showed that an attractive interaction between electrons (even a very weak attractive interaction at low enough temperature) would cause pairs of the electrons (the Cooper pairs) to form bound states near the Fermi energy (see Sect. 8.5.3). Later, we will discuss the BCS theory and show the pairing interaction causes a gap in the density of single-electron states.

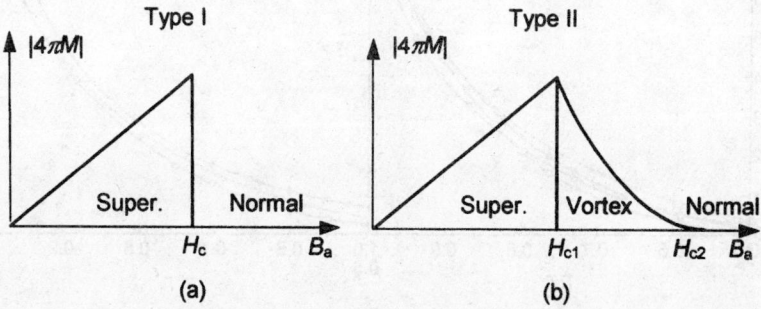

Fig. 8.3. (a) Type I and (b) Type II superconductors

As we have mentioned a distinction is made between type I and type II superconductors. Type I have only one critical field while type II have two critical fields. The idea is shown in Fig. 8.3a and b. $4\pi M$ is the magnetic field produced by the surface superconducting currents induced when the external field is applied. Type I superconductors either have flux penetration (normal state) or flux exclusion (superconductivity state). For type II superconductors, there is no flux penetration below H_{c1}, the lower critical field, and above the upper critical field H_{c2} the material is normal. But, between H_{c1} and H_{c2} the superconductivity regions are threaded by vortex regions of the flux penetration. The idea is shown in Fig. 8.4.

Type I and type II behavior will be discussed in more detail after we discuss the Ginzburg–Landau equations for superconductivity. We now mention some experiments that support the theories of superconductivity.

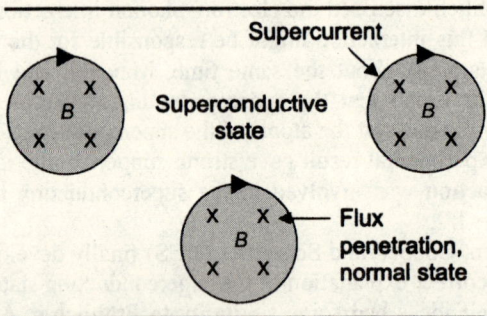

Fig. 8.4. Schematic of flux penetration for type II superconductors. The gray areas represent flux penetration surrounded by supercurrent (vortex). The net effect is that the superconducting regions in between have no flux penetration

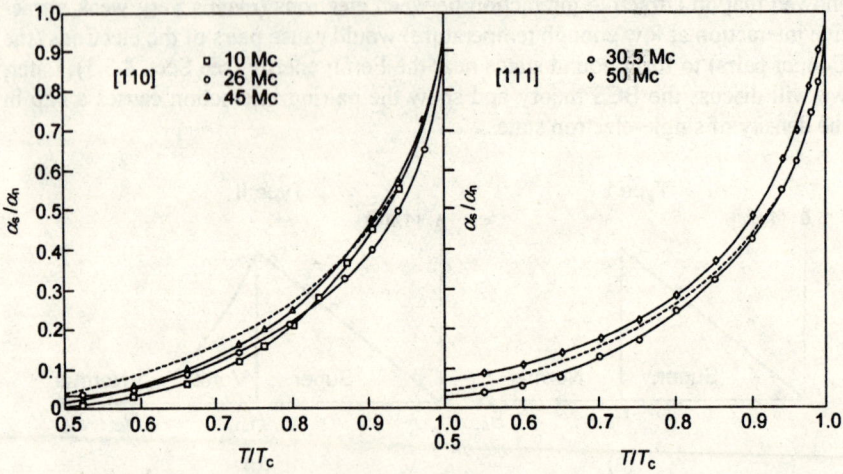

Fig. 8.5. Absorption coefficients ultrasonic attenuation in Pb (α_n refers to the normal state, α_s refers to the superconducting state, and T_c is the transition temperature). The dashed curve is derived from BCS theory and it uses an energy gap of 4.2 kT_c. [Reprinted with permission from Love RE and Shaw RW, *Reviews of Modern Physics* **36**(1) part 1, 260 (1964). Copyright 1964 by the American Physical Society.]

8.1.1 Ultrasonic Attenuation (B)

The BCS theory of the ratio of the normal to the superconducting absorption coefficients (α_n to α_s) as a function of temperature variation of the energy gap (discussed in detail later) can be interpreted in such a way as to give information on the temperature variation of the energy gap. Some experimental results on (α_n/α_s) *versus* temperature are shown in Fig. 8.5. Note the close agreement of experiment and theory, and that the absorption of superconductors is much lower than for the normal case when well below the transition temperature.

8.1.2 Electron Tunneling (B)

There are at least two types of tunneling experiments of interest. One involves tunneling from a superconductor to a superconductor with a thin insulator separating the two superconductors. Here, as will be discussed later, the Josephson effects are caused by the tunneling of pairs of electrons. The other type of tunneling (Giaever) involves tunneling of single quasielectrons from an ordinary metal to a superconducting metal. As will be discussed later, these measurements provide information on the temperature dependence of the energy gap (which is caused by the formation of Cooper pairs in the superconductor), as well as other features.

8.1.3 Infrared Absorption (B)

The measurement of transmission or reflection of infrared radiation through thin films of a superconductor provides direct results for the magnitude of the energy gap in superconductors. The superconductor absorbs a photon when the photon's energy is large enough to raise an electron across the gap.

8.1.4 Flux Quantization (B)

We will discuss this phenomenon in a little more detail later. Flux quantization through superconducting rings of current provides evidence for the existence of paired electrons as predicted by Cooper. It is found that flux is quantized in units of $h/2e$, not h/e.

8.1.5 Nuclear Spin Relaxation (B)

In these experiments, the nuclear spin relaxation time T_1 is measured as a function of temperatures. The time T_1 depends on the exchange of energy between the nuclear spins and the conduction electrons via the hyperfine interaction. The data for T_1 for aluminum looks somewhat as sketched in Fig. 8.6. The rapid change of T_1 near $T = T_c$ can be explained, at least quantitatively, by BCS theory.

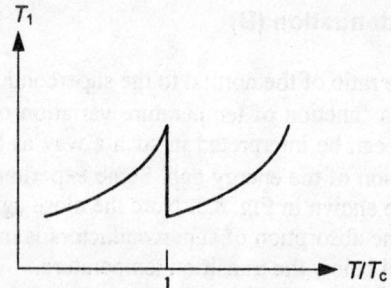

Fig. 8.6. Schematic of nuclear spin relaxation time in a superconductor near T_c

8.1.6 Thermal Conductivity (B)

A sketch of thermal conductivity K versus temperature for a superconductor is shown in Fig. 8.7. Note that if a high enough magnetic field is turned on, the material stays normal—even below T_c. So, a magnetic field can be used to control the thermal conductivity below T_c.

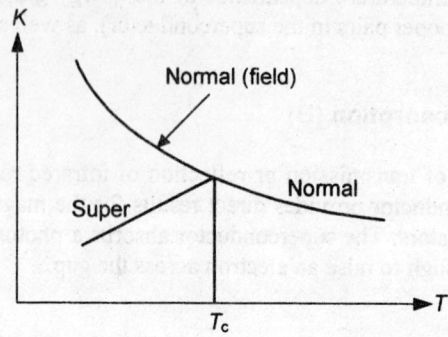

Fig. 8.7. Effect of magnetic field on thermal conductivity K

All of the above experiments have tended to confirm the BCS ideas of the superconducting state. A central topic that needs further elaboration is the criterion for occurrence of superconductivity in any material. We would like to know if the BCS interaction (electrons interacting by the virtual exchange of phonons) is the only interaction. Could there be, for example, superconductivity due to magnetic interactions? Over a thousand superconducting alloys and metals have been found, so superconductivity is not unusual. It is, perhaps, still an open question as to how common it is.

In the chapter on metals, we have mentioned heavy-fermion materials. Superconductivity in these materials seems to involve a pairing mechanism. However, the most probable cause of the pairing is different from the conventional BCS theory. Apparently, the nature of this "exotic" pairing has not been settled as of this writing, and reference needs to be made to the literature (see Sect. 8.7).

For many years, superconducting transition temperatures (well above 20 K) had never been observed. With the discovery of the new classes of high-temperature superconductors, transition temperatures (well above 100 K) have now been observed. We will discuss this later, also. The exact nature of the interaction mechanism is not known for these high-temperature superconductors, either.

8.2 The London and Ginzburg–Landau Equations (B)

We start with a derivation of the Ginzburg–Landau (GL) equations, from which several results will follow, including the London equations. Originally, these equations were proposed on intuitive, phenomenological lines. Later, it was realized they could be derived from the BCS theory. Gor'kov showed the GL theory was a valid description of the BCS theory near T_c. He also showed that the wave function ψ of the GL theory was proportional to the energy gap. Also, the density of superconducting electrons is $|\psi|^2$. Due to spatial inhomogeneities $\psi = \psi(r)$, where $\psi(r)$ is also called the order parameter. This whole theory was developed further by Abrikosov and is often known as the Ginzburg, Landau, Abrikosov, and Gor'kov theory (for further details, see, e.g., Kuper [8.20]

Near the transition temperature, the free energy density in the phenomenological GL theory is assumed to be (gaussian units)

$$F_S(r) = \alpha|\psi|^2 + \frac{1}{2}\beta|\psi|^4 + \frac{1}{2m_S^*}\left|\left(\frac{\hbar}{i}\nabla - \frac{qA}{c}\right)\psi\right|^2 + F_N + \frac{h^2}{8\pi}, \tag{8.2}$$

where N and S refer to normal and superconducting phases. The coefficients α and β are phenomenological coefficients to be discussed. $h^2/8\pi$ is the magnetic energy density ($h = h(r)$ is local and the magnetic induction B is determined by the spatial average of $h(r)$, so A is the vector potential for h, $h = \nabla \times A$). $m^* = 2m$ (for pairs of electrons), $q = 2e$ is the charge and is negative for electrons, and ψ is the complex superconductivity wave function. Requiring (in the usual calculus of variations procedure) $\delta F_S/\delta\psi^*$ to be zero ($\delta F_S/\delta\psi = 0$ would yield the complex conjugate of (8.3)), we obtain the first Ginzburg–Landau equation

$$\left[\frac{1}{2m_S^*}\left(\frac{\hbar}{i}\nabla - \frac{qA}{c}\right)^2 + \alpha + \beta|\psi|^2\right]\psi = 0. \tag{8.3}$$

F_S can be regarded as a functional of ψ and A, so requiring $\delta F_S/\delta A = 0$ we obtain the second GL equation for the current density:

$$\begin{aligned}j &= \frac{c}{4\pi}\nabla \times h = \frac{q\hbar}{2m^*i}(\psi^*\nabla\psi - \psi\nabla\psi^*) - \frac{q^2}{m^*c}\psi^*\psi A \\ &= \frac{q}{m^*}|\psi|^2\left(\hbar\nabla\phi - \frac{q}{c}A\right),\end{aligned} \tag{8.4}$$

where $\psi = |\psi|e^{i\varphi}$. Note (8.3) is similar to the Schrödinger wave equation (except for the term involving β) and (8.4) is like the usual expression for the current density. Writing $n_S = |\psi|^2$ and neglecting, as we have, any spatial variation in $|\psi|$, we find

$$\nabla \times J = -\frac{q^2 n}{m^* c} \nabla \times A = -\frac{q^2 n_S}{m^* c} B,$$

so,

$$\nabla \times J = -\frac{c}{4\pi \lambda_L^2} B, \tag{8.5}$$

where

$$\lambda_L^2 = \frac{m^* c^2}{4\pi n_S q^2}, \tag{8.6}$$

where λ_L is the London penetration depth. Equation (8.5) is London's equation. Note this is the same for a single electron (where $m^* = m$, $q = e$, n_S = ordinary density) or a Cooper pair ($m^* = 2m$, $q = 2e$, $n_S \to n/2$).

Let us show why λ_L is called the London penetration depth. At low frequencies, we can neglect the displacement current in Maxwell's equations and write

$$\nabla \times B = \frac{4\pi}{c} J. \tag{8.7}$$

Combining with (8.5) that we assume to be approximately true, we have

$$\nabla \times (\nabla \times B) = \frac{4\pi}{c} \nabla \times J = -\frac{1}{\lambda_L^2} B, \tag{8.8}$$

or using $\nabla \cdot B = 0$, we have

$$\nabla^2 B = \frac{1}{\lambda_L^2} B. \tag{8.9}$$

For a geometry with a normal material for $x < 0$ and a superconductor for $x > 0$, if the magnetic field at $x = 0$ is B_0, the solution of (8.9) is

$$B(x) = B_0 \exp(-x/\lambda_L). \tag{8.10}$$

Clearly, λ_L is a penetration depth. Thus, if we have a very thin superconducting film (with thickness $\ll \lambda_L$), we really do not have a Meissner effect (flux exclusion). Magnetic flux will penetrate the surface of a superconductor over a distance approximately equal to the London penetration depth $\lambda_L \cong 100$ to 1000 Å. Actually, λ_L is temperature dependent and can be well described by

$$\left(\frac{\lambda_L}{\lambda_{L_0}}\right)^2 = \frac{1}{1-(T/T_c)^4}, \tag{8.11}$$

where

$$\lambda_{L_0} = \sqrt{\frac{m^* c^2}{4\pi n_s q^2}}. \qquad (8.12)$$

8.2.1 The Coherence Length (B)

Consider the Ginzburg–Landau equation in the absence of magnetic fields ($A = 0$). Then, in one dimension from (8.3), we have

$$\left(-\frac{\hbar^2}{2m^*}\frac{d^2}{dx^2} + \alpha + \beta|\psi|^2\right)\psi = 0. \qquad (8.13)$$

We define

$$f = \frac{\psi}{\psi_0} \quad \text{where} \quad \psi_0 = \sqrt{-\frac{\alpha}{\beta}}. \qquad (8.14)$$

Then

$$\left(\frac{-\hbar^2}{2m^*\alpha}\frac{d^2}{dx^2} + 1 - |f|^2\right)f = 0. \qquad (8.15)$$

When f has no gradients $|f| = 1$, which would correspond to being well inside the superconductor. We assume a boundary at $x = 0$ between a normal and a superconductor so $f = 1$, $\psi \to \psi_0$ as $x \to \infty$.

Defining

$$\xi^2(T) = \frac{-\hbar^2}{2m^*\alpha} \qquad (8.16)$$

and letting $f = 1 + g$, where g is small, then

$$\xi^2(T)\frac{d^2g}{dx^2} + 1 + g - (1+g)(1+g)^2 = 0. \qquad (8.17)$$

Keeping only first order in g since it is small,

$$\xi^2(T)\frac{d^2g}{dx^2} - 2g \cong 0 \qquad (8.18)$$

$$g(x) \cong \exp[-\sqrt{2}x/\xi(T)]. \qquad (8.19)$$

Thus, the wave function attains its characteristic value of ψ_0 in a distance $\xi(T)$. $\xi(T)$ is called the coherence length. The coherence length measures the "range" or

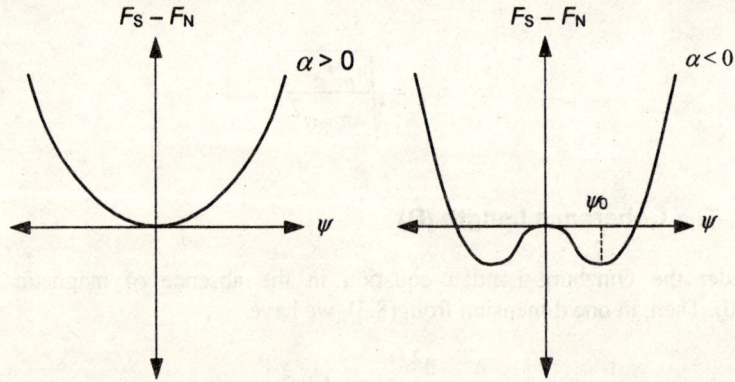

Fig. 8.8. Free energy change at the transition temperature. See (8.20) for how α enters the free energy with no fields or gradients

"size" of Cooper pairs or the distance necessary for the superconducting wave function to change much.

Let us discuss the coherence length further. First, let us review a little about the free energy. The plots for $F_S - F_N$ are shown in Fig. 8.8. The superconducting transition clearly appears at $T = T_c$ or $\alpha = 0$. The free energy for no fields or gradients is

$$F_S - F_N = \alpha\psi^2 + \frac{\beta}{2}\psi^4, \qquad (8.20)$$

so

$$\frac{\partial}{\partial\psi}(F_S - F_N) = 0 \qquad (8.21)$$

gives

$$\psi^2 = \psi_0^2 = -\frac{\alpha}{\beta} \qquad (8.22)$$

at the minimum. Thus at the minimum

$$F_S - F_N = -\frac{\alpha^2}{2\beta}, \qquad (8.23)$$

which would also be the stabilization energy. From thermodynamics, if $F = U - TS$, and $dU = TdS - MdH$, then $dF = -MdH$ at constant T. For perfect diamagnetism,

$$M = \frac{-1}{4\pi}H. \qquad (8.24)$$

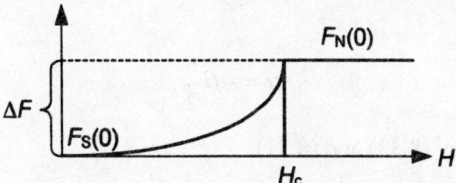

Fig. 8.9. Free energy as a function of field

So,

$$dF_S = \frac{1}{4\pi} H dH, \qquad (8.25)$$

and

$$F_S(H) - F_S(0) = \frac{H^2}{8\pi}. \qquad (8.26)$$

At $H = H_c$, the critical field that destroys superconductivity,

$$F_N(H_c) = F_S(H_c). \qquad (8.27)$$

F_N is almost independent of H so,

$$F_N(H_c) = F_N(0). \qquad (8.28)$$

We show the idea schematically in Fig. 8.9. Therefore,

$$F_S(0) - F_N(0) = -\frac{H_c^2}{8\pi} = -\Delta F \qquad (8.29)$$

would also be the negative of the stabilization energy, or using (8.23)

$$\frac{H_c^2}{8\pi} = \frac{\alpha^2}{2\beta} \qquad (8.30)$$

$$\alpha = \sqrt{\beta \frac{H_c^2}{4\pi}}. \qquad (8.31)$$

Notice deep in the superconductor

$$\psi_S^2 = -\frac{\alpha}{\beta} = n_S = \frac{n}{2}, \qquad (8.32)$$

and

$$\lambda^2 = \frac{m^* c^2}{4\pi q^2 n_S}. \qquad (8.33)$$

So,

$$\alpha = -\beta \frac{n}{2}, \tag{8.34}$$

by (8.32) and thus by (8.33) and (8.31)

$$\alpha = -\frac{2e^2 \lambda^2 H_c^2}{m^* c^2}. \tag{8.35}$$

Combining (8.16) and (8.35), an expression for the coherence length can be derived.

We wish to estimate the upper critical field. In Chap. 2 on electrons, we found the allowed energy levels in a constant **B** field were (in an approximation) free-electron-like parallel to the field and harmonic-oscillator-like in a plane perpendicular to the field. The harmonic energy levels were (dropping * on m)

$$\hbar \omega_c \left(n + \frac{1}{2} \right) = E_{n,k_z} - \frac{\hbar^2 k_z^2}{2m}; \quad \omega_c = \left| \frac{eB}{mc} \right|,$$

or

$$\hbar \frac{|e|B}{mc} \left(n + \frac{1}{2} \right) = -\alpha - \frac{\hbar^2 k_z^2}{2m}, \tag{8.36}$$

for the (linearized) Ginzburg–Landau equation (with $-\alpha$ acting as the eigenvalue in (8.3)). The largest value of B for which solutions of the GL equation exists is ($n = 0$ and letting $k_z = 0$)

$$B_{\max} \equiv H_{c2} = -\frac{2mc\alpha}{\hbar |e|}. \tag{8.37}$$

The two lengths λ and ξ can be defined as a dimensionless ratio K (the GL parameter). H_{c2} can now be described in terms of K and H_c. Using (8.6), (8.16), (8.32), (8.30) and

$$K = \frac{\lambda}{\xi}, \tag{8.38}$$

we find

$$H_{c2} = \sqrt{2} K H_c. \tag{8.39}$$

If $K = \lambda/\xi > 2^{-1/2}$, then $H_{c2} > H_c$. This results in a type II superconductor. The regime of $K > 2^{-1/2}$ is a regime of negative surface energy.

8.2.2 Flux Quantization and Fluxoids (B)

We have for the superconducting current density (by (8.4) with $|\psi|^2$ as spatially constant $= n$)

$$J = \frac{q}{m^*} n \left(\hbar \nabla \varphi - \frac{q}{c} A \right). \tag{8.40}$$

Well inside a superconductor $J = 0$, so

$$\hbar \nabla \varphi = \frac{q}{c} A. \tag{8.41}$$

Applying this to Fig. 8.10 and integrating around the loop gives

$$\hbar \oint \nabla \varphi \cdot dl = \frac{q}{c} \oint A \cdot dl = \frac{q}{c} \int_S B \cdot dS = \frac{q\Phi}{c} \tag{8.42}$$

$$\hbar (2\pi m) = q \frac{\Phi}{c}; \quad m = \text{integer} \tag{8.43}$$

$$\Phi = \frac{hc}{q} m; \quad q = 2e. \tag{8.44}$$

This also applies to Fig. 8.11, so

$$\Phi_0 = \frac{hc}{2e} \equiv \text{the unit of flux of a fluxoid.} \tag{8.45}$$

In the vortex state of the type II superconductors, the minimum current produces the flux Φ_0. In the intermediate state there can be flux tubes threading through the superconductors as shown in Fig. 8.4 and Fig. 8.11. Below, in Fig. 8.12, is a sketch of the penetration depth and coherence length in a superconductor starting with a region of flux penetration. Note $\lambda/\xi < 1$ for type I superconductors.

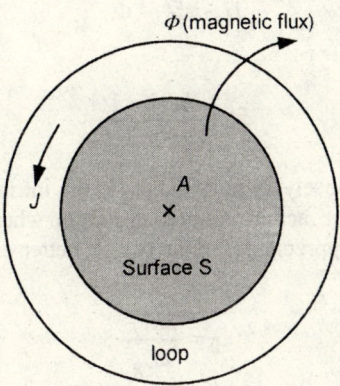

Fig. 8.10. Super current in a ring

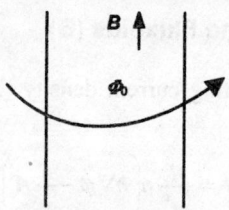

Fig. 8.11. Flux tubes in type II superconductors

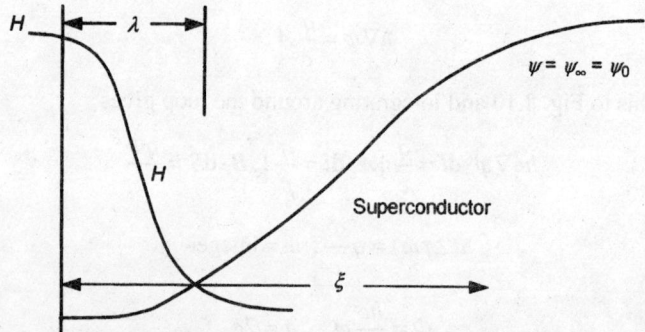

Fig. 8.12. Decay of H and asymptotic value of superconducting wave function

8.2.3 Order of Magnitude for Coherence Length (B)

For type II superconductors, there is a lower critical field H_{c1} for which the flux just begins to penetrate, so

$$H_{c1}\pi\lambda^2 \sim \Phi_0 \tag{8.46}$$

for a single fluxoid. At the upper critical field,

$$H_{c2}\pi\xi_0^2 \sim \Phi_0, \tag{8.47}$$

so that, by (8.44) with $m = 1$,

$$\xi_0^2 \cong \frac{hc}{2e}\frac{1}{\pi}\frac{1}{H_{c2}} \tag{8.48}$$

for fluxoids packed as closely as possible. ξ_0 is the intrinsic coherence length, to be distinguished from the actual coherence length when the superconductor is "dirty" or possessed of appreciable impurities. A better estimate, based on fundamental parameters is[1]

$$\xi_0 = \frac{2\hbar v_F}{\pi E_g}, \tag{8.49}$$

---

[1] See Kuper [8.20 p221].

where v_F is the velocity of the Fermi surface and E_g is the energy gap. The coherence length changes in the presence of scattering. If the electron mean free path is l we have

$$\xi = \xi_0, \tag{8.50}$$

as given by (8.49) for clean superconductors when $\xi_0 < l$ and

$$\xi \cong \sqrt{l\xi_0}, \tag{8.51}$$

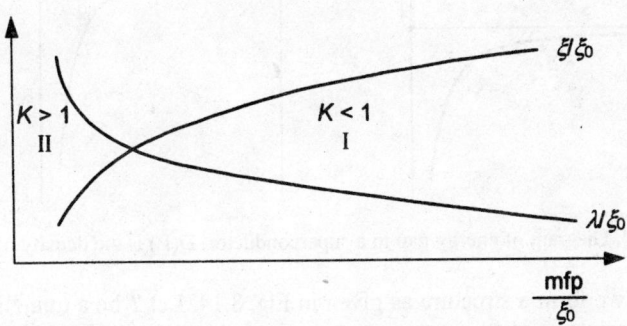

Fig. 8.13. Type I and type II superconductors depending on mfp

for dirty superconductors when $l \ll \xi_0$.[2] That is, dirty superconductors have decreased ξ and increased $K = \lambda/\xi$. The penetration depth can also depend on structure. The idea is schematically shown in Fig. 8.13 where typically the more impure the superconductor the lower the mean free path (mfp) leading to type II behavior.

8.3 Tunneling (B, EE)

8.3.1 Single-Particle or Giaever Tunneling

We anticipate some results of the BCS theory, which we will discuss later. As we will show, when electrons are well separated the electron–lattice interaction can lead to an effective attractive interaction between the electrons. An effective attractive interaction between electrons can cause there to be an energy gap in the single-particle density of states, as we also show later. This energy gap separates the ground state from the excited states and is responsible for most of the unique properties of superconductors.

---

[2] See Saint-James et al [8.27 p. 141]

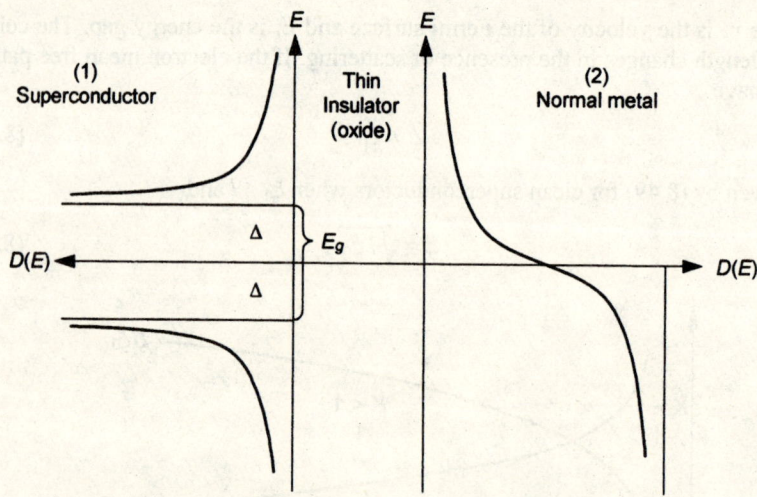

Fig. 8.14. Diagram of energy gap in a superconductor. D(E) is the density of states

Suppose we form a structure as given in Fig. 8.14. Let T be a tunneling matrix element. For the tunneling current we can write (with an applied voltage V)[3]

$$I_{1\to 2} = K' \int_{-\infty}^{\infty} |T|^2 D_1(E) f(E) D_2(E+eV)[1-f(E+eV)] dE \quad (8.52)$$

$$I_{2\to 1} = K' \int_{-\infty}^{\infty} |T|^2 D_1(E) D_2(E+eV) f(E+eV)[1-f(E)] dE \quad (8.53)$$

$$I = I_{1\to 2} - I_{2\to 1} = K' \int_{-\infty}^{\infty} |T|^2 D_1(E) D_2(E+eV)[f(E) - f(E+eV)] dE \quad (8.54)$$

$$I \cong K' D_2(0) |T|^2 \int_{-\infty}^{\infty} D_{1S}(E) \left[-\frac{\partial f}{\partial E} eV \right] dE . \quad (8.55)$$

[4] In the above, K' is a constant, D_i represents density of states, and f is the Fermi function. If we raise the voltage V by $eV = \Delta$, we get the following (see Fig. 8.15) for the net current, and thus, the energy gap can be determined.

---

[3] Note this is actually an oversimplified semiconductor-like picture of a complicated many-body effect [8.14 p. 247], but the picture works well for certain aspects and certainly is the simplest way to get a feel for the experiment.

[4] For the superconducting density of states see Problem 8.2.

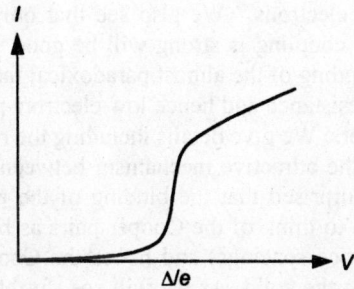

Fig. 8.15. Schematic of Giaever (single-particle) tunneling

8.3.2 Josephson Junction Tunneling

Josephson [8.18] predicted that when two superconductors were separated by an insulator there could be tunneling of Cooper pairs from one to the other provided the insulator was thinner than the coherence length, see Fig. 8.16.

The main concept used to discuss the Josephson effects is that of the phase of the paired electrons. We have already considered this idea in our discussion of flux quantization. F. London had the idea of something like a phase associated with superconducting electrons in that he believed that the motions of electrons in superconductors are correlated over large distances. We now associate the idea of spatial correlation of electrons with the idea of the existence of Cooper pairs. Cooper pairs are sets of two electrons that are attracted to one another (in spite of their Coulomb repulsion) because an electron attracts positive ions. As alluded to earlier, the positive ions in a crystal are much more massive and have, in general, less freedom of movement than the conduction electrons. This means that when an electron has attracted a positive ion to a displaced position, we can imagine the electron as moving out of the area while the positive ion remains displaced for a time. In the region of the crystal where the positive ion(s) is (are) displaced, the crystal has a more positive charge than usual and so this region can attract another electron. We could generalize this argument to consider that the displaced positive ion would be undergoing some sort of motion but still an electron with suitable phase could be attracted to the region of the displaced positive ion. Anyway, the argument seems to make it plausible that there can be an effective attractive interaction between electrons due to the presence of the positive ion lattice. The rather qualitative picture that we have given seems to be the physical content of the statement "Cooper pairs of electrons are formed because of the virtual exchange

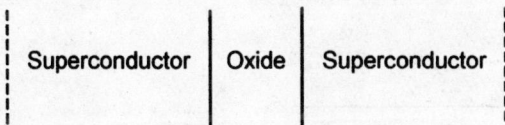

Fig. 8.16. Schematic of Josephson junction

of phonons between the electrons." We also see that only lattices in which the electron–lattice vibration coupling is strong will be good superconductors. Thus, we are led to an understanding of the almost paradoxical fact that good conductors of electricity (with low resistance and hence low electron–phonon coupling) often make poor superconductors. We give details including the role of spin later.

Due to the nature of the attractive mechanism between electrons in a Cooper pair, we should not be surprised that the binding of the electrons is very weak. This means that we have to think of the Cooper pairs as being very large (of the order of many, many lattice spacings) and hence the Cooper pairs overlap with each other a great deal in the solid. As we will see, further analysis of the pairs shows that the electrons in pairs have equal and opposite momentum (in the ground state) and equal and opposite spin. However, the Cooper pairs can accept momentum in such a way that they are still "stable" systems, but so that their center of mass moves. When this happens, the motion of the pairs is influenced by the fact that they are so large many of them must overlap. The Cooper pairs are composed of electrons, and the way electronic wave functions can overlap is limited by the Pauli principle. We now know that overlapping together with the constraint of the Pauli principle causes all Cooper pairs to have the same phase and the same momentum (i.e. the momentum of the center of mass of the Cooper pairs). The pairs are like bosons, in a sense, and condense into a lowest quantum state producing a wave function with phase.

Returning to the coupling of superconductors through an oxide layer, we write a sort of time-dependent "Ginzburg–Landau equations," that allow for coupling,[5]

$$i\hbar \frac{\partial \psi_1}{\partial t} = H_1 \psi_1 + \hbar U \psi_2, \qquad (8.56)$$

$$i\hbar \frac{\partial \psi_2}{\partial t} = H_2 \psi_2 + \hbar U \psi_1. \qquad (8.57)$$

If no voltage or magnetic field is applied, we can assume $H_1 = H_2 = 0$. Then

$$i\hbar \frac{\partial \psi_1}{\partial t} = \hbar U \psi_2, \qquad (8.58)$$

$$i\hbar \frac{\partial \psi_2}{\partial t} = \hbar U \psi_1. \qquad (8.59)$$

---

[5] See, e.g., Feynman et al [8.13], and Josephson [8.18], this was Josephson's Nobel Prize address. See also Dalven [8.11] and Kittel [23 Chap. 12].

We seek solutions of the form (any complex function can always be written as a product of amplitude ρ and $e^{i\varphi}$ where φ is the phase)

$$\psi_1 = \rho_1 \exp(i\varphi_1), \tag{8.60}$$

$$\psi_2 = \rho_2 \exp(i\varphi_2). \tag{8.61}$$

So, using (8.58) and (8.59) we get

$$i\dot{\rho}_1 - \rho_1\dot{\varphi}_1 = U\rho_2 \exp(i\Delta\varphi), \tag{8.62}$$

$$i\dot{\rho}_2 - \rho_2\dot{\varphi}_2 = U\rho_1 \exp(-i\Delta\varphi), \tag{8.63}$$

where

$$\Delta\varphi = (\varphi_2 - \varphi_1) \tag{8.64}$$

is the phase difference between the electrons on the two sides. Separating real and imaginary parts,

$$\dot{\rho}_1 = U\rho_2 \sin\Delta\varphi, \tag{8.65}$$

$$\rho_1\dot{\varphi}_1 = -U\rho_2 \cos\Delta\varphi, \tag{8.66}$$

$$\dot{\rho}_2 = -U\rho_1 \sin\Delta\varphi, \tag{8.67}$$

$$\rho_2\dot{\varphi}_2 = -U\rho_1 \cos\Delta\varphi. \tag{8.68}$$

Assume $\rho_1 \cong \rho_2 \cong \rho$ for identical superconductors, then

$$\frac{d}{dt}(\varphi_2 - \varphi_1) = 0, \tag{8.69}$$

$$\varphi_2 - \varphi_1 \cong \text{constant}, \tag{8.70}$$

$$\dot{\rho}_1 \cong -\dot{\rho}_2. \tag{8.71}$$

The current density J can be written as

$$J \propto \frac{d}{dt}\rho_2^2 = 2\rho_2\dot{\rho}_2, \tag{8.72}$$

so

$$J = J_0 \sin(\varphi_2 - \varphi_1). \tag{8.73}$$

This predicts a dc current with no applied voltage. This is the dc Josephson effect. Another more rigorous derivation of (8.73) is given in Kuper [8.20 p 141]. J_0 is the critical current density or the maximum J that can be carried by Cooper pairs.

The ac Josephson effect occurs if we apply a voltage difference V across the junction, so that $\hbar qV$ with $q = 2e$ is the energy change across the junction. The relevant equations become

$$i\hbar \frac{\partial \psi_1}{\partial t} = \hbar U \psi_2 - eV\hbar \psi_1, \qquad (8.74)$$

$$i\hbar \frac{\partial \psi_2}{\partial t} = \hbar U \psi_1 + eV\hbar \psi_2. \qquad (8.75)$$

Again,

$$i\dot{\rho}_1 - \rho_1 \dot{\varphi}_1 = U\rho_2 \exp(i\Delta\varphi) - eV\rho_1, \qquad (8.76)$$

$$i\dot{\rho}_2 - \rho_2 \dot{\varphi}_2 = U\rho_1 \exp(-i\Delta\varphi) + eV\rho_2. \qquad (8.77)$$

So, separating real and imaginary parts

$$\dot{\rho}_1 = U\rho_2 \sin \Delta\varphi \qquad (8.78)$$

$$\dot{\rho}_2 = -U\rho_1 \sin \Delta\varphi \qquad (8.79)$$

$$\dot{\rho}_1 \cong -\dot{\rho}_2 \qquad (8.80)$$

$$\rho_1 \dot{\varphi}_1 = -U\rho_2 \cos \Delta\varphi + eV \qquad (8.81)$$

$$\rho_2 \dot{\varphi}_2 = -U\rho_1 \cos \Delta\varphi - eV. \qquad (8.82)$$

Remembering $\rho_1 \cong \rho_2 \cong \rho$, so

$$\dot{\varphi}_2 - \dot{\varphi}_1 \cong -2eV. \qquad (8.83)$$

Therefore

$$\Delta\varphi \cong (\Delta\varphi)_0 - 2eVt, \qquad (8.84)$$

and

$$J = J_0 \sin[(\Delta\varphi)_0 - 2eVt]. \qquad (8.85)$$

Again, J_0 is the maximum current carried by Cooper pairs. Additional current is carried by single-particle excitations producing the voltage V. The idea is shown later in Fig. 8.18. Therefore, since V is voltage in units of $\hbar$, the current oscillates with frequency (see (8.85))

$$\omega_J = 2eV = 2e\frac{\text{Voltage}}{\hbar}. \qquad (8.86)$$

For the dc Josephson effect one can say that for low enough currents there is a current across the insulator in the absence of applied voltage. In effect because of the coherence of Cooper pairs, the insulator becomes a superconductor. Above a critical voltage, V_c, one has single electrons and the material becomes ohmic rather than superconducting. The junction then has resistance, but the current also has a component that oscillates with frequency ω_J as above. One understands this by saying that above V_c one has single particles as well as Cooper pairs. The

Cooper pairs change their energy by $2eV = \hbar\omega_J$ as they cross the energy gap causing radiation at this frequency. The ac Josephson effect, which occurs when

$$\omega = \frac{q}{\hbar}\text{Voltage} \qquad (8.87)$$

is satisfied, is even more interesting. With $q = 2e$ (for a Cooper pair), (8.87) is believed to be exact. Thus, the ac Josephson effect can be used for a precise determination of $e/\hbar$. Parker, Taylor, and Langenberg[6] have done this. They used their new value of $e/\hbar$ to determine a new and better value of the fine structure constant α. Their new value of α removed a discrepancy between the quantum-electrodynamics calculation and the experimental value of the hyperfine splitting of atomic hydrogen in the ground state. These experiments have also contributed to better accuracy in the determination of the fundamental constants. There have been many other important developments connected with the Josephson effects, but they will not be presented here. Reference [8.20] is a good source for further discussion. See also Fig. 8.18 for a summary.

Finally, it is worth pointing out another reason why the Josephson effects are so interesting. They represent a quantum effect operating on a macroscopic scale. We can play with words a little, and perhaps convince ourselves that we understand this statement. In order to see quantum effects on a macroscopic scale, we must have many particles in the same state. For example, photons are bosons, and so, we can obtain a large number of them in the same state (which is necessary to see the quantum effects of electrons on a large scale). Electrons are fermions and must obey the Pauli principle. It would appear, then, to be impossible to see the quantum effects of electrons on a macroscopic scale. However, in a certain sense, the Cooper pairs having total spin zero, do act like bosons (but not entirely; the Cooper pairs overlap so much that their motion is highly correlated, and this causes their motion to be different from bosons interacting by a two-boson potential). Hence, we can obtain many electrons in the same state, and we can see the quantum effects of superconductivity on a macroscopic scale.

8.4 SQUID: Superconducting Quantum Interference (EE)

A Josephson junction is shown in Fig. 8.17 below. It is basically a superconductor–insulator–superconductor or a superconducting "sandwich". We now show how flux, due to B, threading the circuit can have profound effects. Using (8.4) with φ the Ginzburg–Landau phase, we have

$$J = \frac{nq}{m}\left[\hbar\nabla\varphi - \frac{q}{c}A\right]. \qquad (8.88)$$

---

[6] See [8.24].

Integrating along the upper path gives

$$\hbar \Delta \varphi_1 = \int_\alpha^\beta \frac{m}{nq} \boldsymbol{J} \cdot d\boldsymbol{l}_1 + \frac{q}{c} \int_\alpha^\beta \boldsymbol{A} \cdot d\boldsymbol{l}_1 , \qquad (8.89)$$

while integrating along the lower path gives

$$\hbar \Delta \varphi_2 = \int_\alpha^\beta \frac{m}{nq} \boldsymbol{J} \cdot d\boldsymbol{l}_2 + \frac{q}{c} \int_\alpha^\beta \boldsymbol{A} \cdot d\boldsymbol{l}_2 . \qquad (8.90)$$

Subtracting, we have

$$\hbar (\Delta \varphi_1 - \Delta \varphi_2) = \frac{m}{nq} \oint \boldsymbol{J} \cdot d\boldsymbol{l} + \frac{q}{c} \oint \boldsymbol{A} \cdot d\boldsymbol{l} , \qquad (8.91)$$

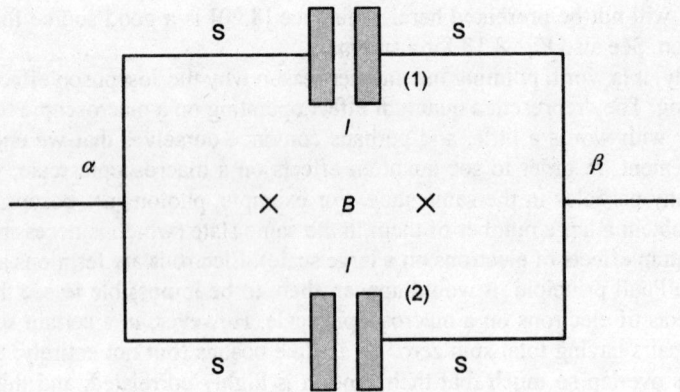

Fig. 8.17. A Josephson junction

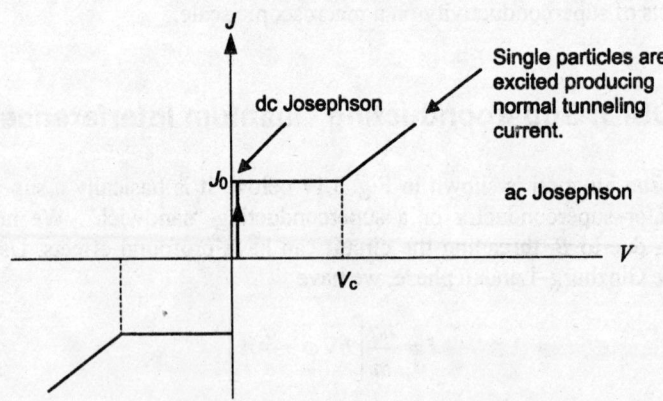

Fig. 8.18. Schematic of current density across junction versus V. The Josephson current $0 < J < J_0$ occurs with no voltage. When $J > J_0$ at $V_c \cong E_g/e$, where E_g is the energy gap, one also has single-particle current

where the first term on the right is zero or negligible. So, using Stokes Theorem and $B = \nabla \times A$ (and choosing a path where $J \cong 0$)

$$\Delta\varphi_1 - \Delta\varphi_2 = \frac{q}{\hbar c}\oint A \cdot dl = \frac{q}{\hbar c}\int B \cdot dA = \frac{q\Phi}{\hbar c}. \tag{8.92}$$

Defining $\Phi_0 = \hbar c/q$ as per (8.45), we have

$$\Delta\varphi_1 - \Delta\varphi_2 = \frac{\Phi}{\Phi_0}, \tag{8.93}$$

so when $\Phi = 0$, and $\Delta\varphi_1 = \Delta\varphi_2$. We assume the junctions are identical so defining $\varphi_0 = (\Delta\varphi_1 + \Delta\varphi_2)/2$, then

$$\Delta\varphi_1 = \varphi_0 + \frac{\Phi}{2\Phi_0}, \tag{8.94}$$

$$\Delta\varphi_2 = \varphi_0 - \frac{\Phi}{2\Phi_0} \tag{8.95}$$

is a solution. By (8.73)

$$J_T = J_1 + J_2 = J_0(\sin\Delta\varphi_1 + \sin\Delta\varphi_2)$$
$$= 2J_0 \sin\left(\frac{\Delta\varphi_1 + \Delta\varphi_2}{2}\right)\cos\left(\frac{\Delta\varphi_1 - \Delta\varphi_2}{2}\right). \tag{8.96}$$

So

$$J_T = 2J_0 \sin(\varphi_0)\cos\left(\frac{\Phi}{2\Phi_0}\right), \tag{8.97}$$

and

$$J_{T_{max}} = 2J_0\left|\cos\left(\frac{\Phi}{2\Phi_0}\right)\right|. \tag{8.98}$$

The maximum occurs when $\Phi = 2n\pi\Phi_0$. Thus, quantum interference can be used to measure small magnetic field changes. The maximum current is a periodic function of Φ and, hence, measures changes in the field. Sensitive magnetometers have been constructed in this way. See the original paper about SQUIDS by Silver and Zimmerman [8.31].

8.4.1 Questions and Answers (B)

Q1. What is the simplest way to understand the dc Josephson effect (a current with no voltage in a super–insulator–super sandwich or SIS)?

A. If the insulator is much thinner than the coherence length, the superconducting pairs of electrons tunnel right through, and the insulator does not interfere with them–it is just one superconductor.

Q2. What is the simplest way to understand the ac Josephson effect (a current with a component of frequency $2eV/\hbar$, where V is the applied voltage)?

A. The Cooper pairs have charge $q = 2e$, and when they tunnel across the insulator, they drop in energy by qV. Thus they radiate with frequency $qV/\hbar$. This radiation is linked to the ac current.

8.5 The Theory of Superconductivity[7] (A)

8.5.1 Assumed Second Quantized Hamiltonian for Electrons and Phonons in Interaction (A)

As has already been mentioned, in many materials the superconducting state can be accounted for by an attractive electron–electron interaction due to the virtual exchange of phonons. See, e.g., Fig. 8.19. Thus, if we are going to try to understand the theory of superconductivity from a microscopic viewpoint, then we must examine, in detail, the nature of the electron–phonon interaction. There is no completely rigorous road to the BCS Hamiltonian. The arguments given below are intended to show how the physical origins of the BCS Hamiltonian could arise. It is not claimed that this is the way it *must* arise. However, given the BCS Hamiltonian, it is fair to say that the way it describes superconductivity is well understood.

One could draw an analogy to the Heisenberg Hamiltonian. The road to this Hamiltonian is also not rigorous for real materials, but there seems to be no doubt that it well describes magnetic phenomena in at least certain materials. The phenomena of superconductivity and ferromagnetism are exact, but the road to a quantitative description is not.

We thus start out with the Hamiltonian, which represents the interaction of electrons and phonons. As before, an intuitive approach suggests

$$\mathcal{H}_{ep} = \sum_{l,b} x_{lb} \cdot [\nabla_{x_{lb}} U(r_i)]_{x=0} \,. \tag{8.99}$$

We have already discussed this Hamiltonian in Chap. 4, which the reader should refer to, if needed. By the theory of lattice vibrations, we also know that (see Chaps. 2 and 4)

$$x_{l,b} = -\sum_{q,p} \sqrt{\frac{\hbar}{2Nm_b \omega_{q,p}}} [\exp(-i q \cdot l) e^*_{q,b,p}(a^\dagger_{q,p} - a_{-q,p})]\,. \tag{8.100}$$

---

[7] See Bardeen et al [8.6].

In the above equation, the a are, of course, phonon creation and annihilation operators.

By a second quantization representation of the terms involving electron coordinates (see Appendix G), we can write

$$\frac{\partial U(r_i)}{\partial x_{l,b}} = \sum_{k,k'} \langle \psi_k | \nabla_{x_{l,b}} U(r_i) | \psi_{k'} \rangle C_k^\dagger C_{k'}, \qquad (8.101)$$

where the C are electron creation and annihilation operators. The only quantities that we will want to calculate involve matrix elements of the operator $\mathcal{H}_{ep}$. As we have already shown, these matrix elements will vanish unless the selection rule $q = k' - k - G_n$ is obeyed. Neglecting umklapp processes (assuming $G_n = 0$ the *first major approximation*), we can write

$$\mathcal{H}_{ep} = -i \sum_{l,b} \sum_{q,p} \sum_{k,k'} \sqrt{\frac{\hbar}{2Nm_b \omega_{q,p}}} \exp(-i q \cdot l) \\ \times \langle \psi_k | e_{q,b,p}^* \nabla_{x_{lb}} U(r_i) | \psi_{k'} \rangle \delta_q^{k'-k} (a_{q,p}^\dagger - a_{-q,p}) C_k^\dagger C_{k'}, \qquad (8.102)$$

or

$$\mathcal{H}_{ep} = -i \sum_{l,b} \sum_{q,p} \sum_{k,k'} \sqrt{\frac{\hbar}{2Nm_b \omega_{q,p}}} \exp(-i q \cdot l) \\ \times \langle \psi_{k'-q} | e_{q,b,p}^* \nabla_{x_{lb}} U(r_i) | \psi_{k'} \rangle (a_{q,p}^\dagger - a_{-q,p}) C_{k'-q}^\dagger C_{k'}. \qquad (8.103)$$

Making the dummy variable changes $k' \to k$, $q \to -q$, and dropping the sum over p (assuming, for example, that only longitudinal acoustic phonons are effective in the interaction—this is the *second major approximation*), we find

$$\mathcal{H}_{ep} = i \sum_{k,q} B_q C_{k+q}^\dagger C_k (a_q - a_{-q}^\dagger) \qquad (8.104)$$

where

$$B_q = \sum_{l,b} \sqrt{\frac{\hbar}{2Nm_b \omega_q}} \exp(i q \cdot l) \langle \psi_{k+q} | e_{-q,b}^* \nabla_{x_{l,b}} U(r_i) | \psi_k \rangle. \qquad (8.105)$$

The only property of B_q that we will use from the above equation is $B_q = B_{-q}^*$. From any reasonable, practical viewpoint, it would be impossible to evaluate the above equation directly and obtain B_q. Thus, B_q will be treated as a parameter to be evaluated from experiment. Note that so far we have not made any approximations that are specifically restricted to superconductivity. The same Hamiltonian could be used in certain electrical-resistivity calculations.

We can now write the total Hamiltonian for interacting electrons and phonons (with $\hbar = 1$, and neglecting the zero-point energy of the lattice vibrations):

$$\mathcal{H} = \mathcal{H}_0 + \mathcal{H}_{ep} = \sum_q \omega_q a_q^\dagger a_q + \sum_k \varepsilon_k C_k^\dagger C_k + i \sum_{q,k} B_q C_{k+q}^\dagger C_k (a_q - a_{-q}^\dagger), \quad (8.106)$$

where the first two terms are the unperturbed Hamiltonian $\mathcal{H}_0$.

The first term is the Hamiltonian for phonons only (with $n_q = a^\dagger_q a_q$ as the phonon occupation number operator). The second term is the Hamiltonian for electrons only (with $n_k = C^\dagger_k C_k$ as the electron occupation number operator). The third term represents the interaction of phonons and electrons. We have in mind that the second term really deals with quasielectrons. We can assign an effective mass to the quasielectrons in such a manner as partially to take into account the electron–electron interactions, electron interactions with the lattice, and at least partially any other interactions that may be important but only lead to a "renormalization" of the electron mass. Compare Sects. 3.1.4, 3.2.2, and 4.3, as well as the introduction in Chap. 4. We should also include a screened Coulomb repulsion between electrons (see Sect. 9.5.3), but we neglect this here (or better, absorb it in $V_{k,k'}$—to be defined later).

Various experiments and calculations indicate that the energy per atom between the normal and superconducting states is of order 10^{-7} eV. This energy is very small compared to the accuracy with which we can hope to calculate the absolute energy. Thus, a frontal attack is doomed to failure. So, we will concentrate on those terms leading to the energy difference. The rest of the terms can then be pushed aside. The results are nonrigorous, and their main justification is the agreement we get with experiment. The method for separating the important terms is by no means obvious. It took many years to find. All that will be done here is to present a technique for doing the separation.

The technique for separating out the important terms involves making a canonical transformation to eliminate off-diagonal terms of $O(B_q)$ in the Hamiltonian. Before doing this, however, it is convenient to prove several useful results. First, we derive an expansion for

$$\mathcal{H}_S \equiv (e^{-S})(\mathcal{H})(e^S), \quad (8.107)$$

where S is an operator.

$$(e^{-S})(\mathcal{H})(e^S) = \left(1 - S + \frac{1}{2}S^2 + \ldots\right)\mathcal{H}\left(1 + S + \frac{1}{2}S^2 + \ldots\right)$$
$$= \mathcal{H} - S\mathcal{H} + \mathcal{H}S + \frac{1}{2}S^2\mathcal{H} - S\mathcal{H}S + \frac{1}{2}\mathcal{H}S^2, \quad (8.108)$$

but

$$[[\mathcal{H}, S], S] = [\mathcal{H}S - S\mathcal{H}, S] = 2\left[\frac{1}{2}\mathcal{H}S^2 + \frac{1}{2}S^2\mathcal{H} - S\mathcal{H}S\right], \quad (8.109)$$

so that

$$\mathcal{H}_S = \mathcal{H} + [\mathcal{H}, S] + \frac{1}{2}[[\mathcal{H}, S], S] + \ldots. \tag{8.110}$$

We can treat the next few terms in a similar way.

The second useful result is obtained by $\mathcal{H} = \mathcal{H}_0 + X\mathcal{H}_{ep}$ where X is eventually going to be set to one. In addition, we choose S so that

$$X\mathcal{H}_{ep} + [\mathcal{H}_0, S] = 0. \tag{8.111}$$

We show that in this case $\mathcal{H}_S$ has no terms of O(X). The result is proved by using (8.110) and substituting $\mathcal{H} = \mathcal{H}_0 + X\mathcal{H}_{ep}$. Then

$$\begin{aligned}\mathcal{H}_S &= \mathcal{H}_0 + X\mathcal{H}_{ep} + [\mathcal{H}_0 + X\mathcal{H}_{ep}, S] + \frac{1}{2}[[\mathcal{H}_0 + X\mathcal{H}_{ep}, S], S] + \ldots \\ &= \mathcal{H}_0 + X\mathcal{H}_{ep} + [\mathcal{H}_0, S] + X[\mathcal{H}_{ep}, S] \\ &\quad + \frac{1}{2}[[\mathcal{H}_0, S], S] + \frac{X}{2}[[\mathcal{H}_{ep}, S], S] + \ldots.\end{aligned} \tag{8.112}$$

Using (8.111), we obtain

$$\mathcal{H}_S = \mathcal{H}_0 + X[\mathcal{H}_{ep}, S] + \frac{X}{2}[[\mathcal{H}_{ep}, S], S] + \frac{1}{2}[[\mathcal{H}_0, S], S] + \ldots. \tag{8.113}$$

Since

$$X\mathcal{H}_{ep} + [\mathcal{H}_0, S] = 0, \tag{8.114}$$

we have

$$X[\mathcal{H}_{ep}, S] = -[[\mathcal{H}_0, S], S], \tag{8.115}$$

so that

$$\mathcal{H}_S = \mathcal{H}_0 + X[\mathcal{H}_{ep}, S] + \frac{X}{2}[[\mathcal{H}_{ep}, S], S] - \frac{X}{2}[\mathcal{H}_{ep}, S], \tag{8.116}$$

or

$$\mathcal{H}_S = \mathcal{H}_0 + \frac{X}{2}[\mathcal{H}_{ep}, S] + O(X^3). \tag{8.117}$$

Since O(S) = X the second term is of order X^2, which was to be proved.

The point of this transformation is to push aside terms responsible for ordinary electrical resistivity (*third major transformation*). In the original Hamiltonian, terms in X contribute to ordinary electrical resistivity in first order.

From $X\mathcal{H}_{ep} + [\mathcal{H}_0, S] = 0$, we can calculate S. This is especially easy if we use a representation in which $\mathcal{H}_0$ is diagonal. In such a representation

$$\langle n|X\mathcal{H}_{ep}|m\rangle + \langle n|\mathcal{H}_0 S - S\mathcal{H}_0|m\rangle = 0 , \qquad (8.118)$$

or

$$\langle n|X\mathcal{H}_{ep}|m\rangle + (E_n - E_m)\langle n|S|m\rangle = 0 , \qquad (8.119)$$

or

$$\langle n|S|m\rangle = \frac{\langle n|X\mathcal{H}_{ep}|m\rangle}{E_m - E_n} . \qquad (8.120)$$

The above equation determines the matrix elements of S and, hence, defines the operator S (for $E_m \neq E_n$).

8.5.2 Elimination of Phonon Variables and Separation of Electron–Electron Attraction Term Due to Virtual Exchange of Phonons (A)

Let us now connect the results we have just derived with the problem of superconductivity. Let $X\mathcal{H}_{ep}$ be the interaction Hamiltonian for the electron–phonon system. Any operator that we present for S that satisfies

$$\langle n|S|m\rangle = \frac{\langle n|X\mathcal{H}_{ep}|m\rangle}{E_m - E_n} \qquad (8.121)$$

is good enough. In the above equation, $|m\rangle$ means both electron and phonon states. However, let us take matrix elements with respect to phonon states only and select S so that if we were to take electronic matrix elements, the above equation would be satisfied. This procedure is done because the behavior of phonons, except insofar as it affects the electrons, is of no interest. The point of this Section is then to find an effective Hamiltonian for the electrons.

We begin with these ideas. Taking phonon matrix elements, we have

$$\langle n_{q'}+1|S|n_{q'}\rangle = \frac{\langle n_{q'}+1|X\mathcal{H}_{ep}|n_{q'}\rangle}{E(\text{total initial state}) - E(\text{total final state})}$$

$$= i\sum_{k,q} B_q \frac{C^\dagger_{k+q} C_k \langle n_{q'}+1|a_q - a^\dagger_{-q}|n_{q'}\rangle}{E_{q'} + \varepsilon_k - (E_{q'} + \omega_{q'}) - \varepsilon_{k+q}}$$

$$= -i\sum_{k,q} B_q \frac{C^\dagger_{k+q} C_k \langle n_{q'}+1|a^\dagger_{-q}|n_{q'}\rangle}{E_{q'} - (E_{q'} + \omega_{q'}) + \varepsilon_k - \varepsilon_{k+q}} , \qquad (8.122)$$

8.5 The Theory of Superconductivity (A)

where ω_q is the energy of the created phonon (with $\hbar = 1$ and $\omega_q = \omega_{-q}$). Using

$$\langle n_q + 1 | a_q^\dagger | n_q \rangle = \sqrt{n_q + 1}, \tag{8.123}$$

we find

$$\begin{aligned}\langle n_q + 1 | S | n_q \rangle &= -i \sum_{k,q'} B_{q'} C_{k+q'}^\dagger C_k \frac{\sqrt{n_{q'}+1}}{\varepsilon_k - \varepsilon_{k+q'} - \omega_{q'}} \delta_{q'}^{-q} \\ &= -i \sum_k B_{-q'} C_{k-q'}^\dagger C_k \frac{\sqrt{n_{q'}+1}}{\varepsilon_k - \varepsilon_{k-q'} - \omega_{q'}}.\end{aligned} \tag{8.124}$$

In a similar way we can show

$$\langle n_{q'} | S | n_{q'} + 1 \rangle = i \sum_k B_{q'} C_{k+q'}^\dagger C_k \frac{\sqrt{n_{q'}+1}}{\varepsilon_k - \varepsilon_{k+q} + \omega_{q'}}. \tag{8.125}$$

Now, using

$$\mathcal{H}_S = \mathcal{H}_0 + \frac{1}{2}[\mathcal{H}_{ep} S - S \mathcal{H}_{ep}] + \ldots, \tag{8.126}$$

with

$$\mathcal{H}_{ep} = i \sum_{k,q} B_q C_{k+q}^\dagger C_k (a_q - a_{-q}^\dagger) \tag{8.127}$$

(X has now been set equal to 1), and taking phonon expectation values for a particular phonon state, we have

$$\begin{aligned}\langle n | \mathcal{H}_S | n \rangle &= \langle n | \mathcal{H}_0 | n \rangle + \frac{1}{2} \sum_m [\langle n | \mathcal{H}_{ep} | m \rangle \langle m | S | n \rangle - \langle n | S | m \rangle \langle m | \mathcal{H}_{ep} | n \rangle] \\ &= \langle n | \mathcal{H}_0 | m \rangle + \frac{1}{2}\big[(\mathcal{H}_{ep})_{n,n-1} S_{n-1,n} + (\mathcal{H}_{ep})_{n,n+1} S_{n+1,n} \\ &\quad - S_{n,n-1} (\mathcal{H}_{ep})_{n-1,n} - S_{n,n+1} (\mathcal{H}_{ep})_{n+1,n}\big].\end{aligned} \tag{8.128}$$

Since we are interested only in electronic coordinates, we will write below $\langle n_q | \mathcal{H}_S | n_q \rangle$ as $\mathcal{H}_S$, and $\langle n_q | \mathcal{H}_0 | n_q \rangle$ as $\mathcal{H}_0$, and hope that no confusion in notation will arise. Using

$$(\mathcal{H}_{ep})_{n_q, n_q - 1} = -i \sum_k B_{-q} C_{k-q}^\dagger C_k \sqrt{n_q}, \tag{8.129}$$

and

$$(\mathcal{H}_{ep})_{n_q, n_q + 1} = i \sum_k B_q C_{k+q}^\dagger C_k \sqrt{n_q + 1}, \tag{8.130}$$

the effective Hamiltonian for electrons is given by combining the above. Thus,

$$\mathcal{H}_S = \mathcal{H}_0 + \frac{1}{2}|B_q|^2 \sum_{k,k'} \Bigg[C_{k-q}^\dagger C_k C_{k'+q}^\dagger C_{k'} n_q \frac{1}{\varepsilon_{k'} - \varepsilon_{k'+q} + \omega_q}$$
$$+ C_{k+q}^\dagger C_k C_{k'-q}^\dagger C_{k'} (n_q + 1) \frac{1}{\varepsilon_{k'} - \varepsilon_{k'-q} - \omega_q}$$
$$- C_{k'-q}^\dagger C_{k'} C_{k+q}^\dagger C_k n_q \frac{1}{\varepsilon_{k'} - \varepsilon_{k'-q} - \omega_q}$$
$$- C_{k'+q}^\dagger C_{k'} C_{k-q}^\dagger C_k (n_q + 1) \frac{1}{\varepsilon_{k'} - \varepsilon_{k'+q} + \omega_q} \Bigg]. \quad (8.131)$$

Making dummy variable changes, dropping terms that do not involve the interaction of electrons (i.e. that do not involve both k and k'), and using the commutation relations for the C, it is possible to write the above in the form

$$\mathcal{H}_S = \mathcal{H}_0 + \frac{1}{2}|B_q|^2 \sum_{k,k'} C_{k'+q}^\dagger C_{k'} C_{k-q}^\dagger C_k$$
$$\times \left(\frac{1}{\varepsilon_k - \varepsilon_{k-q} - \omega_q} - \frac{1}{\varepsilon_{k'} - \varepsilon_{k'+q} + \omega_q} \right). \quad (8.132)$$

In order to properly interpret Hamiltonians such as the above equation, which are expressed in the second quantization notation, it is necessary to keep in mind the appropriate commutation relations of the C. By Appendix G, these are

$$C_k C_{k'}^\dagger + C_{k'}^\dagger C_k = \delta_k^{k'}, \quad (8.133)$$

$$C_k^\dagger C_{k'}^\dagger + C_{k'}^\dagger C_k^\dagger = 0, \quad (8.134)$$

and

$$C_k C_{k'} + C_{k'} C_k = 0. \quad (8.135)$$

The Hamiltonian (8.132) describes a process called a *virtual exchange of a phonon*. It has the diagrammatic representation shown in Fig. 8.19.

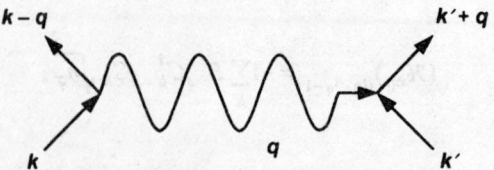

Fig. 8.19. The virtual exchange of a phonon of wave vector q. The k are the wave vectors of the electrons. This is the fundamental process of superconductivity

Note that (8.132) is independent of the number of phonons in mode q, and it is the effective electron Hamiltonian with phonons in the single mode q. To get the effective Hamiltonian with phonons in all modes, we merely have to sum over the modes of q. Thus, the total effective interaction Hamiltonian is given by

$$\mathcal{H}_1 = \frac{1}{2} \sum_q \sum_{k,k'} |B_q|^2 C^\dagger_{k'+q} C_{k'} C^\dagger_{k-q} C_k \\ \times \left(\frac{1}{\varepsilon_k - \varepsilon_{k-q} - \omega_q} - \frac{1}{\varepsilon_{k'} - \varepsilon_{k'+q} + \omega_q} \right). \tag{8.136}$$

By dropping further terms that do not involve the interaction of electrons (terms not involving both k and k') and by making variable changes, we can reduce this Hamiltonian to

$$\mathcal{H}_1 = \sum_q \sum_{k,k'} |B_q|^2 \frac{\omega_q}{(\varepsilon_k - \varepsilon_{k-q})^2 - \omega_q^2} C^\dagger_{k'+q} C^\dagger_{k-q} C_k C_{k'}. \tag{8.137}$$

From the above equation, we see that there is an attractive electron–electron interaction for $|\varepsilon_k - \varepsilon_{k-q}| < |\omega_q|$. We will assume, for appropriate excitation energies, that the main interaction is attractive. In this connection, most of the electron energies of interest are near the Fermi energy ε_f. A typical phonon energy is the Debye energy $\hbar\omega_D$ (or cutoff frequency with $\hbar = 1$). Many approximations have already been made, and so a very simple criterion for the dominance of the attractive interaction will be assumed. It will be assumed that the interaction is attractive when the electronic energies are in the range of

$$\varepsilon_F - \hbar\omega_D < \varepsilon_k < \varepsilon_F + \hbar\omega_D \quad (\hbar \neq 1 \text{ here}). \tag{8.138}$$

The states that do not satisfy this criterion are not directly involved in the superconducting transition, so their properties are of no particular interest. Hence, the effective Hamiltonian can be written in the following form (*fourth major approximation*):

$$\mathcal{H}_1 \equiv -\sum_q \sum_{k,k'} V_q C^\dagger_{k'+q} C^\dagger_{k-q} C_k C_{k'}. \tag{8.139}$$

For simplicity, we will assume that V_q is positive and fitted from experiment, that $V_q = V_{-q}$ and $V_q = 0$, unless q is such that (8.138) is satisfied. We assume that any important interactions not included in the above equation can be included by renormalizing (i.e. changing) the quasiparticle mass.

8.5.3 Cooper Pairs and the BCS Hamiltonian (A)

Let us assume that $\varepsilon_k = 0$ at the Fermi level. The total effective Hamiltonian for the electrons is then

$$\mathcal{H} = \sum_k \varepsilon_k C^\dagger_k C_k - \sum_{k,k',q} V_q C^\dagger_{k'+q} C^\dagger_{k-q} C_k C_{k'}. \tag{8.140}$$

By Appendix G, the Fermion operators satisfy

$$C_j |n_1 ... n_j ...\rangle = (-)^{P_j} n_j |n_1 ...(1-n_j)...\rangle, \qquad (8.141)$$

$$C_j^\dagger |n_1 ... n_j ...\rangle = (-)^{P_j} (1-n_j) |n_1 ...(1+n_j)...\rangle, \qquad (8.142)$$

where

$$P_j = \sum_{P=1}^{j-1} n_P. \qquad (8.143)$$

It is essential to notice the alternation in sign defined by (8.142). This alternation is very important for discovering the nature of the lowest-energy state. When we begin to guess a trial wave function, if we pay no attention to this alternation of sign, the presence of the interaction will result in little lowering of the energy. What we need is a way of selecting the trial wave function so that most of the matrix elements of individual terms in the second sum in (8.138) are negative. The way to do this for the ground state is by grouping the electrons into *Cooper pairs*. (These will be precisely defined below.)

There are several assumptions necessary to construct a minimum energy wave function [60, p. 155ff]. For the ground-state wave function, it will be assumed that the Bloch states are occupied only in pairs. In fact, the superconducting ground state is a coherent superposition of Cooper pairs. The Hamiltonian conserves the wave vector, and only pairs with equal total momentum will be considered, i.e.,

$$\bm{k} + \bm{k}' = \bm{K}, \qquad (8.144)$$

where $\bm{K}$ is the same for each pair. It is reasonable to suppose that $\bm{K}$ is zero for the ground (noncurrent carrying) state of the pairs.

Cooper Pairs[8]

Before proceeding, let us discuss Cooper pairs a little more. A large clue as to the nature of the unusual character of the superconducting state was obtained by L. Cooper in 1956. He showed that the Fermi sea was unstable if electrons interacted by an attractive mechanism—no matter how weak.

Consider the normal Fermi sea of electrons with a well-defined Fermi energy E_F. Now add two more electrons interacting with an attractive interaction $V(1, 2)$ and suppose the only interaction with the other electrons is via the Pauli principle.

We write the Schrödinger wave equation for the two electrons as

$$\left[-\frac{\hbar^2}{2m}\nabla_1^2 - \frac{\hbar^2}{2m}\nabla_2^2 + V(1,2) \right] \psi(1,2) = E\psi(1,2). \qquad (8.145)$$

---

[8] See Cooper [8.10].

We seek a solution of the form

$$\psi(1,2) = \frac{V}{(2\pi)^3} A(1,2) \frac{1}{V} \int e^{i\mathbf{k}\cdot(\mathbf{r}_1-\mathbf{r}_2)} f(\mathbf{k}) d\mathbf{k}, \tag{8.146}$$

where $A(1, 2)$ is the antisymmetric spin zero spin wave function

$$A(1,2) = \frac{1}{\sqrt{2}}[\alpha(1)\beta(2) - \alpha(2)\beta(1)],$$

with α, β being the usual spin-up and -down wave functions (note $A^\dagger A = 1$) and $f(\mathbf{k}) = +f(-\mathbf{k})$ so that the spatial wave function is symmetric (it can be shown that the ψ with spin 1 and antisymmetric wave function yields no energy shift, at least in our approximation, and in any case such wave functions correspond to p-state pairs that we are not considering). Note that the spatial wave function pairs off the electrons into $(\mathbf{k}, -\mathbf{k})$ states.

Inserting (8.146) into (8.145) we have

$$\frac{1}{(2\pi)^3} A(1,2) \int \left[\left(\frac{\hbar^2}{2m} k^2 + \frac{\hbar^2}{2m} k^2 + V \right) f(\mathbf{k}) e^{i\mathbf{k}\cdot(\mathbf{r}_1-\mathbf{r}_2)} \right] d\mathbf{k}$$
$$= \frac{A(1,2)}{(2\pi)^3} \int E f(\mathbf{k}) e^{i\mathbf{k}\cdot(\mathbf{r}_1-\mathbf{r}_2)} d\mathbf{k}. \tag{8.147}$$

Now multiply by

$$A^\dagger(1,2) \frac{1}{V} e^{-i\mathbf{k}'\cdot(\mathbf{r}_1-\mathbf{r}_2)},$$

and integrate over $\mathbf{r}_1$ and $\mathbf{r}_2$ and we obtain ($\mathbf{r} = \mathbf{r}_1 - \mathbf{r}_2$, $V(\mathbf{r}_1, \mathbf{r}_2) = V(\mathbf{r}_1 - \mathbf{r}_2) = V(\mathbf{r})$, and $E_k = \hbar^2 k^2/2m$)

$$\int \int e^{-i\mathbf{k}'\cdot\mathbf{r}} [2E_k + V(\mathbf{r})] f(\mathbf{k}) e^{i\mathbf{k}\cdot\mathbf{r}} d\mathbf{r} d\mathbf{k} = \int E f(\mathbf{k}) e^{i(\mathbf{k}-\mathbf{k}')\cdot\mathbf{r}} d\mathbf{k}. \tag{8.148}$$

Using

$$\frac{1}{(2\pi)^3} \int e^{i\mathbf{k}\cdot\mathbf{r}} d\mathbf{k} = \delta(\mathbf{k}),$$

and

$$V_{\mathbf{k}',\mathbf{k}} = \frac{1}{V} \int e^{-i\mathbf{k}'\cdot\mathbf{r}} V(\mathbf{r}) e^{i\mathbf{k}\cdot\mathbf{r}} d\mathbf{r},$$

we obtain

$$[2E_{k'} - E] f(\mathbf{k}') + \frac{V}{(2\pi)^3} \int f(\mathbf{k}') V_{\mathbf{k}',\mathbf{k}} d\mathbf{k} = 0. \tag{8.149}$$

We suppose

$$V_{k',k} = -V_0 < 0 \quad \text{for} \quad E_F < E_k, \quad E_{k'} < E_F + \hbar\omega_D$$
$$= 0 \quad \text{otherwise.}$$

Notice we are using the ideas that led us to (8.138), divide by $2E_{k'} - E$ and integrate over k' and obtain (after canceling)

$$1 = V_0 \frac{V}{(2\pi)^3} \int \frac{dk'}{2E_{k'} - E}. \tag{8.150}$$

Note that in the limit of large volumes

$$\frac{V/N}{(2\pi)^3} \int dk'(\ ) \leftrightarrow \frac{1}{N} \sum_{k'} \leftrightarrow \int N(E')(\ )dE',$$

where $N(E)$ is the density of state for one spin per unit cell (N unit cells). Thus with $E_{\text{pair}} = E$

$$1 = V_0 \int_{E_F}^{E_F + \hbar\omega_D} \frac{N(E')}{2E' - E_{\text{pair}}} dE'. \tag{8.151}$$

Note we can replace $N(E') \cong N(E_F)$ because $\hbar\omega_D \ll E_F$ so we obtain

$$1 = \frac{V_0 N(E_F)}{2} \ln \frac{2E_F + 2\hbar\omega_D - E_{\text{pair}}}{2E_F - E_{\text{pair}}}. \tag{8.152}$$

Let $\delta = 2E_F - E_{\text{pair}}$ so

$$\delta = \hbar\omega_D \frac{\exp\left(\frac{-1}{V_0 N(E_F)}\right)}{\sinh \frac{1}{V_0 N(E_F)}}, \tag{8.153}$$

and in the weak coupling limit

$$\delta = 2\hbar\omega_D \exp\left(\frac{-2}{V_0 N(E_F)}\right). \tag{8.154}$$

We note in particular, the following points:

1. A pair electron wave function that is independent of the direction of $r_1 - r_2$ is said to be an s wave function, which is consistent with an antisymmetric spin wave function.
2. δ is not an analytic function of V_0 so ordinary perturbation theory would not work.
3. In the BCS theory one considers pairing of all electrons.
4. For $\delta > 0$ then the Fermi sea is unstable with respect to the formation of Cooper pairs.

BCS Hamiltonian

Returning to the mainstream of the BCS argument, the above reasoning can be used to pick out the best wave function to use as a trial wave function for evaluating the ground-state energy by variational principle. For mathematical convenience, it is easier to place these assumptions directly in the Hamiltonian. Also, due to exchange, the spins in the Cooper pairs are usually opposite. Thus, the interaction part of the Hamiltonian is now written (*fifth major approximation*) with $K = 0$,

$$\mathcal{H}_1 = -\sum_{k,q} V_q C^\dagger_{k+q\uparrow} C^\dagger_{-k-q\downarrow} C_{-k\downarrow} C_{k\uparrow}. \tag{8.155}$$

Next, assume a "BCS Hamiltonian" for interacting pairs consistent with (8.155), with $k + q \to k$, $k \to k'$, $V_q = V_{k-k'} = -V_{k,k'}$

$$\mathcal{H} = \sum_{k\sigma} \varepsilon_k C^\dagger_{k\sigma} C_{k\sigma} + \sum_{k,k'} V_{k,k'} C^\dagger_{k\uparrow} C^\dagger_{-k\downarrow} C_{-k'\downarrow} C_{k'\uparrow}, \tag{8.156}$$

where

$$\varepsilon_k = \frac{\hbar^2 k^2}{2m} - \mu, \tag{8.157}$$

and where μ is the chemical potential. Also

$$\mathcal{H} \equiv \mathcal{H}_0 + \mathcal{H}_I, \tag{8.158}$$

and note

$$V_{k,k'} = V_{k',k} = V^*_{k,k'}. \tag{8.159}$$

As before C are Fermion (electron) annihilation operators, and $C^\dagger$ are Fermion (electron) creation operators. Defining the pair creation and annihilation operators

$$b^\dagger_k = C^\dagger_{k\uparrow} C^\dagger_{-k\downarrow}, \tag{8.160}$$

$$b_k = C_{-k\downarrow} C_{k\uparrow}; \tag{8.161}$$

and defining

$$\overline{b}_k = \frac{\mathrm{Tr}(e^{-\beta H} b_k)}{\mathrm{Tr}(e^{-\beta H})}, \tag{8.162}$$

we can show $\overline{b}^\dagger_k = \overline{b}^*_k$ using $\mathrm{Tr}(AB) = \mathrm{Tr}(BA)$. We can also show in the representation we use that $\overline{b}^*_k = \overline{b}_k$. We define

$$\Delta_k = -\sum_{k'} V_{k,k'} \overline{b}_{k'} = \Delta^*_k. \tag{8.163}$$

As we will demonstrate later, this will turn out to be the gap parameter. We can write the interaction term as

$$\mathcal{H}_1 = \sum_{k,k'} V_{k,k'} b_k^\dagger b_{k'}. \tag{8.164}$$

Note

$$b_{k'} = \bar{b}_{k'} + \delta b_{k'} = \bar{b}_{k'} + (b_{k'} - \bar{b}_{k'}), \tag{8.165}$$

and

$$b_{k'}^\dagger = \bar{b}_{k'}^* + \delta b_{k'}^\dagger = \bar{b}_{k'}^* + (b_{k'}^\dagger - \bar{b}_{k'}^*); \tag{8.166}$$

$$b_k^\dagger = \bar{b}_k^* + \delta b_k^\dagger = \bar{b}_k + \delta b_k^\dagger; \tag{8.167}$$

and we will neglect $(\delta b_{k'}) \times (\delta b_k^\dagger)$ terms. (This is sort of a mean-field-like approximation for pairs.) Thus, using (8.166) and (8.167) and neglecting $O(\delta b^2)$ terms, we can write

$$\begin{aligned} b_k^\dagger b_{k'} &= (\bar{b}_k + \delta b_k^\dagger)(\bar{b}_{k'} + \delta b_{k'}) \\ &= \bar{b}_k \bar{b}_{k'} + \bar{b}_{k'} \delta b_k^\dagger + \bar{b}_k \delta b_{k'}, \\ &= \bar{b}_k b_{k'} + \bar{b}_{k'} b_k^\dagger - \bar{b}_k \bar{b}_{k'}, \end{aligned} \tag{8.168}$$

assuming $\bar{b}_k$ is real. Also,

$$\mathcal{H}_1 = \sum_{k,k'} V_{k,k'} (\bar{b}_{k'} b_k^\dagger + \bar{b}_k b_{k'} - \bar{b}_{k'} \bar{b}_k). \tag{8.169}$$

Thus,

$$\mathcal{H} = \sum_{k\sigma} \varepsilon_k C_{k\sigma}^\dagger C_{k\sigma} - \sum_k (\Delta_k b_k^\dagger + \Delta_k b_k - \Delta_k \bar{b}_k). \tag{8.170}$$

We now diagonalize by a Bogoliubov–Valatin transformation:

$$C_{k\uparrow} = u_k \alpha_k + v_k \beta_k^\dagger, \tag{8.171}$$

$$C_{-k\downarrow} = u_k \beta_k - v_k \alpha_k^\dagger; \tag{8.172}$$

where $u_k^2 + v_k^2 = 1$ (to preserve anticommutation relations), u_k and v_k are real, and the α and β given by

$$\alpha_k^\dagger = u_k C_{k\uparrow}^\dagger - v_k C_{-k\downarrow}, \tag{8.173}$$

$$\beta_k^\dagger = u_k C_{-k\downarrow}^\dagger + v_k C_{k\uparrow}, \tag{8.174}$$

are Fermion operators obeying the usual anticommutation relations. The $\alpha_k^\dagger$, and $\beta_k^\dagger$ create "bogolons". The algebra gets a bit detailed here and one can *skip* along unless curious,

$$b_k = C_{-k\downarrow} C_{k\uparrow} = (-v_k \alpha_k^\dagger + u_k \beta_k)(u_k \alpha_k + v_k \beta_k^\dagger) \tag{8.175}$$

$$b_k^\dagger = C_{k\uparrow}^\dagger C_{-k\downarrow}^\dagger = (u_k \alpha_k^\dagger + v_k \beta_k)(-v_k \alpha_k + u_k \beta_k^\dagger) \tag{8.176}$$

$$C_{k\uparrow}^\dagger C_{k\uparrow} = (u_k \alpha_k^\dagger + v_k \beta_k)(u_k \alpha_k + v_k \beta_k^\dagger) \tag{8.177}$$

$$C_{-k\downarrow}^\dagger C_{-k\downarrow} = (-v_k \alpha_k + u_k \beta_k^\dagger)(-v_k \alpha_k^\dagger + u_k \beta_k) \tag{8.178}$$

$$b_k^\dagger = -u_k v_k \alpha_k^\dagger \alpha_k + u_k v_k \beta_k \beta_k^\dagger - v_k^2 \beta_k \alpha_k + u_k^2 \alpha_k^\dagger \beta_k^\dagger \tag{8.179}$$

$$b_k = -v_k u_k \alpha_k^\dagger \alpha_k + u_k v_k \beta_k \beta_k^\dagger - v_k^2 \alpha_k^\dagger \beta_k^\dagger + u_k^2 \beta_k \alpha_k \tag{8.180}$$

$$C_{k\uparrow}^\dagger C_{k\uparrow} = u_k^2 \alpha_k^\dagger \alpha_k + v_k^2 \beta_k \beta_k^\dagger + u_k v_k \alpha_k^\dagger \beta_k^\dagger + u_k v_k \beta_k \alpha_k \tag{8.181}$$

$$C_{-k\downarrow}^\dagger C_{-k\downarrow} = u_k^2 \beta_k^\dagger \beta_k + v_k^2 \alpha_k \alpha_k^\dagger - u_k v_k \beta_k^\dagger \alpha_k^\dagger - v_k u_k \alpha_k \beta_k \tag{8.182}$$

$$\begin{aligned}
\mathcal{H} = \sum_k [&2\varepsilon_k u_k v_k \alpha_k^\dagger \beta_k^\dagger + 2\varepsilon_k \beta_k \alpha_k u_k v_k \\
&+ \varepsilon_k(u_k^2 - v_k^2)(\alpha_k^\dagger \alpha_k + \beta_k^\dagger \beta_k) + 2\varepsilon_k v_k^2 \\
&+ \Delta_k u_k v_k (\alpha_k^\dagger \alpha_k + \beta_k^\dagger \beta_k) \\
&- \Delta_k u_k v_k + \Delta_k v_k^2 \beta_k \alpha_k - \Delta_k u_k^2 \alpha_k^\dagger \beta_k^\dagger \\
&+ \Delta_k^* u_k v_k (\alpha_k^\dagger \alpha_k + \beta_k^\dagger \beta_k) - \Delta_k^* u_k v_k \\
&+ \Delta_k v_k^2 \alpha_k^\dagger \beta_k^\dagger - \Delta_k^* u_k^2 \beta_k \alpha_k] + \sum_k \Delta_k \bar{b}_k.
\end{aligned} \tag{8.183}$$

Rewriting this we get

$$\mathcal{H} = \sum_k \{(2\Delta_k u_k v_k - \varepsilon_k(v_k^2 - u_k^2))(\alpha_k^\dagger \alpha_k + \beta_k^\dagger \beta_k) \\
+ (2\varepsilon_k u_k v_k + \Delta_k(v_k^2 - u_k^2))(\alpha_k^\dagger \beta_k^\dagger + \beta_k \alpha_k)\} + G, \tag{8.184}$$

where

$$G = \sum (2\varepsilon_k v_k^2 - 2\Delta_k u_k v_k + \Delta_k \bar{b}_k).$$

Next, choose

$$2\varepsilon_k u_k v_k + \Delta_k(v_k^2 - u_k^2) = 0, \tag{8.185}$$

so as to diagonalize the Hamiltonian. Also, using $u_k^2 + v_k^2 = 1$ let

$$v_k^2 = \frac{1}{2} - a, \qquad (8.186)$$

$$v_k^2 - u_k^2 = -2a, \qquad (8.187)$$

$$u_k v_k = \sqrt{\frac{1}{4} - a^2}; \qquad (8.188)$$

$$2\varepsilon_k \sqrt{\frac{1}{4} - a^2} = \Delta_k 2a, \qquad (8.189)$$

$$\varepsilon_k^2 \left(\frac{1}{4} - a^2\right) = \Delta_k^2 a^2. \qquad (8.190)$$

Thus

$$a = \frac{\varepsilon_k}{2\sqrt{\varepsilon_k^2 + \Delta_k^2}}. \qquad (8.191)$$

Rewriting,

$$\mathcal{H} = [2\Delta_k u_k v_k - \varepsilon_k (v_k^2 - u_k^2)] \times (\alpha_k^\dagger \alpha_k + \beta_k^\dagger \beta_k) + G. \qquad (8.192)$$

But, define

$$E_k = \sqrt{\varepsilon_k^2 + \Delta_k^2}, \qquad (8.193)$$

$$a = \frac{\varepsilon_k}{2E_k}, \qquad (8.194)$$

and thus

$$2u_k v_k = \frac{\Delta_k}{E_k}. \qquad (8.195)$$

Thus, after a bit of algebra,

$$2\Delta_k u_k v_k - \varepsilon_k (v_k^2 - u_k^2)] = E_k. \qquad (8.196)$$

So

$$\mathcal{H} = \sum_k E_k (\alpha_k^\dagger \alpha_k + \beta_k^\dagger \beta_k) + G, \qquad (8.197)$$

and G can be put in the form

$$G = \sum_k (\varepsilon_k - E_k + \Delta_k \bar{b}_k). \tag{8.198}$$

Note by Fig. 8.20 and (8.193) how E_k predicts a gap, for clearly $E_k \geq \Delta_0$. Continuing

$$b_k = -v_k u_k \alpha_k^\dagger \alpha_k + u_k v_k \beta_k \beta_k^\dagger - v_k^2 \alpha_k^\dagger \beta_k^\dagger + u_k^2 \beta_k \alpha_k. \tag{8.199}$$

But $\bar{b}_k$ involves only diagonal terms, so using an appropriate anticommutation relation

$$\bar{b}_k = u_k v_k \left(1 - \overline{\alpha_k^\dagger \alpha_k} - \overline{\beta_k^\dagger \beta_k}\right), \tag{8.200}$$

so

$$\bar{b}_k = u_k v_k (1 - 2n_k), \tag{8.201}$$

where

$$n_k = \frac{1}{e^{\beta E_k} + 1} = f(E_k). \tag{8.202}$$

$f(E_k)$ is of course the Fermi function but it looks strange without the chemical potential. This is because $\alpha^\dagger, \beta^\dagger$ do not change the particle number. See Marder [8.22]. Therefore,

$$\begin{aligned}\Delta_k &= -\sum V_{k,k'} \bar{b}_{k'} \\ &= -\sum V_{k,k'} u_{k'} v_{k'} [1 - 2f(E_{k'})] \\ &= -\sum_{k'} V_{k,k'} \frac{\Delta_{k'}}{2 E_{k'}} [1 - 2f(E_{k'})]. \end{aligned} \tag{8.203}$$

Now assume that (not using $\hbar = 1$)

$$\Delta_{k'} = \Delta \quad \text{when} \quad |\varepsilon_k| < \hbar \omega_D, \tag{8.204}$$

$$V_{k,k'} = -V \quad \text{when} \quad |\varepsilon_k| < \hbar \omega_D, \tag{8.205}$$

$$\Delta_k = 0 \quad \text{when} \quad |\varepsilon_k| > \hbar \omega_D, \tag{8.206}$$

and

$$V_{k,k'} = 0 \quad \text{when} \quad |\varepsilon_k| > \hbar \omega_D, \tag{8.207}$$

where ω_D is the Debye frequency (see (8.138)) So,

$$\Delta = V \sum_{|\varepsilon_{k'}| < \hbar \omega_D} \frac{\Delta [1 - 2f(E_{k'})]}{2 E_{k'}}. \tag{8.208}$$

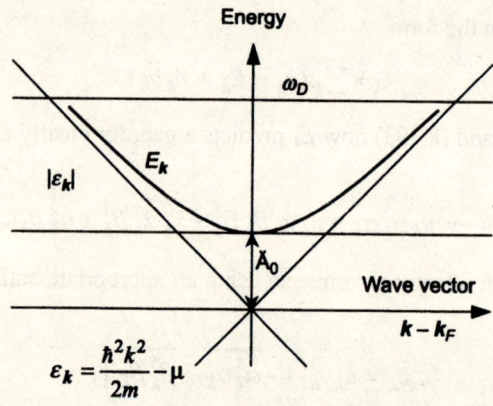

Fig. 8.20. Gap in single-particle excitations near the Fermi energy

For $T = 0$, then

$$\Delta = \sum_{k'} V \frac{\Delta}{2\sqrt{\varepsilon_{k'}^2 + \Delta^2}}, \qquad (8.209)$$

and for $T \neq 0$, then

$$\Delta = \sum_{k'} \frac{V\Delta}{2\sqrt{\varepsilon_{k'}^2 + \Delta^2}} \tanh\left(\frac{E_{k'}}{2kT}\right). \qquad (8.210)$$

We can then write

$$\Delta \cong \int_0^{\hbar\omega_D} N(E) V \frac{\Delta}{\sqrt{E^2 + \Delta^2}} dE. \qquad (8.211)$$

If we further suppose that $N(E) \cong$ constant $\cong N(0) \equiv$ the density of states at the Fermi level, then (8.211) becomes

$$\begin{aligned} \frac{1}{N(0)V} &= \int_0^{\hbar\omega_D} \frac{dE}{\sqrt{E^2 + \Delta^2}} = \ln\left(E + \sqrt{E^2 + \Delta^2}\right)\Big|_0^{\hbar\omega_D} \\ &= \ln\left(\frac{\hbar\omega_D + \sqrt{(\hbar\omega_D)^2 + \Delta^2}}{\Delta}\right). \end{aligned} \qquad (8.212)$$

This equation can be written as

$$\begin{aligned} \exp\left(-\frac{1}{N(0)V}\right) &= \frac{\Delta}{\hbar\omega_D + \sqrt{(\hbar\omega_D)^2 + \Delta^2}} \\ &= \frac{\Delta}{\hbar\omega_0\left(\sqrt{1 + (\Delta/\hbar\omega_D)^2} + 1\right)} = \frac{\Delta}{2\hbar\omega_D} \end{aligned} \qquad (8.213)$$

in the weak coupling limit (when $\Delta \ll \hbar\omega_D$). Thus, in the weak coupling limit, we obtain

$$\Delta \cong 2\hbar\omega_D \exp\left(-\frac{1}{N(0)V}\right). \tag{8.214}$$

From (8.210) by similar reasoning

$$\frac{1}{N(0)V} = \int_0^{\hbar\omega_D} \frac{\tanh\left(\sqrt{\varepsilon^2 + \Delta^2}/2kT\right)}{\sqrt{\varepsilon^2 + \Delta^2}} d\varepsilon, \tag{8.215}$$

where, again, $N(0)$ is the density of states at the Fermi energy.

For T greater than some critical temperature there are no solutions for Δ, i.e. the energy gap no longer exists. We can determine T_c by using the fact that at $T = T_c$, $\Delta = 0$. This says that

$$\frac{1}{N(0)V} = \int_0^{\hbar\omega_D} \frac{\tanh(\varepsilon/2kT_c)}{\varepsilon} d\varepsilon. \tag{8.216}$$

In the weak coupling approximation, when $N(0)V \ll 1$, we obtain from (8.216) that

$$kT_c = 1.14\hbar\omega_D \exp(-1/N(0)V). \tag{8.217}$$

Equation (8.217) is a very important equation. It depends on three material properties:

a) The Debye frequency ω_D

b) V that measures the strength of the electron–phonon coupling and

c) $N(0)$ that measures the number of electrons available at the Fermi energy.

Note that typically $\omega_D \propto (m)^{-1/2}$, where m is the mass of atoms. This leads directly to the isotope effect. Note also the energy gap $E_g = 2\Delta(0)$ at absolute zero.

We can combine this result with our result for the energy gap parameter in the ground state to derive a relation between the energy gap at absolute zero and the critical superconducting transition temperature with no magnetic field. By (8.217) and (8.214), we have that

$$\Delta(0) = 2\hbar\omega_D \exp(-1/N(0)V) = \frac{2}{1.14} kT_c, \tag{8.218}$$

or

$$2\Delta(0) = 3.52 kT_c. \tag{8.219}$$

Note that our expression for $\Delta(0)$ and T_c both involve the factor $\exp(-1/N(0)V)$; that is, a power series (in V) expansion for both $\Delta(0)$ and T_c have an essential singularity in V. We could not have obtained reasonable results if we had tried ordinary perturbation theory because with ordinary perturbation theory, we cannot reproduce the effect of an essential singularity in the perturbation. This is similar to what happened when we discussed a single Cooper pair.

Our discussion has only been valid for weakly coupled superconductors. Roughly speaking, these have $(T_c/\theta_D)^2 \gtrsim (500)^{-2}$. Pb, Hg, and Nb are strongly coupled, and for them $(T_c/\theta_D)^2 \gtrsim (300)^{-2}$. Alternatively, the electron–phonon coupling parameter is about three times larger than is a typical weak coupling superconductor. A result for the strong coupling approximation is given below.

8.5.4 Remarks on the Nambu Formalism and Strong Coupling Superconductivity (A)

The Nambu approach to superconductivity is presented by matrices and diagrams. The Nambu formalism includes the possibility of Cooper pairs in the calculation from the beginning via two component field operators. This approach allows for the treatment of retardation effects that need to be included for the strong (electron lattice) coupling regime. An essential step in the development was taken by Eliashberg and this leads to his equations. The Eliashberg strong coupling calculation of the superconducting transition temperature gives with a computer fit (via McMillan):

$$T_c = \frac{\theta_D}{1.45}\exp\left(\frac{-1.04(1+\lambda)}{\lambda - \mu^*(1+0.62\lambda)}\right).$$

θ_D is the Debye temperature, and for definitions of λ (the coupling constant) and μ^* (the Coulomb pseudopotential term) see Jones and March [8.17]. They also give a nice summary of the calculation. Briefly $\lambda = N(0)V_{phonon}$, $\mu = N(0)V_{coulomb}$ where V in (8.218) is $V_{phonon} - V_{coulomb}$, and

$$\mu^* = \mu\left(1 + \mu\ln\frac{E_F}{k_B\theta_D}\right)^{-1}.$$

Usually λ empirically turns out to be not much larger than 5/4 (or smaller).

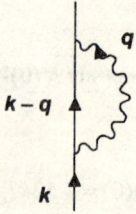

Fig. 8.21. Lowest-order correction to self-energy Feynman diagram (for electrons due to phonons)

The calculation includes the self-energy terms. The lowest-order correction to self-energy for electrons due to phonons is indicated in Fig. 8.21. The BCS theory with the extension of Eliashberg and McMillan has been very successful for many superconductors.

A nice reference to consult is Mattuck [8.23 pp. 267-272].

8.6 Magnesium Diboride (EE, MS, MET)

For a review of the new superconductor magnesium diboride, see, e.g., *Physics Today*, March 2003, p. 34ff. The discovery of the superconductor MgB_2, with a transition temperature of 39 K, was announced by Akimitsu in early 2001. At first sight this might not appear to be a particularly interesting discovery, compared to that of the high-temperature superconductors, but MgB_2 has several interesting properties:

1. It appears to be a conventional BCS superconductor with electron–phonon coupling driving the formation of pairs. It shows a strong isotope effect.

2. It does not appear to have the difficulty that the high-T_c cuprate ceramics have of having grain boundaries that inhibit current.

3. It is a widely available material that comes right off the shelf.

4. MgB_2 is an intermetallic (two metals forming a crystal structure at a well-defined stoichiometry) compound with a transition temperature near double that of Nb_3Ge.

Possibly, the transition temperature can be driven higher by tailoring the properties of magnesium diboride. At this writing, several groups are working intensely on this material, with several interesting results including the fact that it has two superconducting gaps arising from two weakly interacting bands.

8.7 Heavy-Electron Superconductors (EE, MS, MET)

UBe_{13} ($T_c = 0.85$ K), $CeCu_2Si_2$ ($T_c = 0.65$ K), and UPt_3 ($T_c = 0.54$ K) are heavy-electron superconductors. They are characterized by having large low-temperature specific heats due to effective mass being two or three orders of magnitude larger than in normal metals (because of f band electrons). Heavy-electron superconductors do not appear to have a singlet state s-wave pairing, but perhaps can be characterized as d-wave pairing or p-wave pairing (d and p referring to orbital symmetry). It is also questionable whether the pairing is due to the exchange of virtual phonons—it may be due, e.g., to the exchange of virtual magnons. See, e.g., Burns [8.9 p51]. We have already mentioned these in Sect. 5.7.

8.8 High-Temperature Superconductors (EE, MS, MET)

It has been said that Brazil is the country of the future and always has been as well as always will be. A similar comment has been made about superconductors. The problem is that superconductivity applications have been limited by the fact that liquid helium temperatures (of order 4 K) have been necessary to retain superconductivity. Liquid nitrogen (which boils at 77 K) is much cheaper and materials that superconduct at or above the boiling temperature of liquid nitrogen would open a large range of practical applications. Particularly important would be the transport of electrical power.

Just finding a high superconducting transition temperature T_c, however, does not solve all problems. The critical current can be an important limiting factor. Thermally activated creep of fluxoids (due to $J \times B$) can lower J_c (the critical current) as the current interacts with the fluxoid and causes energy loss when the fluxoid becomes unpinned and thus creeps (can move). This is important in the high-T_c superconductors considered in this section.

Until 1986, the highest transition temperature for a superconductor was T_c = 23.2 K for Nb_3Ge. Then Bednortz and Müller found a *ceramic* oxide (product of clay) of lanthanum, barium, and copper became superconducting at about 35 K. For this work they won the Nobel prize for Physics in 1987. Since Bednortz's pioneering work several other high-T_c superconductors have been found.

The "1-2-3" compound $YBa_2Cu_3O_7$, has a T_c of 92 K. The "2-1-4" compound (e.g. $Ba_xLa_{2-x}CuO_{4-y}$) are another class of high-T_c superconductors. $Tl_2Ba_2Ca_2Cu_3O_{10}$ has a remarkably high T_c of 125 K.

The high-T_c materials are type II and typically have a penetration depth to coherence length ratio $K \approx 100$ and typically have a very large upper critical field. As we have mentioned, thermally activated creep of fluxoids due to the $J \times B$ force may cause energy dissipation and limit useable current values. For real materials, the critical current (J_c), critical temperature (T_c), and critical magnetic field (B_c) vary, but can be conveniently represented as shown in Fig. 8.22. As mentioned, the high-temperature superconductors (HTSs) are typically type II and also their J_c parallel to the copper oxide sheets (mentioned below) $\approx 10^7$ A/cm^2, while perpendicular to the sheets J_c can be about 10^7 A/cm^2. A schematic of J, B_c, and T_c is shown in Fig. 8.22 for type I materials. For HTS, the representation of Fig. 8.22 is not complex enough. In Table 8.1 we list selected superconductor elements and compounds along with their transition temperature.

For HTS, we are faced with a puzzle as to what causes some ceramic copper oxide materials to be superconductors at temperatures well above 100 K. In conventional superconductors, we talk about electrons paired into spherically symmetric wave functions (s-waves) due to exchange of virtual phonons. Apparently, lattice vibrations cannot produce a strong enough coupling to produce such high critical temperatures. It appears parallel Cu-O planes in these materials play some very significant but not yet fully understood role. Hole conduction in these planes is important. As mentioned, there is also a strong anisotropy in electrical conduction. Although there seems to be increasing evidence for d-wave pairing, the exchange mechanism necessary to produce the pair is still not clear as of this writing. It could be due to magnetic interactions or there may be new physics. See, e.g., Burns [8.9].

Table 8.1. Superconductors and their transition temperatures

Selected elements*	Transition temperature T_c (K)
Al	1.17
Hg	4.15
Nb	9.25
Sn	3.72
Pb	7.2
Selected compounds*	
Nb_3Ge	23.2
Nb_3Sn	18.
Nb_3Au	10.8
$NbSe_2$	7.2
MgB_2**	39
Copper oxide (HTS)*	
$Bi_2Sr_2Ca_2Cu_3O_{10}$	~110
$YBa_2Cu_3O_7$	~92
$Tl_2Ba_2Ca_3Cu_4O_{11}$	~122
Heavy fermion*	
UBe_{13}	0.85
$CeCu_2Si_2$	0.65
UPt_3	0.54
Fullerenes***	
K_3C_{60}	19.2
$RbCs_2C_{60}$	33

*Reprinted from Burns G, *High Temperature Superconductivity* Table 2-1 p. 8 and Table 3-1 p. 57, Academic Press, Copyright 1992, with permission from Elsevier. On p. 52 Burns also briefly discusses organic superconductors.
**Canfield PC and Crabtree GW, *Physics Today* 56(3), 34 (2003).
***Huffmann DR, "Solid C60," *Physics Today* 41(11), 22 (1991).

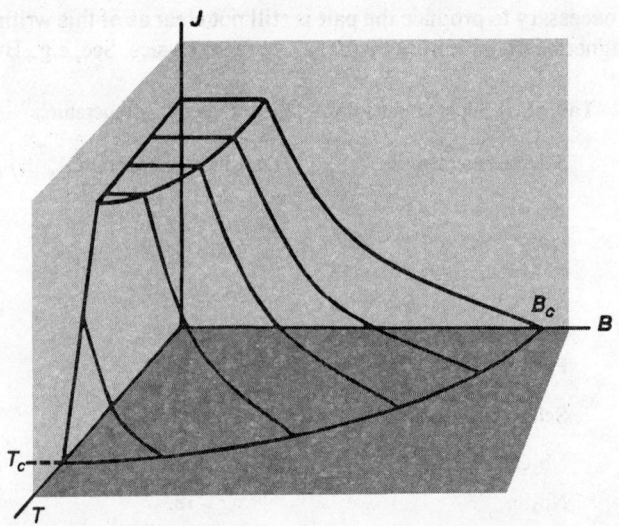

Fig. 8.22. J, B, T surface separating superconducting and normal regions

8.9 Summary Comments on Superconductivity (B)

1. In the superconducting state $E = 0$ (superconductivity implies the resistivity ρ vanishes, $\rho \to 0$).
2. The superconducting state is more than vanishing resistivity since this would imply B was constant, whereas $B = 0$ in the superconducting state (flux is excluded as we drop below the transition temperature).
3. For "normal" BCS theory:

    a) An attractive interaction between electrons can lead to a ground state separated from the excited states by an energy gap. Most of the important properties of superconductors follow from this energy gap.

    b) The electron–lattice interaction, which can lead to an effective attractive interaction, causes the energy gap.

    c) The ideas of the penetration depth (and, hence, the Meissner effect—flux exclusion) and the coherence length follow from the theory of superconductivity.

4. Type II superconductors have upper and lower critical fields and are technically important because of their high upper critical fields. Magnetic flux can penetrate between the upper and lower critical fields, and the penetration is quantized in units of $hc/|2e|$, just as is the magnetic flux through a superconducting ring. Using a unit of charge of $2e$ is consistent with Cooper pairs.

5. In zero magnetic fields, for weak superconductors, superconductivity occurs at the transition temperature:

$$k_B T_c \cong 1.14\hbar\omega_D \exp(-1/N_0 V_0), \qquad (8.220)$$

where N_0 is half the density of single-electron states, V_0 is the effective interaction between electron pairs near the Fermi surface, and $\hbar\omega_D \cong \hbar\theta_D$, where θ_D is the Debye temperature.

6. The energy gap (2Δ) is determined by (weak coupling):

$$\Delta(0) \cong 2\hbar\omega_D \exp(-1/N_0 V_0) = 1.76 k T_c \qquad (8.221)$$

$$\Delta(T) \cong \Delta(0)\left(1 - \frac{T}{T_c}\right)^{1/2}; \quad T \leq T_c. \qquad (8.222)$$

7. The critical field is fairly close to the empirical law (for weak coupling):

$$\frac{H_c(T)}{H_c(0)} \approx 1 - \left(\frac{T}{T_c}\right)^2. \qquad (8.223)$$

8. The coherent motion of the electrons results in a resistanceless flow because a small perturbation cannot disturb one pair of electrons without disturbing all of them. Thus, even a small energy gap can inhibit scattering.

9. The central properties of superconductors are the penetration depth λ (of magnetic fields) and the coherence length ξ (or "size" of Cooper pairs). Small λ/ξ ratios lead to type I superconductors, and large λ/ξ ratios lead to type II behavior. ξ can be decreased by alloying.

10. The Ginzburg–Landau theory is used for superconductors in a magnetic field where one has inhomogeneities in spatial behavior.

11. We should also mention that one way to think about the superconducting transition is a Bose–Einstein condensation, as modified by their interaction, of bosonic Cooper pairs.

12. See the comment on spontaneously broken symmetry in the chapter on magnetism. Superconductivity can be viewed as a broken symmetry.

13. In the paired electrons of superconductivity, in s and d waves, the spins are antiparallel, and so one understands why ferromagnetism and superconductivity don't appear to coexist, at least normally. However, even p-wave superconductors (e.g. Strontium Ruthenate) with parallel spins the magnetic fields are commonly expelled in the superconducting state. Recently, however, two materials have been discovered in which ferromagnetism and superconductivity coexist. They are UGe_2 (under pressure) and $ZrZn_2$ (at ambient pressure). One idea is that these two materials are p-wave superconductors. The issues about these materials are far from settled, however. See *Physics Today*, p. 16, Sept. 2001.

14. Also, high-T_c (over 100 K) superconductors have been discovered and much work remains to understand them.

In Table 8.2 we give a subjective "Top Ten" of superconductivity research.

Table 8.2 Top 10 of superconductivity (subjective)

Person	Achievement	Date/comments
1. H. Kammerlingh Onnes	Liquefied He Found resistance of Hg $\rightarrow$ 0 at 4.19 K	1908 – Started low-T physics 1911 – Discovered superconducting state 1911 – Nobel Prize
2. W. Meissner and R. Ochsenfeld	Perfect diamagnetism	1933 – Flux exclusion
3. F. and H. London	London equations and flux expulsion	1935 – B proportional to curl of J
4. V.L. Ginzburg and L.D. Landau	Phenomenological equations	1950 – Eventually GLAG equations 1962 – Nobel Prize, Landau 2003 – Nobel Prize, Ginzburg
A. A. Abrikosov	Improvement to GL equations, Type II	1957 – Negative surface energy 2003 – Nobel Prize
L. P. Gor'kov	GL limit of BCS and order parameter	1959 – Order parameter proportional to gap parameter
5. A. B. Pippard	Nonlocal electrodynamics	1953 – x and l dependent on mean free path in alloys
6. J. Bardeen, L. Cooper, and J. Schrieffer	Theory of superconductivity	1957 – e.g. see (8.217) 1972 – Nobel Prize (all three)
7. I. Giaver	Single-particle tunneling	1960 – Get gap energy 1973 – Nobel Prize
8. B. D. Josephson	Pair tunneling	1962 – SQUIDS and metrology 1973 – Nobel Prize
9. Z. Fisk, et al	Heavy fermion "exotic" superconductors	1985 – Pairing different than BCS, probably
10. J.G. Bednorz and K. A. Muller	High-temperature superconductivity	1986 – Now, T_cs are over 100 K 1987 – Nobel Prize (both)

Problems

8.1 Show that the flux in a superconducting ring is quantized in units of h/q, where $q = |2e|$.

8.2 Derive an expression for the single-particle tunneling current between two superconductors separated by an insulator at absolute zero. If E^T is measured from the Fermi energy, you can calculate a density of states as below.

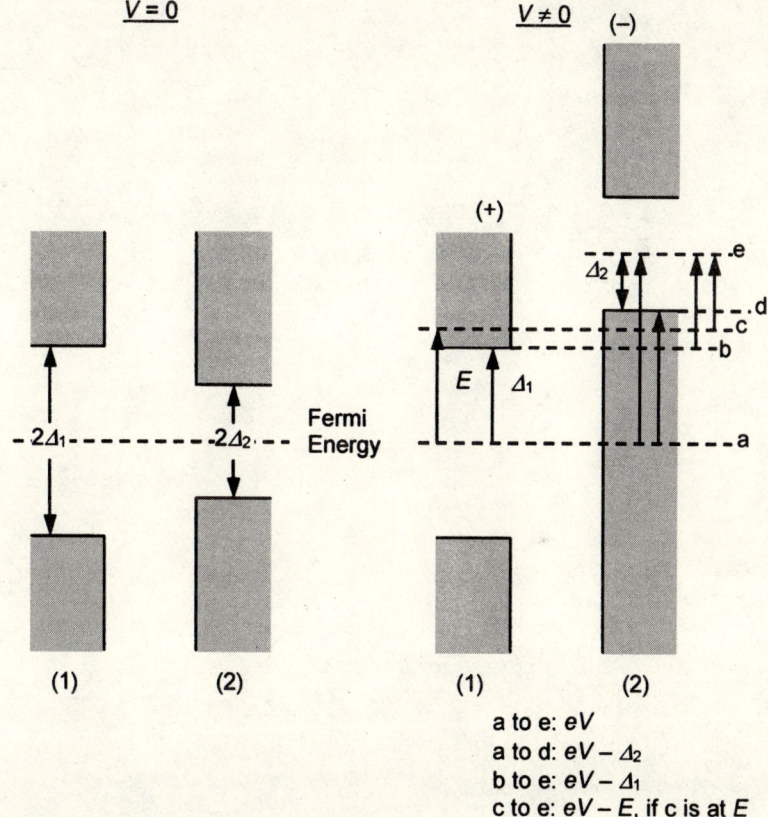

a to e: eV
a to d: $eV - \Delta_2$
b to e: $eV - \Delta_1$
c to e: $eV - E$, if c is at E

Note:
$$E^T = \sqrt{\varepsilon^2 + \Delta^2} \quad \text{(compare (8.93))},$$

$$D_S(E^T) = \frac{dn(E^T)}{dE^T} = \frac{\frac{dn(E^T)}{d\varepsilon}}{\frac{dE^T}{d\varepsilon}} \cong D(0)\frac{E^T}{\sqrt{(E^T)^2 - \Delta^2}},$$

where $D(0)$ is the number of states per unit energy without pairing.

9 Dielectrics and Ferroelectrics

Despite the fact that the concept of the dielectric constant is often taught in introductory physics – because, e.g., of its applications to capacitors – the concept involves subtle physics. The purpose of this chapter is to review the important dielectric properties of solids without glossing over the intrinsic difficulties.

Dielectric properties are important for insulators and semiconductors. When a dielectric insulator is placed in an external field, the field (if weak) induces a polarization that varies linearly with the field. The constant of proportionality determines the dielectric constant. Both static and time-varying external fields are of interest, and the dielectric constant may depend on the frequency of the external field. For typical dielectrics at optical frequencies, there is a simple relation between the index of refraction and the dielectric constant. Thus, there is a close relation between optical and dielectric properties. This will be discussed in more detail in the next chapter.

In some solids, below a critical temperature, the polarization may "freeze in." This is the phenomena of ferroelectricity, which we will also discuss in this chapter. In some ways ferroelectric and ferromagnetic behavior are analogous.

Dielectric behavior also relates to metals particularly by the idea of "dielectric screening" in a quasifree-electron gas. In metals, a generalized definition of the dielectric constant allows us to discuss important aspects of the many-body properties of conduction electrons. We will discuss this in some detail.

Thus, we wish to describe the ways that solids exhibit dielectric behavior. This has practical as well as intrinsic interest and is needed as a basis for the next chapter on optical properties.

9.1 The Four Types of Dielectric Behavior (B)

1. *The polarization of the electronic cloud around the atoms*: When an external electric field is applied, the electronic charge clouds are distorted. The resulting polarization is directly related to the dielectric constant. There are "anomalies" in the dielectric constant or refractive index at frequencies in which the atoms can absorb energies (resonance frequencies, or in the case of solids, interband frequencies). These often occur in the visible or ultraviolet. At lower frequencies, the dielectric constant is practically independent of frequency.

2. *The motion of the charged ions*: This effect is primarily of interest in ionic crystals in which the positive and negative ions can move with respect to one another

and thus polarize the crystal. In an ionic crystal, the resonant frequencies associated with the relative motion of the positive and negative ions are in the infrared and will be discussed in the optics chapter in connection with the restrahlen effect.

3. *The rotation of molecules with permanent dipole moments*: This is perhaps the easiest type of dielectric behavior to understand. In an electric field, the dipoles tend to line up with the electric fields, while thermal effects tend to oppose this alignment, and so, the phenomenon is temperature dependent. This type of dielectric behavior is mostly relevant for liquids and gases.

4. *The dielectric screening of a quasifree electron gas*: This is a many-body problem of a gas of electrons interacting via the Coulomb interaction. The technique of using the dielectric constant with frequency and wave-vector dependence will be discussed. This phenomena is of interest for metals.

Perhaps we should mention *electrets* here as a fifth type of dielectric behavior in which the polarization may remain, at least for a very long time after the removal of an electric field. In some ways an electret is analogous to a magnet. The behavior of electrets appears to be complex and as yet they have not found wide applications. Electrets occur in organic waxes due to frozen in disorder that is long lived but probably metastable.[1]

9.2 Electronic Polarization and the Dielectric Constant (B)

The ideas in this Section link up closely with optical properties of solids. In the chapter on the optical properties of solids, we will relate the complex index of refraction to the absorption and reflection of electromagnetic radiation. Now, we remind the reader of a simple picture, which relates the complex index of refraction to the dynamics of electron motion. We will include damping.

Our model considers matters only from a classical point of view. We limit discussion to electrons in bound states, but for some solids we may want to consider quasifree electrons or both bound and quasifree electrons. For electrons bound by Hooke's law forces, the equation describing their motion in an alternating electric field $E = E_0 \exp(-i\omega t)$ may be written ($e > 0$)

$$m\frac{d^2x}{dx^2} + \frac{m}{\tau}\frac{dx}{dt} + m\omega_0^2 x = -eE_0 \exp(-i\omega t). \qquad (9.1)$$

---

[1] See Gutmann [9.9]. See also Bauer et al [9.1].

9.2 Electronic Polarization and the Dielectric Constant (B)

The term containing τ is the damping term, which can be due to the emission of radiation or the other frictional processes. ω_0 is the natural oscillation frequency of the elastically bound electron of charge e and mass m. The steady-state solution is

$$x(t) = -\frac{e}{m} \frac{E_0 \exp(-i\omega t)}{\omega_0^2 - \omega^2 - i\omega/\tau}. \tag{9.2}$$

Below, we will assume that the field at the electronic site is the same as the average internal field. This completely neglects local field effects. However, we will follow this discussion with a discussion of local field effects, and in any case, much of the basic physics can be done without them. In effect, we are looking at atomic effects while excluding some interactions.

If N is the number of charges per unit volume, with the above assumptions, we write:

$$P = -Nex = \left(\frac{\varepsilon}{\varepsilon_0} - 1\right)\varepsilon_0 E = N\alpha E, \tag{9.3}$$

where ε is the dielectric constant and α is the polarizability. Using $E = E_0 \exp(-i\omega t)$,

$$\alpha = -\frac{ex}{E} = \frac{e^2}{m} \frac{1}{\omega_0^2 - \omega^2 - i\omega/\tau}. \tag{9.4}$$

The complex dielectric constant is then given by

$$\frac{\varepsilon}{\varepsilon_0} = 1 + \frac{N}{\varepsilon_0} \frac{e^2}{m} \frac{1}{\omega_0^2 - \omega^2 - i\omega/\tau} \equiv \varepsilon_r + i\varepsilon_i, \tag{9.5}$$

where we have absorbed the ε_0 into ε_r and ε_i for convenience. The real and the imaginary parts of the dielectric constant are then given by:

$$\varepsilon_r = 1 + \frac{Ne^2}{m\varepsilon_0} \frac{\omega_0^2 - \omega^2}{(\omega_0^2 - \omega^2)^2 + \omega^2/\tau^2}, \tag{9.6}$$

$$\varepsilon_i = \frac{Ne^2}{m\varepsilon_0} \frac{\omega/\tau}{(\omega_0^2 - \omega^2)^2 + \omega^2/\tau^2}. \tag{9.7}$$

In the chapter on optical properties, we will note that the connection (10.8) between the complex refractive index and the complex dielectric constant is:

$$n_c^2 = (n + in_i)^2 = (\varepsilon_r + i\varepsilon_i). \tag{9.8}$$

Therefore,

$$n^2 - n_i^2 = \varepsilon_r, \tag{9.9}$$

$$2nn_i = \varepsilon_i. \tag{9.10}$$

Thus, explicit equations for fundamental optical constants n and n_i are:

$$n^2 - n_i^2 = 1 + \frac{Ne^2}{m\varepsilon_0} \frac{\omega_0^2 - \omega^2}{(\omega_0^2 - \omega^2)^2 + \omega^2/\tau^2} \tag{9.11}$$

$$2nn_i = \frac{Ne^2}{m\varepsilon_0} \frac{\omega/\tau}{(\omega_0^2 - \omega^2)^2 + \omega^2/\tau^2}. \tag{9.12}$$

Quantum mechanics produces very similar equations. The results as given by Moss[2] are

$$n^2 - n_i^2 = 1 + \sum_j \frac{(Ne^2 f_{ij}/m\varepsilon_o)(\omega_{ij}^2 - \omega^2)}{(\omega_{ij}^2 - \omega^2) + \omega^2/\tau_j^2}, \tag{9.13}$$

$$2nn_i = \sum_j \frac{(Ne^2 f_{ij}/m\varepsilon_o)\omega/\tau_j}{(\omega_{ij}^2 - \omega^2) + \omega^2/\tau_j^2}, \tag{9.14}$$

where the f_{ij} are called oscillator strengths and are defined by

$$f_{ij} = 2\omega_{ij} \frac{m|\langle\psi_i|x|\psi_i\rangle|^2}{\hbar}, \tag{9.15}$$

where

$$\omega_{ij} = \frac{E_i - E_j}{\hbar}, \tag{9.16}$$

with E_i and E_j being the energies corresponding to the wave functions ψ_i and ψ_j. In a solid, because of the presence of neighboring dipoles, the local electric field does not equal the applied electric field.

Clearly, dielectric and optical properties are not easy to separate. Further discussion of optical-related dielectric properties comes in the next chapter.

We now want to examine some consequences of local fields. We also want to keep in mind that we will be talking about total dielectric constants and total polarizability. Thus in an ionic crystal, there are contributions to the polarizabilities and dielectric constants from both electronic and ionic motion.

The first question we must answer is, "If an external field, E, is applied to a crystal, what electric field acts on an atom in the crystal?" See Fig. 9.1. The slab is maintained between two plates that are connected to a battery of constant voltage V. Fringing fields are neglected. Thus, the electric field, E_0, between the plates before the slab is inserted, is the same as the electric field in the solid-state after insertion

---

[2] See Moss [9.13]. Note n_i refers to the imaginary part of the dielectric constant on the left of these equations and in f_{ij}, i refers to the initial state, while j refers to the final state.

(so, $E_0 d = V$). This is also the same as the electric field in a needle-shaped cavity in the slab. The electric field acting on the atom is

$$E_{\text{loc}} = E'_0 + E_a + E_b + E_c, \tag{9.17}$$

where, E'_0 is the electric field due to charge on the plates after the slab is inserted, E_a is the electric field due to the polarization charges on the faces of the slab, and E_b is the electric field due to polarization charges on the surface of the spherical cavity (which exists in our imagination), and E_c is the polarization due to charges interior to the cavity that we assume (in total) sums to zero.

It is, of course, an approximation to write E_{loc} in the above form. Strictly speaking, to find the field at any particular atom, we should sum over the contributions to this field from all other atoms. Since this is an impossible task, we treat macroscopically all atoms that are sufficiently far from A (and outside the cavity).

By Gauss' law, we know the electric field due to two plates with a uniform charge density ($\pm\sigma$) is $E = \sigma/\varepsilon$. Further, σ due to P ending on the boundary of a slab is $\sigma = P$ (from electrostatics). Since the polarization charges on the surface of the slabs will oppose the electric field of the plate and since charge will flow to maintain constant voltage.

$$\varepsilon_0 E_0 = \varepsilon_0 E'_0 - P, \tag{9.18}$$

or

$$E'_0 = E_0 + \frac{P}{\varepsilon_0}. \tag{9.19}$$

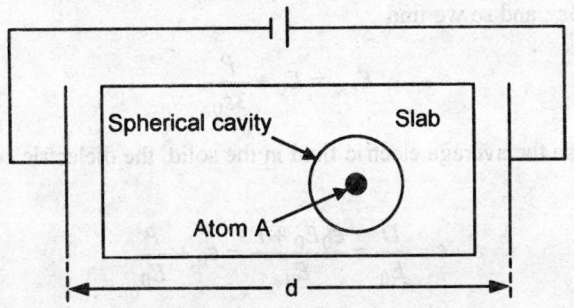

Fig. 9.1. Geometry for local field

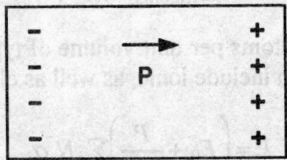

Fig. 9.2. The polarized slab

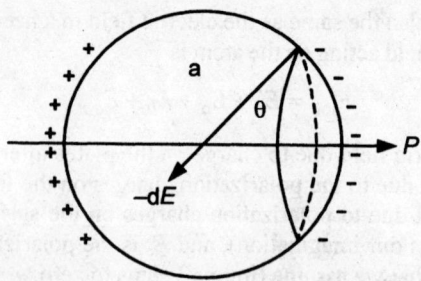

Fig. 9.3. Polarized charges around the cavity

Clearly, $E_a = -P/\varepsilon_0$ (see Fig. 9.2), and for all cubic crystals, $E_c = 0$. So,

$$E_{\text{loc}} = E_0 + E_b . \tag{9.20}$$

Using Fig. 9.3, since $\sigma_p = \boldsymbol{P}\cdot\boldsymbol{n}$ (n is outward normal), the charge on an annular region of the surface of the cavity is

$$dq = -P\cos\theta \cdot 2\pi a\sin\theta \cdot a d\theta , \tag{9.21}$$

$$dE_b = \frac{1}{4\pi\varepsilon_0} \frac{dq}{a^2} \cdot \cos\theta , \tag{9.22}$$

$$E_b = -\frac{P}{2\varepsilon_0} \int_0^\pi \cos^2\theta \cdot d\cos\theta . \tag{9.23}$$

Thus $E_b = P/3\varepsilon_0$, and so we find

$$E_{\text{loc}} = E_0 + \frac{P}{3\varepsilon_0} . \tag{9.24}$$

Since E_0 is also the average electric field in the solid, the dielectric constant is defined as

$$\varepsilon = \frac{D}{E_0} = \frac{\varepsilon_0 E_0 + P}{E_0} = \varepsilon_0 + \frac{P}{E_0} . \tag{9.25}$$

The polarization is the dipole moment per unit volume, and so, it is given by

$$P = \sum_{i(\text{atoms})} E_{\text{loc}}^i N_i \alpha_i , \tag{9.26}$$

where N_i is the number of atoms per unit volume of type i, and α_i is the appropriate polarizability (which can include ionic, as well as electronic motions). Thus,

$$P = \left(E_0 + \frac{P}{3\varepsilon_0} \right) \sum_i N_i \alpha_i , \tag{9.27}$$

or

$$\frac{P}{E_0} = \frac{\sum_i N_i \alpha_i}{1 - \frac{1}{3\varepsilon_0} \sum_i N_i \alpha_i},$$ (9.28)

or

$$\frac{\varepsilon}{\varepsilon_0} = 1 + \frac{1}{\varepsilon_0} \frac{\sum_i N_i \alpha_i}{\left(1 - \frac{1}{3\varepsilon_0} \sum_i N_i \alpha_i\right)},$$ (9.29)

which can be arranged to give the Clausius–Mossotti equation

$$\frac{(\varepsilon/\varepsilon_0) - 1}{(\varepsilon/\varepsilon_0) + 2} = \frac{1}{3\varepsilon_0} \sum_i N_i \alpha_i .$$ (9.30)

In the optical range of frequencies (the order of but less than 10^{15} cps), $n^2 = \varepsilon/\varepsilon_0$, and the equation becomes the Lorentz–Lorenz equation

$$\frac{n^2 - 1}{n^2 + 2} = \frac{1}{3\varepsilon_0} \sum_i N_i \alpha_i .$$ (9.31)

Finally, we show that when one resonant peak dominates, the only effect of the local field is to shift the dormant resonant (natural) frequency. From

$$\frac{\varepsilon}{\varepsilon_0} = 1 + \frac{1}{\varepsilon_0} \frac{N\alpha}{(1 - N\alpha/3\varepsilon_0)},$$ (9.32)

and

$$\alpha = \frac{e^2}{m} \frac{1}{\omega_0^2 - \omega^2 - i\omega/\tau},$$ (9.33)

we have

$$\frac{(\varepsilon/\varepsilon_0) - 1}{(\varepsilon/\varepsilon_0) + 2} = \frac{N\alpha}{3\varepsilon_0} = \frac{\omega_p^2}{3} \frac{1}{\omega_o^2 - \omega^2 - i\omega/\tau},$$ (9.34)

where

$$\omega_p = \sqrt{Ne^2/m\varepsilon_0}$$ (9.35)

is the plasma frequency. From this, we easily show

$$\frac{\varepsilon}{\varepsilon_0} = 1 + \frac{\omega_p^2}{\omega_0'^2 - \omega^2 - i\omega/\tau},$$ (9.36)

where

$$\omega_0'^2 = \omega_0^2 - \frac{1}{3}\omega_p^2, \qquad (9.37)$$

which is exactly what we would have obtained in the beginning (from (9.32) and (9.33)) if $\omega_0 \to \omega'_0$, and if the term $N\alpha/3\varepsilon_0$ had been neglected.

9.3 Ferroelectric Crystals (B)

All ferroelectric crystals are polar crystals.[3] Because of their structure, polar crystals have a permanent electric dipole moment. If $\rho(r)$ is the total charge density, we know for polar crystals

$$\int r\rho(r)dV \neq 0. \qquad (9.38)$$

Pyroelectric crystals have a polarization that changes with temperature. All polar crystals are pyroelectric, but not all polar crystals are ferroelectric. Ferroelectric crystals are polar crystals whose polarization can be reversed by an electric field. All ferroelectric crystals are also piezoelectric, in which stress changes the polarization. Piezoelectric crystals are suited for making electromechanical transducers with a variety of applications.

Ferroelectric crystals often have unusual properties. Rochelle salt $NaKC_4H_4O_6 \, 4H_2O$, which was the first ferroelectric crystal discovered, has both an upper and lower transition temperature. The crystal is only polarized between the two transition temperatures. The "TGS" type of ferroelectric, including triglycine sulfate and triglycine selenate, is another common class of ferroelectrics and has found application to IR detectors due to its pyroelectric properties. Ferroelectric crystals with hydrogen bonds (e.g. KH_2PO_4, which was the second ferroelectric crystal discovered) undergo an appreciable change in transition temperature when the crystal is deuterated (with deuterons replacing the H nuclei). $BaTiO_3$ was the first mechanically hard ferroelectric crystal that was discovered. Ferroelectric crystals are often classified as *displacive*, involving a lattice distortion (i.e. barium titinate, $BaTiO_3$, see Fig. 9.4), or *order–disorder* (i.e. potassium dihydrogen phosphate, KH_2PO_4, which involves the ordering of protons).

---

[3] Ferroelectrics: The term ferro is used but iron has nothing to do with it. Low symmetry causes spontaneous polarization.

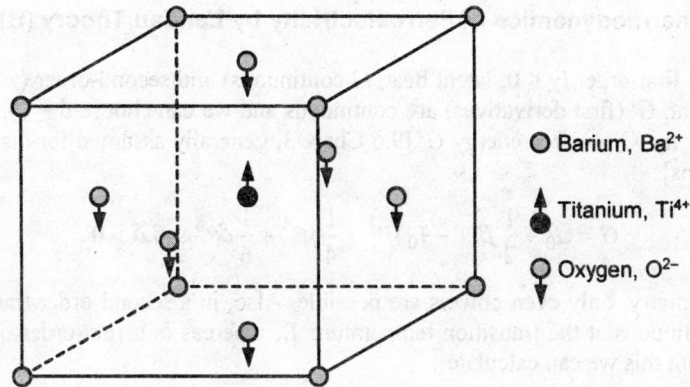

Fig. 9.4. Unit cell of barium titanate. The displacive transition is indicated by the direction of the arrows

In a little more detail, displacive ferroelectrics involve transitions associated with the displacement of a whole sublattice. How this could arises is discussed in Sect. 9.3.3 where we talk about the soft mode model. The soft mode theory, introduced in 1960, has turned out to be a unifying principle in ferroelectricity (see Lines and Glass [9.12]). Order–disorder ferroelectrics have transitions associated with the ordering of ions. We have mentioned in this regard KH_2PO_4 as a crystal with hydrogen bonds in which the motion of protons is important. Ferroelectrics have found application as memories, their high dielectric constant is exploited in making capacitors, and ferroelectric cooling is another area of application.

Other examples include ferroelectric cubic perovskite (PZT) $PbZr_{(x)}Ti_{(1-x)}O_3$, $T_c = 670$ K. The ferroelectric $BaMgF_4$ (BMF) does not show a Curie T even up to melting. These are other familiar ferroelectrics as given below.

The central problem of ferroelectricity is to be able to describe the onset of spontaneous polarization. Spontaneous polarization is said to exist if, in the absence of an electric field, the free energy is minimum for a finite value of the polarization. There may be some ordering involved in a ferroelectric transition, as in a ferromagnetic transition, but the two differ by the fact that the ferroelectric transition in a solid always involves the creation of dipoles.

Just as for ferromagnets, a ferroelectric crystal undergoes a phase transition from the paraelectric phase to the ferroelectric phase, typically, as the temperature is lowered. The transition can be either first order (with a latent heat, i.e. $BaTiO_3$) or second order (without latent heat, i.e. $LiTaO_3$). Just as for ferromagnets, the ferroelectric will typically split into domains of varying size and orientation of polarization. The domain structure forms to reduce the energy. Ferroelectrics show hysteresis effects just like ferromagnets. Although we will not discuss it here, it is also possible to have antiferroelectrics that one can think of as arising from antiparallel orientation of neighboring unit cells. A simple model of spontaneous polarization is obtained if we use the Clausius–Mossotti equation and assume (unrealistically for solids) that polarization arises from orientation effects. This is discussed briefly in a later section.

9.3.1 Thermodynamics of Ferroelectricity by Landau Theory (B)

For both first-order ($\gamma < 0$, latent heat, G continuous) and second-order ($\gamma > 0$, no latent heat, G' (first derivatives) are continuous and we can choose $\delta = 0$), we assume for the Gibbs free energy G' [9.6 Chap. 3, generally assumed for displacive transitions],

$$G = G_0 + \frac{1}{2}\beta(T-T_0)P^2 + \frac{1}{4}\gamma P^4 + \frac{1}{6}\delta P^6 \; ; \; \beta, \delta > 0. \tag{9.39}$$

(By symmetry, only even powers are possible. Also, in a second-order transition, P is continuous at the transition temperature T_c, whereas in a first-order one it is not.) From this we can calculate

$$E = \frac{\partial G}{\partial P} = \beta(T-T_0)P + \gamma P^3 + \delta P^5, \tag{9.40}$$

$$\frac{1}{\chi} = \frac{\partial E}{\partial P} = \beta(T-T_0) + 3\gamma P^2 + 5\delta P^4. \tag{9.41}$$

Notice in the paraelectric phase, $P = 0$ so $E = 0$ and $\chi = 1/\beta(T-T_0)$, and therefore Curie–Weiss behavior is included in (9.39). For $T < T_c$ and $E = 0$ for second order where $\delta = 0$, $\beta(T-T_0)P + \gamma P^3 = 0$, so

$$P^2 = -\frac{\beta}{\gamma}(T-T_0), \tag{9.42}$$

or

$$P = \pm\sqrt{\frac{\beta}{\gamma}(T_0-T)}, \tag{9.43}$$

which again is Curie–Weiss behavior (we assume $\gamma > 0$). For $T = T_c = T_0$, we can show the stable solution is the polarized one.

For first order set $E = 0$, solve for P and exclude the solution for which the free energy is a maximum. We find (where we assume $\gamma < 0$)

$$P_S = \pm\left[-\frac{\gamma}{2\delta}\left(1+\sqrt{1-\frac{4\delta\beta}{\gamma^2}(T-T_0)}\right)\right]^{1/2}.$$

Now, $G(P_{SC}) = G_{\text{polar}} = G_{\text{nonpolar}} = G_0$ at the transition temperature. Using the expression for G (9.39) and the expression that results from setting $E = 0$ (9.40), we find

$$T_c = T_0 - \frac{\gamma}{4\beta}P_{SC}^2. \tag{9.44}$$

By $E = 0$, we find (using (9.44))

$$\frac{3\gamma}{4} P_{SC}^3 + \delta P_{SC}^5 = 0, \qquad (9.45)$$

so

$$P_{SC}^2 = -\frac{3\gamma}{4\delta}. \qquad (9.46)$$

Putting (9.46) into (9.44) gives

$$T_c = T_0 + \frac{3\gamma^2}{16\beta\delta}. \qquad (9.47)$$

Figures 9.5, 9.6, and 9.7 give further insight into first- and second-order transitions.

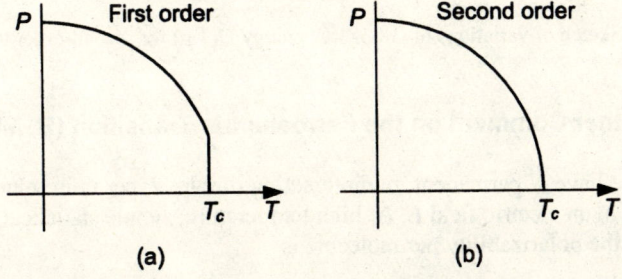

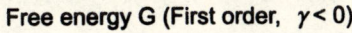

Fig. 9.5. Sketch of (**a**) first-order and (**b**) second-order ferroelectric transitions

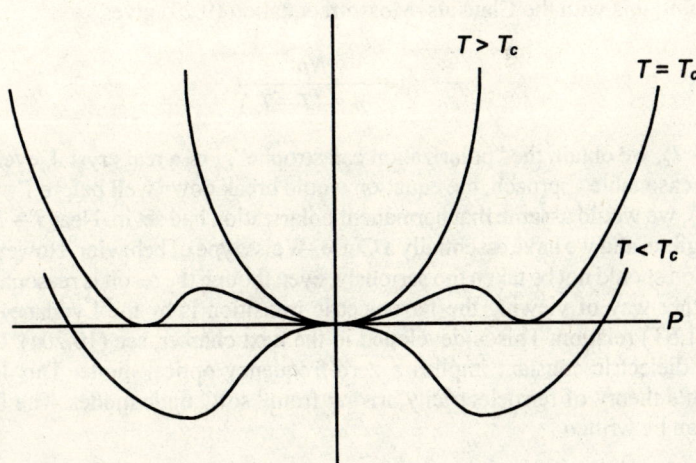

Fig. 9.6. Sketch of variation of Gibbs free energy $G(T, p)$ for first-order transitions

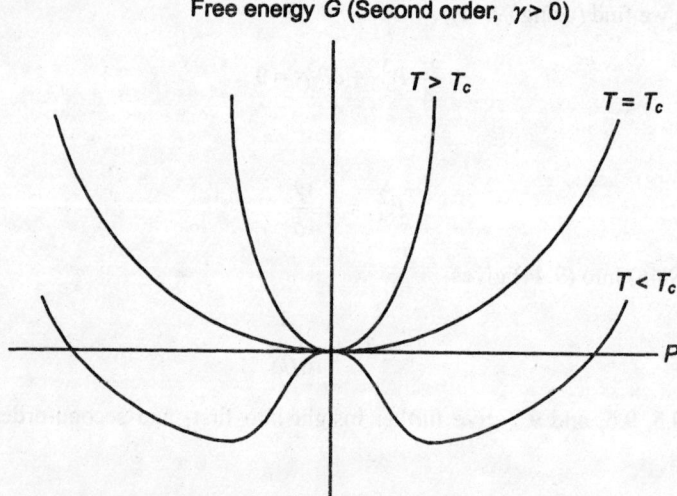

Fig. 9.7. Sketch of variations of Gibbs free energy $G(T,p)$ for second-order transitions

9.3.2 Further Comment on the Ferroelectric Transition (B, ME)

Suppose we have N permanent, noninteracting dipoles P per unit volume, at temperature T, in an electric field E. At high temperature, simple statistical mechanics shows that the polarizability per molecule is

$$\alpha = \frac{p^2}{3kT}. \tag{9.48}$$

Combining this with the Clausius–Mossotti equation (9.29) gives

$$\frac{\varepsilon}{\varepsilon_0} = 1 + \frac{Np^2}{3k\varepsilon_0(T-T_c)}. \tag{9.49}$$

As $T \to T_c$, we obtain the "polarization catastrophe". For a real crystal, even if this were a reasonable approach, the equation would break down well before $T = T_c$, and at $T = T_c$, we would assume that permanent polarization had set in. Near $T = T_c$, the 1 is negligible, and we have essentially a Curie–Weiss type of behavior. However, this derivation should not be taken too seriously, even though the result is reasonable.

Another way of viewing the ferroelectric transition is by the Lyddane–Sachs–Teller (LST) relation. This is developed in the next chapter, see (10.204). Here an infinite dielectric constant implies a zero-frequency optical mode. This leads to Cochran's theory of ferroelectricity arising from "soft" optic modes. The LST relation can be written

$$\frac{\omega_T^2}{\omega_L^2} = \frac{\varepsilon(\infty)}{\varepsilon(0)}, \tag{9.50}$$

where ω_T is the transverse optical frequency, ω_L is the longitudinal optical frequency (both at low wave vector), $\varepsilon(\infty)$ is the high-frequency limit of the dielectric constant and $\varepsilon(0)$ is the low-frequency (static) limit. Thus a Curie–Weiss behavior for $\varepsilon(0)$ as

$$\frac{1}{\varepsilon(0)} \propto (T - T_c) \tag{9.51}$$

is consistent with

$$\omega_T^2 \propto (T - T_c). \tag{9.52}$$

Table 9.1. Selected ferroelectric crystals

Type	Crystal	T_c (K)
KDP	KH_2PO_4	123
TGS	Triglycine sulfate	322
Perovskites	$BaTiO_3$	406
	$PbTiO_3$	765
	$LiNbO_3$	1483

From Anderson HL (ed), *A Physicists Desk Reference* 2nd edn, American Institute of Physics, Article 20: Frederikse HPR, p.314, Table 20.02.C.1., 1989, with permission of Springer-Verlag. Original data from Kittel C, *Introduction to Solid State Physics*, 4th edn, p.476, Wiley, NY, 1971.

Cochran has pioneered the approach to a microscopic theory of the onset of spontaneous polarization by the soft mode or "freezing out" (frequency going to zero) of an optic mode of zero wave vector. The vanishing frequency appears to result from a canceling of short-range and long-range (Coulomb) forces between ions. Not all ferroelectric transitions are easily associated with phonon modes. For example, the order–disorder transition is associated with the ordering of protons in potential wells with double minima above the transition. Transition temperatures for some typical ferroelectrics are given in Table 9.1.

9.3.3 One-Dimensional Model of the Soft Mode of Ferroelectric Transitions (A)

In order to get a better picture of what the soft mode theory involves, we present a one-dimensional model below that is designed to show ferroelectric behavior.

Anderson and Cochran have suggested that the phase transition in certain ferroelectrics results from an instability of one of the normal vibrational modes of the lattice. Suppose that at some temperature T_c

a) An infinite-wavelength optical mode is accompanied by the condition that the vibrational frequency ω for that mode is zero.

b) The effective restoring force for this mode for the ion displacements equals zero. This condition has prompted the terminology, "soft" mode ferroelectrics.

If these conditions are satisfied, it is seen that the static ion displacements would give rise to a "frozen-in" electric dipole moment–that is, spontaneous polarization. The idea is shown in Fig. 9.8.

We now consider a one-dimensional lattice consisting of two atoms per unit cell, see Fig. 9.9. The atoms (ions) have, respectively, mass m_1 and m_2 with charge $e_1 = e$ and $e_2 = -e$. The equilibrium separation distance between atoms is the distance $a/2$.

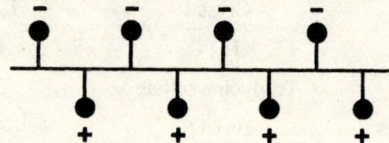

Fig. 9.8. Schematic for ferroelectric mode in one dimension

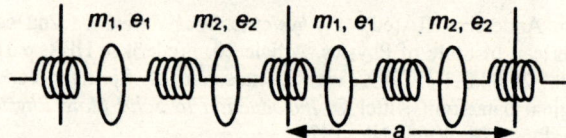

Fig. 9.9. One-dimensional model for ferroelectric transition (masses m_i, charges e_i)

It should be pointed out that in an ionic, one-dimensional model, a unit cell exhibits a nonzero electric polarization—even when the ions are in their equilibrium positions. However, in three dimensions, one can find a unit cell that possesses zero polarization when the atoms are in equilibrium positions. Since our interest is to present a model that reflects important features of the more complicated three-dimensional model, we are interested only in the electric polarization that arises because of displacements away from equilibrium positions. We could propose for the one-dimensional model the existence of fixed charges that will cancel the equilibrium position polarization but that have no other effect. At any rate, we will disregard equilibrium position polarization.

9.3 Ferroelectric Crystals (B)

We define x_{kb} as the displacement from its equilibrium position of the bth atom ($b = 1, 2$) in the kth unit cell. For N atoms, we assume that the displacements of the atoms from equilibrium give rise to a polarization, P, where

$$P = \frac{1}{N}\sum_{k',b'} x_{k'b'} e_b . \tag{9.53}$$

The equation of motion of the bth atom in the kth unit cell can be written

$$m_b \ddot{x}_{kb} + \sum_{k',b'} J_{bb'}(k-k') x_{k'b'} = ce_b P , \tag{9.54}$$

where

$$J_{bb'}(k-k') = \frac{\partial^2 V}{\partial x_{kb} \partial x_{k'b'}} . \tag{9.55}$$

This equation is, of course, Newton's second law, $F = ma$, applied to a particular ion. The second term on the left-hand side represents a "spring-like" interaction obtained from a power series expansion to the second order of the potential energy, V, of the crystal. The right-hand side represents a long-range electrical force represented by a local electric field that is proportional to the local electric field $E_{loc} = cP$, where c is a constant.

As a further approximation, we assume the spring-like interactions are nearest neighbors, so

$$V = \frac{\gamma}{2}\sum_{k''}(x_{k''2} - x_{k''1})^2 + \frac{\gamma}{2}\sum_{k''}(x_{k''+1,1} - x_{k''2})^2 , \tag{9.56}$$

where γ is the spring constant. By direct calculation, we find for the $J_{bb'}$

$$\begin{aligned} J_{11}(k'-k) &= 2\gamma \delta_k^{k'} = J_{22}(k'-k), \\ J_{12}(k'-k) &= -\gamma(\delta_k^{k'} + \delta_k^{k'+1}), \\ J_{21}(k'-k) &= -\gamma(\delta_k^{k'} + \delta_k^{k'-1}). \end{aligned} \tag{9.57}$$

We rewrite our dynamical equation in terms of $h = k' - k$

$$m_b \ddot{x}_{kb} + \sum_{h,b'} J_{bb'}(h) x_{h+k,b'} = \frac{ce_b}{N}\sum_{h,b'} x_{h+k,b'} e_{b'} . \tag{9.58}$$

Since this equation is translationally invariant, it has solutions that satisfy Bloch's theorem. Thus, there exists a wave vector k such that

$$x_{kb} = \exp(ikqa) x_{ob} , \tag{9.59}$$

where x_{ob} is the displacement of the bth atom in the cell chosen as the origin for the lattice vectors. Substituting, we find

$$m_b \ddot{x}_{kb} + \sum_{h,b'} J_{bb'}(k) \exp(ihqa) x_{ob'} = \frac{ce_b}{N}\sum_{h,b'} \exp(ihqa) x_{ob'} e_{b'} . \tag{9.60}$$

We simplify by defining

$$G_{bb'}(q) = \sum_h J_{bb'}(h)\exp(ihqa).\qquad(9.61)$$

Using the results for $J_{bb'}$, we find

$$\begin{aligned}G_{11} &= 2\gamma = G_{22},\\ G_{12} &= -\gamma[1+\exp(iqa)],\\ G_{21} &= -\gamma[1-\exp(-iqa)].\end{aligned}\qquad(9.62)$$

In addition, since

$$\sum_h \exp(ihqa) = N\delta_q^0,\qquad(9.63)$$

we finally obtain,

$$m_b \ddot{x}_{ob} + \sum_{b'} G_{bb'}(q) x_{ob'} = ce_b \sum_{b'} \delta_q^0 x_{ob'} e_{b'}.\qquad(9.64)$$

As in the ordinary theory of vibrations, we assume x_{ob} contains a time factor $\exp(i\omega t)$, so

$$\ddot{x}_{ob} = -\omega^2 x_{ob}.\qquad(9.65)$$

The polarization term only affects the $q \to 0$ solution, which we look at now. Letting $q = 0$, and $e_1 = -e_2 = e$, we obtain the following two equations:

$$-m_1\omega^2 x_{o1} + 2\gamma x_{o1} - 2\gamma x_{o2} = ce(x_{o1}e - x_{o2}e),\qquad(9.66)$$

and

$$-m_2\omega^2 x_{o2} - 2\gamma x_{o1} + 2\gamma x_{o2} = -ce(x_{o1}e - x_{o2}e).\qquad(9.67)$$

These two equations can be written in matrix form:

$$\begin{bmatrix} -m_1\omega^2 + d & -d \\ -d & -m_2\omega^2 + d \end{bmatrix}\begin{bmatrix} x_{o1} \\ x_{o2} \end{bmatrix} = 0,\qquad(9.68)$$

where $d = 2\gamma - ce^2$. From the secular equation, we obtain the following:

$$\omega^2[m_1 m_2 \omega^2 - (m_1 + m_2)d] = 0.\qquad(9.69)$$

The solution $\omega = 0$ is the long-wavelength acoustic mode frequency. The other solution, $\omega^2 = d/\mu$ with $1/\mu = 1/m_1 + 1/m_2$, is the optic mode long-wavelength frequency. For this frequency

$$-m_1 x_{o1} = m_2 x_{o2}.\qquad(9.70)$$

So,

$$P = x_{o1}e\left(1+\frac{m_1}{m_2}\right), \tag{9.71}$$

and $P \neq 0$ if $x_{o1} \neq 0$. Suppose

$$\lim_{T \to T_c}[2\gamma(T) - ce^2] = 0, \tag{9.72}$$

then

$$\omega^2 = \frac{d}{\mu} \to 0 \quad \text{at} \quad T = T_c, \tag{9.73}$$

and

$$F_1 = m_1 \ddot{x}_{o1} = d(x_{o1} + x_{o2}) \to 0 \text{ as } T \to T_c. \tag{9.74}$$

So, a solution is x_{o1} = constant $\neq 0$. That is, the model shows a ferroelectric solution for $T \to T_c$.

9.4 Dielectric Screening and Plasma Oscillations (B)

We begin now to discuss more complex issues. We want to discuss the nature of a gas of interacting electrons. This topic is closely related to the occurrence of oscillations in gas-discharge plasmas and is linked to earlier work of Langmuir and Tonks.[4] We begin by considering the subject of plasma oscillations. The general idea can be presented from a classical viewpoint, so we start by assuming the simultaneous validity of Newton's laws and Maxwell's equations.

Let n_0 be the number density of electrons in equilibrium. We assume an equal distribution of positive charge that remains uniform and, thus, supplies a constant background. We will consider one dimension only and, thus, consider only longitudinal plasma oscillations.

---

[4] See Tonks and Langmuir [9.19].

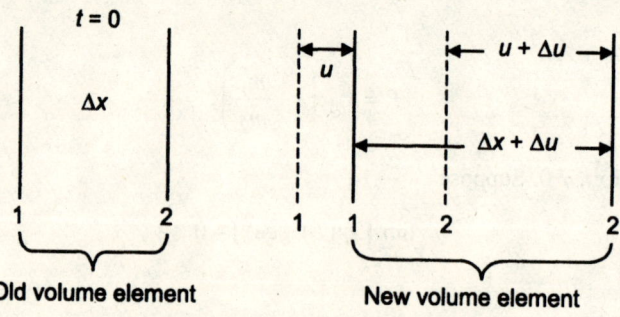

Fig. 9.10. Schematic used to discuss plasma vibration

Let $u(x, t)$ represent the displacement of electrons whose equilibrium position is x and refer to Fig. 9.10 to compute the change in density Let e represent the magnitude of the electronic charge. Since the positive charge remains at rest, the total charge density is given by $\rho = -(n - n_0)e$. Since the same number of electrons is contained in the new volume as the old volume.

$$n = \frac{n_0 \Delta x}{(\Delta x + \Delta u)} \cong n_0\left(1 - \frac{du}{dx}\right). \tag{9.75}$$

Thus,

$$\rho = n_0 e \frac{du}{dx}. \tag{9.76}$$

In one-dimension, Gauss' law is

$$\frac{dE_x}{dx} = \frac{\rho}{\varepsilon_0} = \frac{n_0 e}{\varepsilon_0}\frac{du}{dx}. \tag{9.77}$$

Integrating and using the boundary condition that $(E_x)_{n=0} = 0$, we have

$$E_x = \frac{n_0 e}{\varepsilon_0} u. \tag{9.78}$$

A simpler derivation is discussed in the optics chapter (see Sect. 10.9). Using Newton's second law with force $-eE_x$, we have

$$m\frac{d^2u}{dt^2} = -\frac{n_0 e^2}{\varepsilon_0} u, \tag{9.79}$$

with solution

$$u = u_0 \cos(\omega_p t + \text{const.}), \tag{9.80}$$

where

$$\omega_p = \sqrt{n_0 e^2 / m\varepsilon_0} \qquad (9.81)$$

is the plasma frequency of electron oscillation. The quanta associated with this type of excitation are called plasmons. For a typical gas in a discharge tube, $\omega_P \cong 10^{10}$ s^{-1}, while for a typical metal, $\omega_P \cong 10^{16}$ s^{-1}.

More detailed discussions of plasma effects and electrons can be made by using frequency- and wave-vector-dependent dielectric constants. See Sect. 9.5.3 for further details where we will discuss screening in some detail. We define $\varepsilon(q, \omega)$ as the proportionality constant between the space and time Fourier transform components of the electric field and electric displacement vectors. We generally assume $\varepsilon(\omega) = \varepsilon(q = 0, \omega)$ provides an adequate description of dielectric properties when $q^{-1} \gg a$, where a is the lattice spacing. It is necessary to use $\varepsilon(q, \omega)$ when spatial variations not too much larger than the lattice constant are important.

The basic idea is contained in (9.82) and (9.83). For electrical interactions, if the actual perturbation of the potential is of the form

$$V' = \iint v'(q,\omega) \exp(i\mathbf{q} \cdot \mathbf{r}) \exp(i\omega t) dq \cdot d\omega . \qquad (9.82)$$

Then, the perturbation of the energy is given by

$$\varepsilon' = \iint \frac{v'(q,\omega)}{\varepsilon(q,\omega)} \exp(i\mathbf{q} \cdot \mathbf{r}) \exp(i\omega t) dq \cdot d\omega . \qquad (9.83)$$

$\varepsilon(q,\omega)$ is used to discuss (a) plasmons, (b) the ground-state energy of a many-electron system, (c) screening and Friedel oscillation in charge around a charged impurity in a sea of electrons, (d) the Kohn effect (a singularity in the dielectric constant that implies a change in phonon frequency.), and (e) even other elementary energy excitations, provided enough physics is included in $\varepsilon(q,\omega)$. Some of this is elaborated in Sect. 9.5.

We now discuss two kinds of waves that can occur in plasmas. The first kind concerns waves that propagate in a region with only one type of charge carrier, and in the second we consider both signs of charge carrier. In both cases we assume overall charge neutrality. Both cases deal with electromagnetic waves propagating in a charged media in the direction of a constant magnetic field. Both cases only relate somewhat indirectly to dielectric properties through the Coulomb interaction. They seem to be worth discussing as an aside.

9.4.1 Helicons (EE)

Here we consider electrons as the charge carriers. The helicons are low-frequency (much lower than the cyclotron frequency) waves of circularly polarized electromagnetic radiation that propagate, with little attenuation, along the direction of the external magnetic field. They have been observed in sodium at high field (~ 2.5 T) and low temperatures (~ 4 K). The existence of these waves was predicted by

P. Aigrain in 1960. Since their frequency depends on the Hall coefficient, they have been used to measure it in solids. Their dispersion relation shows that lower frequencies have lower velocities. When high-frequency helicons are observed in the ionosphere, they are called whistlers (because of the way their signal sounds when converted to audio).

For electrons (charge $-e$) in E and B fields with drift velocity v, relaxation time τ, and effective mass m, we have

$$m\left(\frac{d}{dt} + \frac{1}{\tau}\right)v = -e(E + v \times B). \tag{9.84}$$

Assuming $B = B\mathbf{k}$ and low frequencies so $\omega\tau \ll 1$, we can neglect the time derivatives and so

$$v_x = -\frac{e\tau E_x}{m} - \omega_c \tau v_y,$$
$$v_y = -\frac{e\tau E_y}{m} + \omega_c \tau v_x, \tag{9.85}$$
$$v_z = -\frac{e\tau E_z}{m},$$

where $\omega_c = eB/m$ is the cyclotron frequency. Letting, $\sigma_0 = m/ne^2\tau$, where n is the number of charges per unit volume, and the Hall coefficient $R_H = -1/ne$, we can write (noting $j = -nev$, $j = v/R_H$):

$$v_x = \sigma_0 R_H (E_x + B v_y), \tag{9.86}$$
$$v_y = \sigma_0 R_H (E_y - B v_x). \tag{9.87}$$

Neglecting the displacement current, from Maxwell's equations we have:

$$\nabla \times B = \mu_0 j,$$
$$\nabla \times E = -\frac{\partial B}{\partial t}.$$

Assuming $\nabla \cdot E = 0$ (overall neutrality), these give

$$\nabla^2 E = \mu_0 \frac{\partial j}{\partial t}. \tag{9.88}$$

If solutions of the form $E = E_0 \exp[i(kx - \omega t)]$ and $v = v_0 \exp[i(kx - \omega t)]$ are sought, we require:

$$-k^2 E = -i\omega\mu_0 \frac{v}{R_H},$$

$$E_x = i\frac{\omega\mu_0}{k^2} \frac{v_x}{R_H},$$

9.4 Dielectric Screening and Plasma Oscillations (B)

$$E_y = i\frac{\omega\mu_0}{k^2}\frac{v_y}{R_H}.$$

Thus

$$\left(1 - i\sigma_0\frac{\omega\mu_0}{k^2}\right)v_x - \sigma_0 R_H B v_y = 0,$$

$$\sigma_0 R_H B v_x + \left(1 - i\sigma_0\frac{\omega\mu_0}{k^2}\right)v_y = 0. \tag{9.89}$$

Assuming large conductivity, $\sigma_0\omega\mu_0/k^2 \gg 1$, and large B, we find:

$$\omega = \frac{k^2}{\mu_0}|R_H|B = \frac{k^2}{\mu_0 n e}B, \tag{9.90}$$

or the phase velocity is

$$v_p = \frac{\omega}{k} = \sqrt{\frac{\omega B}{\mu_0 n e}}, \tag{9.91}$$

independent of m. Note the group velocity is just twice the phase velocity. Since the plasma frequency ω_p is $(ne^2/m\varepsilon_0)^{1/2}$, we can write also

$$v_p = c\sqrt{\frac{\omega\omega_c}{\omega_p^2}}. \tag{9.92}$$

Typically v_p is of the order of sound velocities.

9.4.2 Alfvén Waves (EE)

Alfvén waves occur in a material with two kinds of charge carriers (say electrons and holes). As for helicon waves, we assume a large magnetic field with electromagnetic radiation propagating along the field. Alfvén waves have been observed in Bi, a semimetal at 4 K. The basic assumptions and equations are:

1. $\nabla \times \boldsymbol{B} = \mu_0 \boldsymbol{j}$, neglecting displacement current.
2. $\nabla \times \boldsymbol{E} = -\partial \boldsymbol{B}/\partial t$, Faraday's law.
3. $\rho\dot{\boldsymbol{v}} = \boldsymbol{j} \times \boldsymbol{B}$, where $\boldsymbol{v}$ is the fluid velocity, and the force per unit volume is dominated by magnetic forces.
4. $\boldsymbol{E} = -(\boldsymbol{v} \times \boldsymbol{B})$, from the generalized Ohm's law $\boldsymbol{j}/\sigma = \boldsymbol{E} + \boldsymbol{v} \times \boldsymbol{B}$ with infinite conductivity.
5. $\boldsymbol{B} = B_x \hat{\boldsymbol{i}} + B_y \hat{\boldsymbol{j}}$, where $B_x = B_0$ and is constant while $B_y = B_1(t)$.

6. Only the j_x, E_x, and v_y components need be considered (v_y is the velocity of the plasma in the y direction and oscillates with time).
7. $\dot{v} = \partial v/\partial t$, as we neglect $(v\cdot\nabla)v$ by assuming small hydrodynamic motion. Also we assume the density ρ is constant in time.

Combining (1), (3), and (7) we have

$$\mu_0 \rho \frac{\partial v_y}{\partial t} = [(\nabla \times B) \times B]_y \cong \frac{\partial B_1}{\partial x} B_0. \tag{9.93}$$

By (4)

$$E_z = -B_0 v_y,$$

so

$$-\frac{\mu_0 \rho}{B_0} \frac{\partial E_z}{\partial t} = B_0 \frac{\partial B_1}{\partial x}. \tag{9.94}$$

By (2)

$$\frac{\partial E_z}{\partial x} = -\frac{\partial B_1}{\partial t},$$

so

$$\frac{\partial^2 E_z}{\partial t^2} = -\frac{B_0^2}{\mu_0 \rho} \frac{\partial^2 B_1}{\partial x \partial t} = +\frac{B_0^2}{\mu_0 \rho} \frac{\partial^2 E_z}{\partial x^2}. \tag{9.95}$$

This is the equation of a wave with velocity

$$v_A = \frac{B}{\sqrt{\mu_0 \rho}}, \tag{9.96}$$

the Alfvén velocity. For electrons and holes of equal number density n and effective masses m_e and m_h,

$$v_A = \frac{B}{\sqrt{\mu_0 n(m_e + m_h)}}, \tag{9.97}$$

Notice that $v_A = (B^2/\mu_0 \rho)^{1/2}$ is the velocity in a string of tension B^2/μ_0 and density ρ. In some sense, the media behaves as if the charges and magnetic flux lines move together.

A unified treatment of helicon and Alfvén waves can be found in Elliot and Gibson [9.5] and Platzman and Wolff [9.15]. Alfvén waves are also discussed in space physics, e.g. in connection with the solar wind.

9.5 Free-Electron Screening

9.5.1 Introduction (B)

If you place one charge in the midst of other charges, they will redistribute themselves in such a way as to "damp out" the long-range effects of the original charge. This long-range damping is an aspect of screening. Its origin resides in the Coulomb interactions of charges. This phenomenon was originally treated classically by the Debye–Huckel theory. A semiclassical form is called the Thomas–Fermi Approximation, which also assumes a free-electron gas. Neither the Debye–Huckel Theory nor the Thomas–Fermi model treats screening accurately at small distances. To do this, it is necessary to use the Lindhard theory.

We begin with the linearized Thomas–Fermi and Debye–Huckel methods and show how to use them to calculate the screening due to a single charged impurity. Perhaps the best way to derive this material is through the dielectric function and derive the Lindhard expression for it for a free-electron gas. The Lindhard expression for $\varepsilon(\omega \to 0, q)$ for small q then gives us the Thomas–Fermi expression. Generalization of the dielectric function to band electrons can also be made. The Lindhard approach follows in Sect. 9.5.3.

9.5.2 The Thomas–Fermi and Debye–Huckel Methods (A, EE)

We assume an electron gas with a uniform background charge (jellium). We assume a point charge of charge Ze ($e > 0$) is placed in the jellium. This will produce a potential $\varphi(r)$, which we assume to be weak and to vary slowly over a distance of order $1/k_F$ where k_F is the wave vector of the electrons whose energy equals the Fermi energy. For distances close to the impurity, where the potential is neither weak nor slowly varying our results will not be a very good approximation. Consistent with the slowly varying potential approximation, we assume it is valid to think of the electron energy as a function of position.

$$E_k = \frac{\hbar^2 k^2}{2m} - e\varphi(r), \qquad (9.98)$$

where $\hbar$ is Planck's constant (divided by 2π), k is the wave vector, and m is the electronic effective mass.

In order to exhibit the effects of screening, we need to solve for the potential φ. We assume the static dielectric constant is ε and ρ is the charge density. Poisson's equation is

$$\nabla^2 \varphi = \frac{-\rho}{\varepsilon}, \qquad (9.99)$$

where the charge density is

$$\rho = eZ\delta(\mathbf{r}) + n_0 e - ne, \tag{9.100}$$

where $eZ\delta(\mathbf{r})$ is the charge density of the added charge. For the spin $1/2$ electrons obeying Fermi–Dirac statistics, the number density (assuming local spatial equilibrium) is

$$n = \int \frac{1}{\exp[\beta(E_k - \mu)] + 1} \frac{d\mathbf{k}}{4\pi^3}, \tag{9.101}$$

where $\beta = 1/k_B T$ and k_B is the Boltzmann constant. When $\varphi = 0$, then $n = n_0$, so

$$n_0 = n_0(\mu) = \int \frac{1}{\exp[\beta((\hbar^2 k^2 / 2m) - \mu)] + 1} \frac{d\mathbf{k}}{4\pi^3}. \tag{9.102}$$

Note by (9.98) and (9.102), we also have

$$n = n_0[\mu + e\varphi(\mathbf{r})]. \tag{9.103}$$

This means the charge density can be written

$$\rho = eZ\delta(\mathbf{r}) + \rho^{\text{ind}}(\mathbf{r}), \tag{9.104}$$

where

$$\rho^{\text{ind}}(\mathbf{r}) = -e[n_0(\mu + e\varphi(\mathbf{r})) - n_0(\mu)]. \tag{9.105}$$

We limit ourselves to weak potentials. We can then expand n_0 in powers of φ and obtain:

$$\rho^{\text{ind}}(\mathbf{r}) = -e^2 \frac{\partial n_0}{\partial \mu} \varphi(\mathbf{r}). \tag{9.106}$$

The Poisson equation then becomes

$$\nabla^2 \varphi = -\frac{1}{\varepsilon} \left[Ze\delta(\mathbf{r}) - e^2 \frac{\partial n_0}{\partial \mu} \varphi(\mathbf{r}) \right]. \tag{9.107}$$

A convenient way to solve this equation is by the use of Fourier transforms. The Fourier transform of the potential can be written

$$\varphi(\mathbf{q}) = \int \varphi(\mathbf{r}) \exp(-i\mathbf{q} \cdot \mathbf{r}) d\mathbf{r}, \tag{9.108}$$

with inverse

$$\varphi(r) = \frac{1}{(2\pi)^3} \int \varphi(q) \exp(i\mathbf{q} \cdot \mathbf{r}) \mathrm{d}q , \qquad (9.109)$$

and the Dirac delta function can be represented by

$$\delta(r) = \frac{1}{(2\pi)^3} \int \exp(i\mathbf{q} \cdot \mathbf{r}) \mathrm{d}q . \qquad (9.110)$$

Taking the Fourier transform of (9.107), we have

$$q^2 \varphi(q) = \frac{1}{\varepsilon}\left[Ze - e^2 \frac{\partial n_0}{\partial \mu} \varphi(q) \right]. \qquad (9.111)$$

Defining the screening parameter as

$$k_S^2 = \frac{e^2}{\varepsilon} \frac{\partial n_0}{\partial \mu}, \qquad (9.112)$$

we find from (9.111) that

$$\varphi(q) = \frac{Ze}{\varepsilon} \frac{1}{q^2 + k_S^2}. \qquad (9.113)$$

Then, using (9.109), we find from (9.113) that

$$\varphi(r) = \frac{Ze}{4\pi\varepsilon r} \exp(-k_S r). \qquad (9.114)$$

Equations (9.112) and (9.114) are the basic equations for screening.
For the classical nondegenerate case, we have from (9.102)

$$n_0(\mu) = \exp(\beta\mu) \int \exp(-\beta\hbar^2 k^2 / 2m) \frac{\mathrm{d}k}{4\pi^3}, \qquad (9.115)$$

so that by (9.112)

$$k_S^2 = \frac{e^2}{\varepsilon} \frac{n_0}{k_B T}, \qquad (9.116)$$

we get the classical Debye–Huckel result. For the degenerate case, it is convenient to rewrite (9.102) as

$$n_0(\mu) = \int D(E) f(E) \mathrm{d}E , \qquad (9.117)$$

so

$$\frac{\partial n_0}{\partial \mu} = \int D(E) \frac{\partial f}{\partial \mu} dE, \qquad (9.118)$$

where $D(E)$ is the density of states per unit volume and $f(E)$ is the Fermi function

$$f(E) = \frac{1}{\exp[\beta(E-\mu)] + 1}. \qquad (9.119)$$

since

$$\frac{\partial f(E)}{\partial \mu} \cong \delta(E-\mu), \qquad (9.120)$$

at low temperatures when compared with the Fermi temperature; so we have

$$\frac{\partial n_0}{\partial \mu} \cong D(\mu). \qquad (9.121)$$

Since the free-electron density of states per unit volume is

$$D(E) = \frac{1}{2\pi^2}\left(\frac{2m}{\hbar^2}\right)^{3/2}\sqrt{E}, \qquad (9.122)$$

and the Fermi energy at absolute zero is

$$\mu = \frac{\hbar^2}{2m}(3\pi^2 n_0)^{2/3}, \qquad (9.123)$$

where $n_0 = N/V$, we find

$$D(\mu) = \frac{3n_0}{2\mu}, \qquad (9.124)$$

which by (9.121) and (9.112) gives the linearized Thomas–Fermi approximation. If we further use

$$\mu = \frac{3}{2}k_B T_F, \qquad (9.125)$$

we find

$$k_S^2 = \frac{e^2}{\varepsilon}\frac{n_0}{k_B T_F}, \qquad (9.126)$$

which looks just like the Debye–Huckel result except T is replaced by the Fermi temperature T_F. In general, by (9.112), (9.118), (9.119), and (9.122), we have for free-electrons,

$$k_S^2 = \frac{e^2 n_0}{\epsilon k_B T_F} \frac{F'_{1/2}(\eta)}{F_{1/2}(\eta)}, \tag{9.127}$$

where $\eta = \mu/k_B T$ and

$$F_{1/2}(\eta) = \int_0^\infty \frac{\sqrt{x}\,dx}{\exp(x-\eta)+1}, \tag{9.128}$$

is the Fermi integral. Typical screening lengths $1/k_S$ for good metals are of order 1 Å, whereas for typical semiconductors 60 Å is more appropriate. For $\eta \ll -1$, $F'_{1/2}(\eta)/F_{1/2}(\eta) \approx 1$, which corresponds to the classical Debye–Huckel theory, and for $\eta \gg 1$, $F'_{1/2}(\eta)/F_{1/2}(\eta) = 3/(2\eta)$ is the Thomas–Fermi result.

9.5.3 The Lindhard Theory of Screening (A)

Here we do a more general discussion that is self-consistent.[5] We start with the idea of an external potential that determines a set of electronic states. Electronic states in turn give rise to a charge density from which a potential can be determined. We wish to show how we can determine a charge density and a potential in a self-consistent way by using the concept of a frequency- and wave-vector-dependent dielectric constant.

The specific problem we wish to solve is that of the self-consistent response to an applied field. We will assume small applied fields and linear responses. The electronic response to the applied field is called screening, and it arises from the interaction of the electrons with each other and with the external field. Only screening by a free-electron gas will be considered.

Let a charge ρ^{ext} be placed in jellium, and let it produce a potential φ^{ext} (by itself). Let φ be the potential caused by the extra charge, the free-electrons, and the uniform background charge (i.e. extra charge plus jellium). We also let ρ be the corresponding charge density. Then

$$\nabla^2 \varphi^{ext} = -\frac{\rho^{ext}}{\epsilon}, \tag{9.129}$$

$$\nabla^2 \varphi = -\frac{\rho}{\epsilon}. \tag{9.130}$$

The induced charge density ρ^{ind} is then defined by

$$\rho^{ind} = \rho - \rho^{ext}. \tag{9.131}$$

---

[5] This topic is also treated in Ziman JM [25, Chap. 5], and Grosso and Paravicini [55 p245ff].

9 Dielectrics and Ferroelectrics

We Fourier analyze the equations in both the space and time domains:

$$q^2 \varphi^{\text{ext}}(q,\omega) = \frac{\rho^{\text{ext}}(q,\omega)}{\varepsilon}, \qquad (9.132a)$$

$$q^2 \varphi(q,\omega) = \frac{\rho(q,\omega)}{\varepsilon}, \qquad (9.132b)$$

$$\rho(q,\omega) = \rho^{\text{ext}}(q,\omega) + \rho^{\text{ind}}(q,\omega). \qquad (9.132c)$$

Subtracting (9.132a) from (9.132b) and using (9.132c) yields:

$$\varepsilon q^2 [\varphi(q,\omega) - \varphi^{\text{ext}}(q,\omega)] = \rho^{\text{ind}}(q,\omega). \qquad (9.133)$$

We have assumed weak field and linear responses, so we write

$$\rho^{\text{ind}}(q,\omega) = g(q,\omega)\varphi(q,\omega), \qquad (9.134)$$

which defines $g(q, \omega)$. Thus, (9.133) and (9.134) give this as

$$\varepsilon q^2 [\varphi(q,\omega) - \varphi^{\text{ext}}(q,\omega)] = g(q,\omega)\varphi(q,\omega). \qquad (9.135)$$

Thus,

$$\varphi(q,\omega) = \frac{\varphi^{\text{ext}}(q,\omega)}{\varepsilon(q,\omega)}, \qquad (9.136)$$

where

$$\varepsilon(q,\omega) = 1 - \frac{g(q,\omega)}{\varepsilon q^2}. \qquad (9.137)$$

To proceed further, we need to calculate $\varepsilon(q, \omega)$ directly. In the process of doing this, we will verify the correctness of the linear response assumption. We write the Schrödinger equation as

$$\mathcal{H}_0 |k\rangle = E_k |k\rangle. \qquad (9.138)$$

We assume an external perturbation of the form

$$\delta V(r,t) = [V \exp(\mathrm{i}(q \cdot r + \omega t)) + V \exp(-\mathrm{i}(q \cdot r + \omega t))]\exp(\alpha t). \qquad (9.139)$$

The factor $\exp(\alpha t)$ has been introduced so that the perturbation vanishes as $t = -\infty$, or in other words, as the perturbation is slowly turned on. V is assumed real. Let

$$\mathcal{H} = \mathcal{H}_0 + \delta V. \qquad (9.140)$$

9.5 Free-Electron Screening

We then seek an approximate solution of the time-dependent Schrödinger wave equation

$$\mathcal{H}\psi = i\hbar \frac{\partial \psi}{\partial t}. \qquad (9.141)$$

We seek solutions of the form

$$|\psi\rangle = \sum_{\mathbf{k}'} C_{\mathbf{k}'}(t) \exp(-iE_{\mathbf{k}'}t/\hbar)|\mathbf{k}'\rangle. \qquad (9.142)$$

Substituting,

$$\begin{aligned}\sum_{\mathbf{k}'}(\mathcal{H}_0 + \delta V) C_{\mathbf{k}'}(t) \exp(-iE_{\mathbf{k}'}t/\hbar)|k'\rangle \\ = i\hbar \frac{\partial}{\partial t} \sum_{\mathbf{k}'} C_{\mathbf{k}'}(t) \exp(-iE_{\mathbf{k}'}t/\hbar)|k'\rangle. \end{aligned} \qquad (9.143)$$

Using (9.138) to cancel two terms in (9.143), we have

$$\sum_{\mathbf{k}'} \delta V C_{\mathbf{k}'}(t) \exp(-iE_{\mathbf{k}'}t/\hbar)|k'\rangle = i\hbar \sum_{\mathbf{k}'} \dot{C}_{\mathbf{k}'}(t) \exp(-iE_{\mathbf{k}'}t/\hbar)|k'\rangle. \qquad (9.144)$$

Using

$$\langle k''|k'\rangle = \delta_{k'}^{k''}, \qquad (9.145)$$

$$\langle k''|\delta V|k'\rangle = \langle k''|\delta V|k' \pm q\rangle \delta_{k'}^{k'' \pm q}, \qquad (9.146)$$

$$\begin{aligned}\dot{C}_{k''}(t) = \frac{1}{i\hbar} C_{k''+q} \exp(-iE_{k''+q}t/\hbar)\langle k''|\delta V|k''+q\rangle \exp(iE_{k''}t/\hbar) \\ + \frac{1}{i\hbar} C_{k''-q} \exp(-iE_{k''-q}t/\hbar)\langle k''|\delta V|k''-q\rangle \exp(iE_{k''}t/\hbar). \end{aligned} \qquad (9.147)$$

Using (9.139), we have

$$\begin{aligned}\dot{C}_{k''}(t) = \frac{1}{i\hbar} C_{k''+q} \exp(-i(E_{k''+q} - E_{k''})t/\hbar) V \exp(-i\omega t)\exp(\alpha t) \\ + \frac{1}{i\hbar} C_{k''-q} \exp(-i(E_{k''-q} - E_{k''})t/\hbar) V \exp(i\omega t)\exp(\alpha t). \end{aligned} \qquad (9.148)$$

We assume a weak perturbation, and we begin in the state k with probability $f_0(k)$, so we have

$$C_{k''}(t) = \sqrt{f_0(k)} \delta_{k'',k} + \lambda C_{k''}^{(1)}(t). \qquad (9.149)$$

We write out (9.147) to first order for two interesting cases:

$$\dot{C}_{k+q}(t) = \lambda \dot{C}^{(1)}_{k+q}(t)$$
$$= \left(\frac{1}{i\hbar}\right)\sqrt{f_0(k)} \exp(-i(E_k - E_{k+q})t/\hbar)V \exp(i\omega t)\exp(\alpha t), \quad (9.150)$$

$$\dot{C}_{k-q}(t) = \lambda \dot{C}^{(1)}_{k-q}(t)$$
$$= \left(\frac{1}{i\hbar}\right)\sqrt{f_0(k)} \exp(-i(E_k - E_{k-q})t/\hbar)V \exp(-i\omega t)\exp(\alpha t). \quad (9.151)$$

Integrating, we find, since $C_{k\pm q}(\infty) = 0$

$$C_{k+q}(t) = \sqrt{f_0(k)}\frac{\exp(-i(E_k - E_{k+q})t/\hbar)V \exp(i\omega t)\exp(\alpha t)}{E_k - E_{k+q} - \hbar\omega + i\hbar\alpha}, \quad (9.152)$$

$$C_{k-q}(t) = \sqrt{f_0(k)}\frac{\exp(-i(E_k - E_{k-q})t/\hbar)V \exp(-i\omega t)\exp(\alpha t)}{E_k - E_{k-q} + \hbar\omega + i\hbar\alpha}. \quad (9.153)$$

We write (9.142) as

$$\psi^{(k)} = \sum_{k'} C_{k'}(t)\exp(-iE_{k'}t/\hbar)\psi_{k'}, \quad (9.154)$$

where

$$\psi_{k'}(r) = \frac{1}{\sqrt{\Omega}}e^{ik'\cdot r}, \quad (9.155)$$

and Ω is the volume. We put a superscript on ψ because we assume we start in the state k. More specifically, (9.153) can be written as

$$\psi^{(k)} = \exp(-iE_k t/\hbar)\sqrt{f_0(k)}\psi_k$$
$$+ C_{k+q}(t)\exp(-iE_{k+q}t/\hbar)\psi_{k+q} + C_{k-q}(t)\exp(-iE_{k-q}t/\hbar)\psi_{k-q}. \quad (9.156)$$

Any charge density in jellium is an induced charge density (in equilibrium, jellium is uniform and has a net density of zero). Thus,

$$\rho^{\text{ind}} = \frac{eN}{\Omega} - e\sum_k \left|\psi^{(k)}\right|^2. \quad (9.157)$$

Now, note

$$\left|\psi^{(k)}\right|^2 = \frac{1}{\Omega} \quad \text{and} \quad \sum_{\text{all } k} f_0(k) = N, \quad (9.158)$$

9.5 Free-Electron Screening

so putting (9.155) into (9.156) and retaining no terms beyond first order,

$$\rho^{ind} = \frac{eN}{\Omega} - \frac{e}{\Omega}\sum_k f_0(k)\left\{1 + \frac{V \exp(i q \cdot r)\exp(i\omega t)\exp(\alpha t)}{E_k - E_{k+q} - \hbar\omega + i\hbar\alpha} \right. $$
$$\left. + \frac{V \exp(-i q \cdot r)\exp(-i\omega t)\exp(\alpha t)}{E_k - E_{k+q} + \hbar\omega + i\hbar\alpha} + c.c.\right\}, \quad (9.159)$$

or

$$\rho^{ind} = -\frac{e}{\Omega}\sum_k \left\{\frac{[f_0(k) - f_0(k+q)]V \exp(iq\cdot r)\exp(i\omega t)\exp(\alpha t)}{E_k - E_{k+q} - \hbar\omega + i\hbar\alpha} + c.c.\right\}. \quad (9.160)$$

Using

$$V(q,\omega) = -e\varphi(q,\omega), \quad (9.161)$$

and identifying $\rho^{ind}(q,\omega)$ as the coefficient of $\exp(iq\cdot r)\exp(i\omega t)$, we have

$$\rho^{ind}(q,\omega) = -\frac{e^2}{\Omega}\sum_k \left\{\frac{f_0(k) - f_0(k-q)}{E_{k-q} - E_k - \hbar\omega + i\hbar\alpha}\right\}\varphi(q,\omega). \quad (9.162)$$

By (9.134) we find $g(q,\omega)$ and by (9.137), we thus find

$$\varepsilon(q,\omega) = 1 + \frac{e^2}{\varepsilon\Omega q^2}\sum_k \frac{f_0(k) - f_0(k-q)}{E_{k-q} - E_k - \hbar\omega + i\hbar\alpha}. \quad (9.163)$$

Finally, a few notes are provided on notation. We can redefine the Fourier components so as to change the sign of q. For example, we can say

$$\varphi(r) = \frac{1}{(2\pi)^2}\int \exp(-iq\cdot r)\varphi(q)dq. \quad (9.164)$$

Then defining

$$v_q = \frac{e^2}{\varepsilon\Omega q^2}, \quad (9.165)$$

gives $\varepsilon(q,\omega)$ in the form given in many textbooks:

$$\varepsilon(q,\omega) = 1 - v_q\sum_k \frac{f_0(k+q) - f_0(k)}{E_{k+q} - E_k - \hbar\omega + i\hbar\alpha}. \quad (9.166)$$

The limit as $\alpha \to 0$ is tacitly implied in (9.166). In the limit as q becomes small, (9.165) gives, as we will show below, the Thomas–Fermi approximation (when $\omega = 0$). Two notable effects follow from (9.165), but they are not included in the small q limit. An expression for $\varepsilon(q,0)$ at large q is readily obtained for our free-electron case. The result for $\omega = 0$ is

$$\varepsilon(q,\omega) = 1 + (\text{constant}) D(E_F) \left[\frac{1}{2} + \frac{1-x^2}{4x} \ln\left|\frac{1+x}{1-x}\right| \right], \quad (9.167)$$

where $D(E_F)$ is the density of states at the Fermi energy and $x = q/2k_F$ with k_F being the wave vector at the Fermi energy. This expression has a singularity at $q = 2k_F$, which causes the screening of a charged impurity to have a weakly decaying oscillating term (beyond the Fermi–Thomas potential). This is the origin of *Friedel oscillations*. The Friedel oscillations damp out with distance due to electron scattering. At finite temperature, the singularity disappears causing the Friedel oscillation to damp out.

Further, since ion–ion interactions are screening by $\varepsilon(q)$, the singularity at $q = 2k_F$ is reflected in the phonon spectrum. Kinks in the phonon spectrum due to the singularity in $\varepsilon(q)$ are called *Kohn anomalies*.

Finally, we look at (9.165) for small q, $\omega = 0$ and $\alpha = 0$. We find

$$\begin{aligned}\varepsilon(q,\omega) &= 1 - \frac{e^2}{\varepsilon \Omega q^2} \sum_k \frac{\partial f_0/\partial k}{\partial E_k/\partial k} \\ &= 1 - \frac{e^2}{\varepsilon q^2} \sum_k D(E) \frac{\partial f}{\partial E} dE = 1 + \frac{k_S^2}{q^2}.\end{aligned} \quad (9.168)$$

and hence comparing to previous work, we get exactly the Thomas–Fermi approximation.

Problems

9.1 Show that $E'_0 = E_0 + P/\varepsilon_0$, where E_0 is the electric field between the plates before the slab is inserted (9.19).

9.2 Show that $E_1 = -P/\varepsilon_0$ (see Fig. 9.2).

9.3 Show that $E_2 = P/3\varepsilon_0$ (9.23).

9.4 Show for cubic crystals that $E_3 = 0$ (chapter notation is used).

9.5 If we have N permanent free dipoles p per unit volume in an electric field E, find an expression for the polarization. At high temperatures show that the polarizability (per molecule) is $\alpha = p^2/3kT$. What magnetic situation is this analogous to?

9.6 Use (9.30) and (9.48) to show (9.49)

$$\frac{\varepsilon}{\varepsilon_0} = 1 + \frac{Np^2}{3k\varepsilon_0(T-T_c)}.$$

Find T_c. How likely is this to apply to any real material?

9.7 Use the trial wave function $\psi = \psi_{100}(1 + pz)$ (where p is the variational parameter) for a hydrogen atom (in an external electric field in the z direction) to show that we obtain for the polarizability $16\pi\varepsilon_0 a_0^3$. (ψ_{100} is the ground-state wave function of the unperturbed hydrogen atom, a_0 is the radius of the first Bohr orbit of the hydrogen atom, and the exact polarizability is $18\pi\varepsilon_0 a_0^3$.)

9.8 (a) Given the Gibbs free energy

$$G = G_0 + \frac{1}{2}\beta(T-T_0)P^2 + \frac{1}{4}\gamma P^4 + \frac{1}{6}\delta P^6;$$
$$\beta, \delta > 0, \gamma < 0 \quad \text{(first order)},$$

derive an expression for T_c in terms of P_{sc} where $G(P_{sc}) = G_0$ and $E = 0$.

(b) Put the expression for T_c in terms without P_{sc}. That is, fill in the details of Sect. 9.3.1.

10 Optical Properties of Solids

10.1 Introduction (B)

The organization of a solid-state course may vary towards its middle or end. Logical beginnings are fairly easy. One defines the solid-state universe, and this is done with a Section on crystal structures and how they are determined. Then one introduces the main players, and so there are sections on lattice vibrations, phonons, band structure, and electrons. Following this, one can present topics based on the interaction of electrons and phonons and hence discuss, for example, transport. After that come specific materials (semiconductors, magnetic materials, metals, and superconductors) and properties (dielectric, optic, defect, surface, etc). The problem is that some of these categories overlap so that a clean separation is not possible. Optical properties, in particular, seem to spread into many areas, so a well-focused segment on the optical properties of solids can be somewhat tricky to put together.[1]

By optical properties of solids, we mean those properties that relate to the interaction of solids with electromagnetic radiation whose wavelength is in the infrared to the ultraviolet. There are several aspects to optical properties of solids and looking at the subject in full generality can often lead to complexity, whereas treating each part as a separate case often leads to confusion. We will try to keep to a middle ground between these, by emphasizing only one topic (absorption) but treating it in some detail. Although we will concentrate on absorption, we will mention other optical phenomena including emission, reflection, scattering, and photoemission of electrons.

There are several processes involved in absorption, but the main five seem to be:

(a) Absorption due to electronic transitions between bands that involve wavelengths typically less than ten micrometers;

(b) Absorption by excitons at wavelengths with energies just below the absorption edge due to valence–conduction band transitions (in semiconductors);

(c) Excitation and ionization of impurities that involve wavelengths ranging from about one micrometer to one thousand micrometers;

---

[1] A good treatment is Fox [10.12].

(d) Excitation of lattice vibrations (optical phonons) in polar solids for which the usual wavelengths are ten to fifty micrometers;

(e) Free-carrier absorption for frequencies up to the plasma edge. Free-carrier absorption is particularly important in metals, of course. By gathering data about any optical process, we can gain information about the inner workings of the solid.

10.2 Macroscopic Properties (B)

We start by relating the dielectric properties to optical properties, particularly those involving absorption and reflection. The complex dielectric constant, and the relation of its two components by the Kronig–Kramers relation, is particularly important. The imaginary part relates to the absorption coefficient. We assume the total charge density $\rho_{total} = 0$, $j = \sigma E$, and $\mu = \mu_0$ (no internal magnetic effects, all in the usual notation). We assume a wavelength large compared with atomic dimensions but small compared with the dimensions of the sample. We start with Maxwell's equations and the constitutive relations in SI in the usual notation:

$$\nabla \cdot E = 0 \qquad \nabla \times E = -\mu_0 \frac{\partial H}{\partial t} = -\frac{\partial B}{\partial t}$$

$$\nabla \cdot B = 0 \qquad \nabla \times H = j + \frac{\partial D}{\partial t} \qquad (10.1)$$

$$D = \varepsilon_0 E + P = \varepsilon E$$

$$B = \mu_0(H + M) = \mu H.$$

One then finds

$$\nabla^2 E = \mu_0 \sigma \frac{\partial E}{\partial t} + \mu_0 \frac{\partial^2}{\partial t^2}(\varepsilon E). \qquad (10.2)$$

We look for solutions for each Fourier component

$$E(k, \omega) = E_0 \exp[i(k \cdot r - \omega t)], \qquad (10.3)$$

and keep in mind that ε should be written $\varepsilon(k, \omega)$. Substituting, one finds

$$k^2 = \mu_0(\varepsilon \omega^2 + i\sigma \omega). \qquad (10.4)$$

Or, since $c = 1/(\mu_0 \varepsilon_0)^{1/2}$,

$$k = \frac{\omega}{c}\left(\frac{\varepsilon}{\varepsilon_0} + i\frac{\sigma}{\varepsilon_0 \omega}\right)^{1/2}. \qquad (10.5)$$

For an insulator, $\sigma = 0$ so,

$$k = \frac{\omega}{v} = \frac{\omega n}{c} = \frac{\omega}{c}\sqrt{\frac{\varepsilon}{\varepsilon_0}}. \qquad (10.6)$$

10.2 Macroscopic Properties (B)

where n is the index of refraction. It is then natural to define a complex dielectric constant ε_c and a complex index of refraction n_c so,

$$k = \frac{\omega}{c} n_c, \qquad (10.7)$$

where

$$n_c = \sqrt{\frac{\varepsilon}{\varepsilon_0} + i\frac{\sigma}{\varepsilon_0 \omega}} = n + i n_i. \qquad (10.8)$$

Letting

$$\varepsilon_c = \varepsilon_r + i\varepsilon_i = \frac{\varepsilon(k,\omega)}{\varepsilon_0} + i\frac{\sigma}{\varepsilon_0 \omega}, \qquad (10.9)$$

squaring both sides and equating real and imaginary parts, we find

$$\varepsilon_r(k,\omega) = n^2 - n_i^2, \qquad (10.10)$$

and

$$\varepsilon_i(k,\omega) = 2 n n_i. \qquad (10.11)$$

Now, assuming the wave propagates in the z direction, if we substitute

$$k = \frac{\omega}{c}(n + i n_i), \qquad (10.12)$$

we have

$$E = E_0 \exp\left[i\omega\left(\frac{nz}{c} - t\right)\right] \exp\left(-\frac{\omega}{c} n_i z\right). \qquad (10.13)$$

So, since energy in the wave is proportional to $|E|^2$, we have that the absorption coefficient is given by

$$\alpha = \frac{2 n_i \omega}{c}. \qquad (10.14)$$

Another readily measured quantity can be related to n and n_i. If we apply appropriate boundary conditions to a solid surface, we can show as noted below that the reflection coefficient for normal incidence is given by

$$R = \frac{(n-1)^2 + n_i^2}{(n+1)^2 + n_i^2}. \qquad (10.15)$$

This relation follows directly from the Maxwell relations. From Faraday's law, we can show that the tangential component of E is continuous, and from Ampere's Law we can show the tangential component of H is continuous. Further manipulation leads to the desired relation. Let us work this out. For normal incidence from the vacuum on a surface at $z = 0$, the incident and reflected waves can be written as

$$E_{i+r} = E_1 \exp\left[i\omega\left(\frac{z}{c} - t\right)\right] + E_2 \exp\left[-i\omega\left(\frac{z}{c} + t\right)\right], \tag{10.16}$$

and the refracted wave is given by

$$E_{rf} = E_0 \exp\left[i\omega\left(n_c \frac{z}{c} - t\right)\right], \tag{10.17}$$

where n_c is the complex index of refraction. Since

$$\nabla \times E + \frac{\partial B}{\partial t} = 0, \tag{10.18}$$

we can use the loop of Fig. 10.1 to write

$$\int_A (\nabla \times E) \cdot d\mathbf{s} + \int \frac{\partial B}{\partial t} \cdot d\mathbf{s} = 0, \tag{10.19}$$

or

$$\oint_C E \cdot d\mathbf{r} = -\frac{d}{dt} \int B \cdot d\mathbf{s}, \tag{10.20}$$

as

$$(E_T^1 - E_T^2)l + O(\varepsilon) = -\frac{d}{dt}(B_\perp 2l\varepsilon), \tag{10.21}$$

where the subscript T means the tangential component of the electric field, and the subscript $\perp$ means perpendicular to the page of the paper. Taking the limit as $\varepsilon \to 0$, we obtain

$$E_T^1 - E_T^2 = 0, \tag{10.22}$$

or the tangential component of E is continuous. In a similar way we can use

$$\nabla \times H = j + \frac{\partial D}{\partial t} \tag{10.23}$$

to show that

$$(H_T^1 - H_T^2)l + O(\varepsilon) = \int_A j \cdot d\mathbf{s} + \frac{d}{dt}(D_\perp 2l\varepsilon) = j_\perp(2l\varepsilon) + \frac{d}{dt}(D_\perp 2l\varepsilon). \tag{10.24}$$

Again taking the limit as $\varepsilon \to 0$, we find

$$(H_T^1 - H_T^2) = 0, \tag{10.25}$$

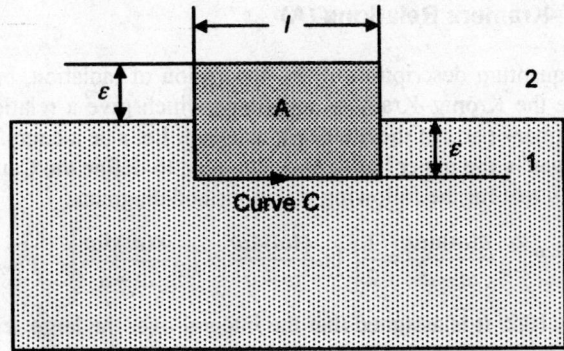

Fig. 10.1. Loop used for deriving field boundary conditions (notice this ε is a distance)

or that the tangential component of H is also continuous. Continuity of the tangential component of H requires (using $\nabla \times E = -\partial B/\partial t$, proper constitutive relations, and (10.16) and (10.17))

$$n_c E_0 = E_1 - E_2. \tag{10.26}$$

Continuity of the tangential component of E requires ((10.16) and (10.17))

$$E_0 = E_1 + E_2. \tag{10.27}$$

Adding these two equations gives

$$E_1 = \frac{E_0(n_c+1)}{2}. \tag{10.28}$$

Subtracting these equations gives

$$E_2 = \frac{E_0(-n_c+1)}{2}. \tag{10.29}$$

Thus, the reflection coefficient is given by

$$R = \left|\frac{E_2}{E_1}\right|^2 = \left|\frac{1-n_c}{1+n_c}\right|^2 = \left|\frac{(n-1)^2 + n_i^2}{(n+1)^2 + n_i^2}\right|. \tag{10.30}$$

Enough has been said to indicate that the theory of the optical properties of solids is intimately related to the complex index of refraction of solids. The complex dielectric constant equals the square of the complex index of refraction. Thus, the optical properties of solids are intimately related to the study of the dielectric properties of solids, and the measurement of the absorptivity and reflectivity determine n and n_i, and hence, ε_r and ε_i.

10.2.1 Kronig–Kramers Relations (A)

We will give a quantum description of the absorption of radiation, but first it is helpful to derive the Kronig–Kramers equations, which give a relation between the real and imaginary parts of the dielectric constant. Let ε be a complex function of ω that converges in the upper half-plane. We need to define the Cauchy principal value P with a real for the following equations and diagrams:

$$P\int_{-\infty}^{\infty}\frac{\varepsilon(\omega)d\omega}{\omega-a}=\frac{1}{2}\left[\int_{C'}\frac{\varepsilon(\omega)d\omega}{\omega-a}+\int_{C''}\frac{\varepsilon(\omega)d\omega}{\omega-a}\right], \quad (10.31)$$

as shown by Fig. 10.2. It is assumed that the integral over the large semicircles is zero. From complex variables, we know that if C encloses a and if f has no singularity in C, then

$$\oint_C \frac{f(Z)dZ}{Z-a}=2\pi_i f(a). \quad (10.32)$$

Using the definition of Cauchy principal value, since we have the integral

$$\int_{C''}=0, \quad P\to\frac{1}{2}\left(\int_{C'}-\int_{C''}\right). \quad (10.33)$$

Thus,

$$P\int\frac{\varepsilon(\omega)d\omega}{\omega-a}=\frac{1}{2}\int_{\substack{\text{small circle}\\\rho\to 0}}\frac{\varepsilon(\omega)d\omega}{\omega-a}=i\pi\varepsilon(a), \quad (10.34)$$

and we have used that $\varepsilon(\omega)$ on the big circle is zero (actually, to achieve this we should use that $\varepsilon(\omega) = [\varepsilon_r(\omega) - 1] + i\varepsilon_i(\omega)$, which we will put in explicitly at the end). Taking real and imaginary parts we then have,

$$P\int_{-\infty}^{\infty}\frac{\text{Re}[\varepsilon(\omega)]d\omega}{\omega-a}=-\pi\,\text{Im}[\varepsilon(a)], \quad (10.35)$$

and

$$P\int_{-\infty}^{\infty}\frac{\text{Im}[\varepsilon(\omega)]d\omega}{\omega-a}=+\pi\,\text{Re}[\varepsilon(a)]. \quad (10.36)$$

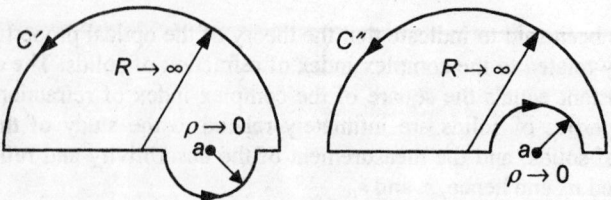

Fig. 10.2. Contours used for Cauchy principal value

There are some other ways to write these relationships,

$$\varepsilon_r(\omega) = \frac{1}{\pi} P \int_{-\infty}^{\infty} \frac{\varepsilon_i(\omega)}{\omega - a} d\omega = \frac{1}{\pi} P \left[\int_0^{\infty} \frac{\varepsilon_i(\omega)}{\omega - a} d\omega + \int_{-\infty}^{0} \frac{\varepsilon_i(\omega)}{\omega - a} d\omega \right]. \quad (10.37)$$

But, the second term can be written

$$-\int_0^{-\infty} \frac{\varepsilon_i(\omega)}{\omega - a} d\omega = \int_0^{-\infty} \frac{[-\varepsilon_i(\omega)]}{-\omega + a} d(-\omega) = \int_0^{\infty} \frac{[-\varepsilon_i(-\omega)]}{\omega + a} d\omega, \quad (10.38)$$

and $\varepsilon^*(r, t) = \varepsilon(r, t)$, so $\varepsilon(-q,-\omega) = \varepsilon^*(q, \omega)$. Therefore,

$$\varepsilon^*(-\omega) = \varepsilon(\omega); \quad (10.39)$$

or

$$\varepsilon_r(-\omega) = \varepsilon_r(\omega), \quad (10.40)$$

and

$$\varepsilon_i(-\omega) = -\varepsilon_i(\omega). \quad (10.41)$$

We get

$$\frac{1}{\pi} P \int_{-\infty}^{0} \frac{\varepsilon_i(\omega)}{\omega - a} d\omega = \int_0^{\infty} \frac{\varepsilon_i(\omega)}{\omega + a} d\omega. \quad (10.42)$$

We can thus write the real component of the dielectric constant as

$$\varepsilon_r(a) = \frac{P}{\pi} \left[\int_0^{\infty} \frac{\varepsilon_i(\omega) d\omega}{\omega - a} + \int_0^{\infty} \frac{\varepsilon_i(\omega) d\omega}{\omega + a} \right] = \frac{2P}{\pi} \int_0^{\infty} \frac{\omega \varepsilon_i(\omega) d\omega}{\omega^2 - a^2}, \quad (10.43)$$

and similarly the imaginary component can be written

$$\begin{aligned}
\varepsilon_i(a) &= -\frac{P}{\pi} \int_{-\infty}^{\infty} \frac{\varepsilon_r d\omega}{\omega - a} = -\frac{P}{\pi} \left[\int_0^{\infty} \frac{\varepsilon_r d\omega}{\omega - a} + \int_{-\infty}^{0} \frac{\varepsilon_r d\omega}{\omega - a} \right] \\
&= -\frac{P}{\pi} \left[\int_0^{\infty} \frac{\varepsilon_r d\omega}{\omega - a} + \int_{-\infty}^{0} \frac{\varepsilon_r(-\omega) d(-\omega)}{-\omega + a} \right] \\
&= -\frac{P}{\pi} \left[\int_0^{\infty} \frac{\varepsilon_r(\omega) d\omega}{\omega - a} + \int_{-\infty}^{0} \frac{\varepsilon_r(\omega) d\omega}{\omega + a} \right] \\
&= -\frac{P}{\pi} \left[\int_0^{\infty} \frac{\varepsilon_r(\omega) d\omega}{\omega - a} - \int_0^{\infty} \frac{\varepsilon_r(\omega) d\omega}{\omega + a} \right] = -\frac{2aP}{\pi} \int_0^{\infty} \frac{\varepsilon_r(\omega) d\omega}{\omega^2 - a^2}.
\end{aligned} \quad (10.44)$$

In summary, the Kronig–Kramers relations can be written, where $\varepsilon_r(\omega) \to \varepsilon_r(\omega) - 1$ should be substituted

$$\varepsilon_r(a) = \frac{P}{\pi} \int_{-\infty}^{\infty} \frac{\text{Im}[\varepsilon(\omega)] d\omega}{\omega - a} = \frac{2P}{\pi} \int_0^{\infty} \frac{\omega \varepsilon_i(\omega) d\omega}{\omega^2 - a^2}, \quad (10.45)$$

$$\varepsilon_i(a) = -\frac{P}{\pi} \int_{-\infty}^{\infty} \frac{\text{Re}[\varepsilon(\omega)] d\omega}{\omega - a} = -\frac{2Pa}{\pi} \int_0^{\infty} \frac{\varepsilon_r(\omega) d\omega}{\omega^2 - a^2}. \quad (10.46)$$

10.3 Absorption of Electromagnetic Radiation—General (B)

We now give a fairly general discussion of the absorption process by quantum mechanics (see also Yu and Cardona [10.27 Chap. 6] as well as Fox op. cit. Chap. 3). Although much of the discussion is more general, we have in mind the absorption due to transitions between the valence and conduction bands of semiconductors. If $-e$ is the electronic charge, and if we assume the electromagnetic field is described by a vector potential A and a scalar potential ϕ, the Hamiltonian describing the electron in the field is in SI

$$\mathcal{H} = \frac{1}{2m}[\mathbf{p} + e\mathbf{A}]^2 - e(\phi + V), \tag{10.47}$$

where V is the potential in the absence of an electromagnetic field; V would be a periodic potential if the electron were in a solid. We will use the Coulomb gauge to describe the electromagnetic field so $\phi = 0$, $\nabla \cdot A = 0$ and the fields are given by

$$E = -\frac{\partial A}{\partial t}, \quad B = \nabla \times A. \tag{10.48}$$

The Hamiltonian can then be written

$$\mathcal{H} = \frac{1}{2m}[p^2 + eA \cdot p + ep \cdot A + e^2 A^2] - eV. \tag{10.49}$$

The terms quadratic in A will be ignored as they are normally small compared to the terms linear in A. Further in the Coulomb gauge, we can write

$$p \cdot A\psi \propto \nabla \cdot (A\psi) = (\nabla \cdot A)\psi + (A \cdot \nabla)\psi = A \cdot \nabla \psi, \tag{10.50}$$

so that the Hamiltonian can be written

$$\mathcal{H} = \mathcal{H}_0 + \mathcal{H}', \quad \mathcal{H}_0 = \frac{p^2}{2m} - eV;$$

where the perturbation is

$$\mathcal{H}' = \frac{e}{m}A \cdot p. \tag{10.51}$$

We assume the matrix element responsible for electronic transitions will be in the form $\langle f|H|i \rangle$, where i and f refer to the initial and final electron states and $\mathcal{H}'$ is the perturbing Hamiltonian. We assume the vector potential is given by

$$A(r,t) = ae\{\exp[i(k \cdot r - \omega t)] + \exp[-i(k \cdot r - \omega t)]\}, \tag{10.52}$$

where $e \cdot k = 0$ and a^2 is given by

$$a = \sqrt{E^2/2\omega^2}, \tag{10.53}$$

where $\overline{E^2}$ is the averaged squared electric field. Then,

$$P_{i \to f}^{\text{absorption}} = \frac{2\pi}{\hbar} a^2 \frac{e^2}{m^2} |\langle f|\exp(i\mathbf{k} \cdot \mathbf{r})\mathbf{e} \cdot \mathbf{p}|i\rangle|^2, \qquad (10.54)$$

and for emission

$$P_{i \to f}^{\text{emission}} = \frac{2\pi}{\hbar} a^2 \frac{e^2}{m^2} |\langle f|\exp(-i\mathbf{k} \cdot \mathbf{r})\mathbf{e} \cdot \mathbf{p}|i\rangle|^2. \qquad (10.55)$$

10.4 Direct and Indirect Absorption Coefficients (B)

Let us now look at the absorption coefficient. Using Bloch wave functions ($\psi_k = e^{i\mathbf{k}\cdot\mathbf{r}} u(\mathbf{r})$), we have

$$\langle f|\exp(i\mathbf{k}\cdot\mathbf{r})\mathbf{e}\cdot\mathbf{p}|i\rangle = \int u_f^* \exp[i(\mathbf{k}-\mathbf{k}_f+\mathbf{k}_i)\cdot\mathbf{r}]\hbar\mathbf{e}\cdot\mathbf{k}_i u_i d\Omega \\ + \int u_f^* \exp[i(\mathbf{k}-\mathbf{k}_f+\mathbf{k}_i)\cdot\mathbf{r}]\hbar\mathbf{e}\cdot\mathbf{p} u_i d\Omega. \qquad (10.56)$$

The first integral can be written as proportional to

$$\int \psi_f^* \psi_i \exp(i\mathbf{k}\cdot\mathbf{r}) d\Omega = \sum_j \exp[i(\mathbf{k}-\mathbf{k}_f+\mathbf{k}_i)\cdot\mathbf{R}_j] \\ \times \int_{\Omega_c} u_f^* \exp[i(\mathbf{k}-\mathbf{k}_f+\mathbf{k}_i)\cdot\mathbf{r}] u_i d\Omega \qquad (10.57) \\ \cong N \int_{\Omega_c} \psi_f \psi_i d\Omega \cong 0,$$

by orthogonality and assuming k is approximately zero, where we have also used

$$\sum_j \exp[i(\mathbf{k}-\mathbf{k}_f+\mathbf{k}_i)\cdot\mathbf{R}_j] = \delta_k^{k_f - k_i}(N), \qquad (10.58)$$

and Ω_c is the volume of a unit cell. The neglect of all terms but the $k = 0$ terms (called the electric dipole approximation) allows a similar description of the emission term. Following a similar procedure for the second term in (10.56), we obtain for absorption,

$$\langle f|\exp(i\mathbf{k}\cdot\mathbf{r})\mathbf{e}\cdot\mathbf{p}|i\rangle = N\int_{\Omega_c} u_f^* \mathbf{e}\cdot\mathbf{p} u_i d\Omega, \qquad (10.59)$$

with $k = 0$ and $k_i = k_f$.

Notice in the electric dipole approximation since $k_i = k_f$, we have what are called *direct optical transitions*. If something else such as phonons is involved, direct transitions are not required but the whole discussion must be modified to include this new physical ingredient. The electric dipole transition probability for photon absorption per unit time is

$$P_{i \to f}^{\text{abs}} = \frac{2\pi}{\hbar} \sum_k \frac{\overline{E^2}}{2\omega^2} \frac{e^2 N^2}{m^2} \left| \int_{\Omega_c} u_f^* \mathbf{e}\cdot\mathbf{p} u_i d\Omega \right|^2 \delta[E_c(k) - E_v(k) - \hbar\omega]. \qquad (10.60)$$

The power (per unit volume) lost by the field due to absorption in the medium is the transition probability per unit volume P multiplied by the energy of each photon (where in carrying out the sum over k in (10.60), we will assume we are summing over k states per unit volume). Carrying out the manipulations below, we finally find an expression for the absorption coefficient and, hence, the imaginary part of the dielectric constant. The power lost equals

$$P\hbar\omega = -\frac{dI}{dt}, \qquad (10.61)$$

where I is the energy/volume. But,

$$-\frac{dI}{dt} = -\frac{dI}{dx}\frac{dx}{dt} = \alpha I\left(\frac{c}{n}\right), \qquad (10.62)$$

where $\alpha = 2n_i\omega/c$, and $n_i = \varepsilon_i/2n$. Thus,

$$-\frac{dI}{dt} = \frac{\varepsilon_i \omega I}{n^2} = P\hbar\omega. \qquad (10.63)$$

Using

$$I = \frac{1}{2}n^2\varepsilon_0 \overline{E^2} \times 2, \qquad (10.64)$$

where $n = (\varepsilon/\varepsilon_0)^{1/2}$ if $\mu = \mu_0$ and the factor of 2 comes from both magnetic and electric fields carrying current, we find

$$\varepsilon_i(\omega) = \frac{P\hbar}{\varepsilon_0}\frac{1}{\overline{E^2}}. \qquad (10.65)$$

Using the Kronig–Kramers relations, we can also derive an expression for the real part of the dielectric constant. Defining

$$|M_{vc}| = \left|\int_\Omega u_f^* e \cdot p u_i d\Omega\right|, \qquad (10.66)$$

we have (using (10.65), (10.66), and (10.60))

$$\varepsilon_i = \frac{\pi}{\varepsilon_0}\left(\frac{e}{m\omega}\right)^2 \sum_k |M_{vc}|^2 \delta(E_c - E_v - \hbar\omega), \qquad (10.67)$$

and by (10.45)

$$\varepsilon_r = 1 + \frac{e^2}{m\varepsilon_0}\sum_k \left(\frac{2}{m\hbar\omega_{cv}}\frac{|M_{vc}|^2}{\omega_{cv}^2 - \omega^2}\right) \qquad (10.68)$$

10.4 Direct and Indirect Absorption Coefficients (B) 553

(where $E_c - E_v \equiv \hbar\omega_{cv}$ and $\delta(ax) = \delta(x)/a$ has been used). Recall that the Σ_k has to be per unit volume and the oscillator strength is defined by

$$f_{vc} = \frac{2|M_{vc}|^2}{m\hbar\omega_{cv}}. \tag{10.69}$$

Classically, the oscillator strength is the number of oscillators per unit volume with frequency ω_{cv}. Thus, the real part of the dielectric constant can be written

$$\varepsilon_r = 1 + \frac{e^2}{m\varepsilon_0}\sum_k \left(\frac{f_{vc}}{\omega_{cv}^2 - \omega^2}\right). \tag{10.70}$$

We want to work this out in a little more detail for direct absorption edges. For direct transitions between parabolic valence and conduction bands, effective mass concepts enter because one has to deal with both the valence band and conduction band. For parabolic bands we write

$$E_{vc} = E_g + \frac{\hbar^2 k^2}{2\mu}, \tag{10.71}$$

where

$$\frac{1}{\mu} = \frac{1}{m_c} + \frac{1}{m_v}. \tag{10.72}$$

The joint density of states per unit volume (see (10.94)) is then given by

$$D_j = \left[\frac{\sqrt{2}\mu^{3/2}}{\pi^2\hbar^3}\right]\sqrt{E_{vc} - E_g} \quad \text{where} \quad E_{vc} > E_g, \tag{10.73}$$

and

$$D_j = 0 \quad \text{where} \quad E_{vc} < E_g. \tag{10.74}$$

Thus, we obtain that the imaginary part of the dielectric constant is given by

$$\varepsilon_i(\omega) = \begin{cases} K\sqrt{\Omega - 1}, & \Omega > 1 \\ 0, & \Omega < 1 \end{cases}, \tag{10.75}$$

where

$$K = \frac{2e^2(2\mu)^{3/2}|M_{vc}|^2\sqrt{E_g}}{m^2\omega^2\hbar^3}, \tag{10.76}$$

and

$$\Omega = \frac{\hbar\omega}{E_g}. \tag{10.77}$$

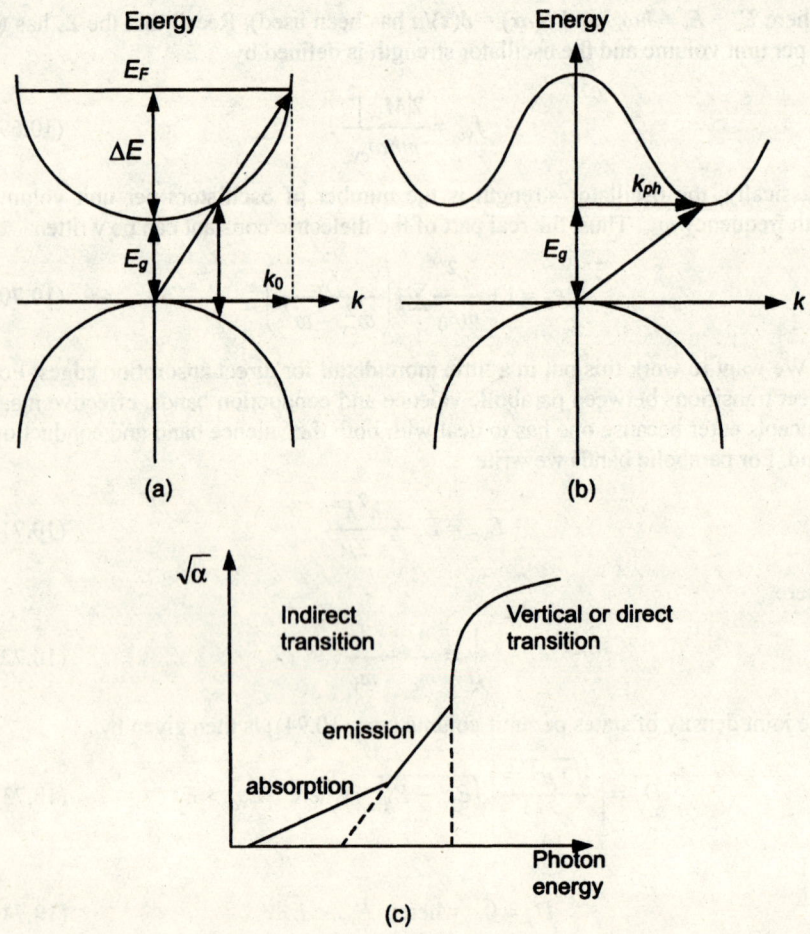

Fig. 10.3. (a) Direct transitions and indirect transitions due to band filling; (b) Indirect transitions, where k_{ph} is the phonon wave vector; (c) Vertical transitions dominate indirect transitions when energy is sufficient to cause them. Emission and absorption refer to phonons in all sketches

From this, one then has an expression for the absorption coefficient (since $\alpha = \omega \varepsilon_i / nc$). Thus for direct transitions and parabolic bands, a plot of the square of the absorption coefficient as a function of the photon energy should be a straight line, at least over a limited frequency. Figure 10.3 illustrates direct and indirect transitions and absorption. Indirect transitions are discussed below.

The fundamental absorption edge due to the bandgap determines the apparent color of semiconductors as seen by transmission.

We now want to discuss indirect transitions. So far, our analysis has assumed a direct bandgap. This means that the k of the initial and final electronic states defining the absorption edge are almost the same (as has been mentioned, the k of the photon causing the absorption is negligible, compared to the Brillouin zone width, for visible wavelengths). This is not true for the two most common semiconductors Si and Ge. For these semiconductors, the maximum energy of the valence band and the minimum energy of the conduction band do not occur at the same k vectors, one has what is called an indirect bandgap semiconductor. For a minimum energy transition across the bandgap, something else, typically a phonon, must be involved in order to conserve wave vector. The requirement of having, for example, a phonon being involved reduces the probability of the event; see Fig. 10.3b, c, Fig. 10.4, (10.82), and consider also Fermi's Golden Rule.

Even in a direct bandgap semiconductor, processes can cause the fundamental absorption edge to shift from direct to indirect, see Fig. 10.3a. For degenerate semiconductors, the optical absorption edge may be a function of the carrier density. In simple models, the location of the Fermi energy in the conduction band can be estimated on the free-electron model. When the Fermi energy is above the bottom of the conduction band, the k vector of the minimum energy that can cause a transition has also shifted from the k of the conduction band minimum. Now direct transitions will originate from deeper states in the valence band, they will be stronger than the threshold energy transitions, but of higher energy.

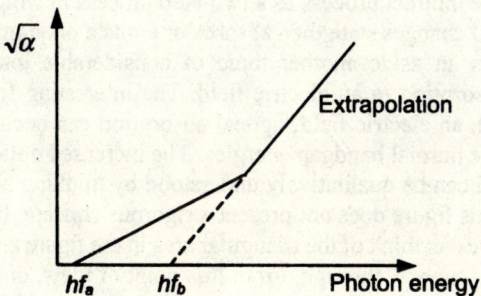

Fig. 10.4. Indirect transitions: $hf_a = E_g - E_{phonon}$, $hf_b = E_g + E_{phonon}$, $E_g = (hf_a + hf_b)/2$, sketch

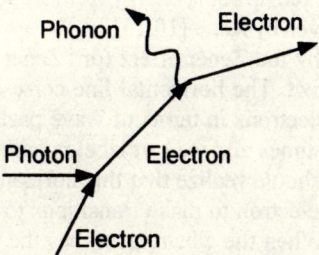

Fig. 10.5. An indirect process viewed in two steps

For indirect transitions, we can write the energy and momentum conservation conditions as follows:

$$k' = k + K \pm q, \tag{10.78}$$

where K = photon $\cong 0$ and, q = phonon (=k_{ph} in Fig. 10.3b). Also

$$E(k') = E(k) + \hbar\omega \pm \hbar\omega_q, \qquad (10.79)$$

where $\hbar\omega$ = photon, and $\hbar\omega_q$ = phonon. Note: although the photon makes the main contribution to the transition energy, the phonon carries the burden of insuring that momentum is conserved. Now the Hamiltonian for the process would look like

$$\mathcal{H}' = \mathcal{H}'_{\text{photon}} + \mathcal{H}'_{\text{phonon}}, \qquad (10.80)$$

where

$$\mathcal{H}'_{\text{photon}} = \frac{e}{m} p \cdot A, \qquad (10.81)$$

and

$$\mathcal{H}'_{\text{phonon}} = \sum_{\sigma k q} M_{kq} (a^+_{-q} + a_q) c^+_{k+q,\sigma} c_{k,\sigma}. \qquad (10.82)$$

One can sketch the indirect process as a two-step process in which the electron absorbs a photon and changes state then absorbs or emits a phonon. See Fig. 10.5.

We mention as an aside another topic of considerable interest. We discuss briefly optical absorption in an electric field. The interesting feature of this phenomenon is that in an electric field, optical absorption can occur for photon energies lower than the normal bandgap energies. The increased optical absorption due to an electric field can be qualitatively understood by thinking about pictures such as in Fig. 10.6. This figure does not present a rigorous concept, but it is helpful.

Very simply, we can think of the triangular area in the figure as a potential barrier that electrons can "tunnel" through. From this point of view, one perhaps believes than an electric field can cause electronic transitions from band 2 to band 1 (This is called the Zener effect). Obviously, the process of tunneling would be greatly enhanced if the electron "picked up some energy from a photon before it began to tunnel." Further details are given by Kane [10.15].

It is not hard to see why the Zener effect (or "Zener breakdown") can be considered as a tunneling effect. The horizontal line corresponds to the motion of an electron (if we describe electrons in terms of wave packets, then we can speak of where they are at various times and we can label positions in terms of distances in the bands). Actually, we should realize that this horizontal line corresponds to the electric field causing the electron to make transitions to higher and higher stationary states in the crystal. When the electron reaches the top of the lower band, we normally think of the electron as being Bragg reflected. However, we should remember what we mean by the energy gap.

The energy gap, E_g, does not represent an absolutely forbidden gap. It simply represents energies corresponding to attenuated, nonpropagating wave functions. The attenuation will be of the form e^{-Kx}, where x represents the distance traveled (K is real and greater than zero) and K is actually a function of x, but this will be ignored here. The electron gains energy from the electric field E as $|eEx|$. When

10.4 Direct and Indirect Absorption Coefficients (B)

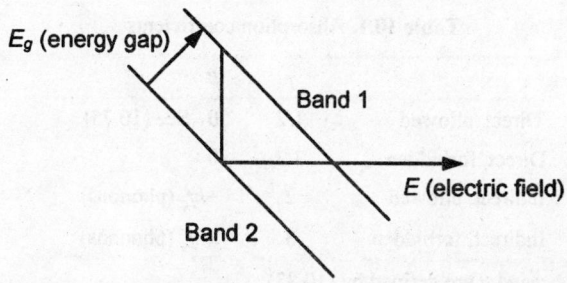

Fig. 10.6. Qualitative effect of an electric field on the energy bands in a solid

the electron has traveled $x = |E_g/eE|$, it has gained sufficient energy to get into the bottom of the upper band if it started at the top of the lower band. In order for the process to occur, we must require that the electron's wave function not be too strongly attenuated, i.e. Zener breakdown will occur if $1/K \gg |E_g/eE|$. To see the analogy to tunneling, we observe that the electron's wave function in the triangular region also behaves as e^{-Kx} from a tunneling viewpoint (also with K a function of x), and that the larger we make the electric field, the thinner the area we have to tunnel across, so the greater a band-to-band transition. A more quantitative discussion of this effect is obtained by evaluating K not from the picture, but directly from the Schrödinger equation. The x dependence on K turns out to be fairly easy to handle in the WKB approximation.

Finally, we can summarize the results for many cases in Table 10.1. Absorption coefficients α for various cases (parabolic bands) can be written

$$\alpha = \left(\frac{A}{hf}\right)(hf + \beta - E_g)^\gamma, \qquad (10.83)$$

where γ, β depend on the process as shown in the table. When phonons are involved we need to add both the absorption and emission (±) possibilities to get the total absorption coefficient.[2] A very clean example of optical absorption is given in Fig. 10.7. Good optical absorption experiments on InSb were done in the early days by Gobeli and Fan [10.15]. In general, one also needs to take into account the effect of temperature. For example, the indirect allowed term should be written

$$\alpha = \left(\frac{A'}{hf}\right)\left[\frac{(hf + \beta - E_g)^2}{\exp(\beta/kT) - 1} + \frac{(hf - \beta - E_g)^2}{\exp(\beta/kT) - 1}\exp(\beta/kT)\right], \qquad (10.84)$$

where A' is a constant independent of the temperature, see, e.g., Bube [10.4] and Pankove [10.22].

---

[2] An additional very useful reference is Greenaway and Harbeke [10.16]. See also Yu and Cardona [10.27].

Table 10.1. Absorption coefficients

	γ	β
Direct, allowed	1/2	0 See (10.75)
Direct, forbidden	3/2	0
Indirect, allowed	2	$\pm hf_q$ (phonons)
Indirect, forbidden	3	$\pm hf_q$ (phonons)

γ and β are defined by (10.83).

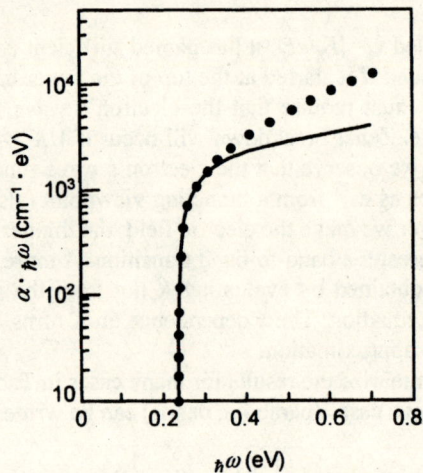

Fig. 10.7. Optical absorption in indium antimonide, InSb at 5 K. The transition is direct because both conduction and valence band edges are at the center of the Brillouin zone, $k = 0$. Notice the sharp threshold. The dots are measurements and the solid line is $(\hbar\omega - E_g)^{1/2}$. (Reprinted with permission from Sapoval B and Hermann C, *Physics of Semiconductors*, Fig. 6.3 p. 154, Copyright 1988 Springer Verlag, New York.)

10.5 Oscillator Strengths and Sum Rules (A)

Let us define the oscillator strength by

$$f_{ij} = b\omega_{ij} |\langle i|e \cdot r|j\rangle|^2 .$$

We will show this is equivalent to the previous definition with the proper choice of b by using commutation relations to cast it in another form. From $[x, p_x] = i\hbar$ we can show

$$[\mathcal{H}, x] = -\frac{i\hbar}{m} p_x . \tag{10.85}$$

Also,

$$[\mathcal{H}, e\cdot r] = -\left(\frac{i\hbar}{m}\right)e\cdot p, \qquad (10.86)$$

therefore

$$\langle i|e\cdot p|j\rangle = im\omega_{ij}\langle i|e\cdot r|j\rangle. \qquad (10.87)$$

Thus we can write the oscillator strength as,

$$f_{ij} = b\frac{|\langle i|e\cdot p|j\rangle|^2}{m^2\omega_{ij}}, \qquad (10.88)$$

which is consistent with how we wrote it before, if $b = -2m/\hbar$ (see (10.69), (10.66)). It is also interesting to show that the oscillator strength obeys a sum rule. If $e = i$, then

$$\begin{aligned}
\sum_j f_{ij} &= b\sum_j \omega_{ij}|\langle i|e\cdot r|j\rangle|^2 \\
&= \frac{b}{2}\sum_j \frac{1}{im}\{\langle i|e\cdot p|j\rangle\langle j|e\cdot r|i\rangle - \langle i|e\cdot r|j\rangle\langle j|e\cdot p|i\rangle\} \qquad (10.89) \\
&= \frac{b}{2}\frac{1}{im}[\langle i|e\cdot p, e\cdot r|i\rangle] = \frac{b}{2im}[-i\hbar] = 1.
\end{aligned}$$

Classically, for bound states with no damping, we can derive the dielectric constant. Assume N states with frequency ω_0. The result is

$$\frac{\varepsilon}{\varepsilon_0} = 1 + \frac{Ne^2}{m\varepsilon_0}\frac{1}{\omega_0^2 - \omega^2}, \qquad (10.90)$$

which follows from (9.6) with $\tau \to \infty$ and ω_0 used for several bound states labeled with i. Note that it is just the same as the quantum result (10.70) provided the oscillator strength from one oscillator is one. From this we have the index of refraction, and it is given by $n^2 = \varepsilon/\varepsilon_0$, since ε is real with $\tau \to \infty$. When $\varepsilon/\varepsilon_0$ as the preceding, the resulting equation is often called Sellmeier's equation.

10.6 Critical Points and Joint Density of States (A)

Optical absorption spectra give many details about the band structure. This can be explained by the Van Hove singularities, which appear in the joint density of states as mentioned below. In the integral for the imaginary part of the dielectric constant, we had an expression of the form (10.67):

$$\varepsilon_i \propto \frac{2}{\omega^2}\int\frac{d^3k}{(2\pi)^3}|M_{vc}|^2\delta(E_c - E_v). \qquad (10.91)$$

A property of delta functions can be written as

$$\int_a^b g(x)\delta[f(x)]dx = \sum_{x_p} g(x_p) \frac{1}{\left.\frac{\partial f}{\partial x}\right|_{x=x_p}}, \quad (10.92)$$

where x_p are the zeros of $f(x)$. From which we conclude that the imaginary part of the dielectric constant can be written as

$$\varepsilon_i \propto \frac{2}{\omega^2} \frac{1}{(2\pi)^3} \int_s \frac{dS |M_{vc}|^2}{\left.|\nabla_k (E_c - E_v)|\right|_{E_c-E_v=\hbar\omega}}, \quad (10.93)$$

where dS is a surface of constant $\hbar\omega = E_c - E_v$. The joint density of states is defined as (Yu and Cardona [10.27 p251])

$$J_{vc} = \int \frac{2}{(2\pi)^3} \frac{dS}{\left.|\nabla_k (E_c - E_v)|\right|_{E_c-E_v=\hbar\omega}}, \quad (10.94)$$

and typically the matrix element M_{vc} is a slowly varying function compared with the joint density of states. Now the joint density of states is a strongly varying function of k where the denominator is zero, i.e. where

$$\nabla_k (E_c - E_v) = 0. \quad (10.95)$$

Both valence and conduction band energies must be periodic functions in reciprocal space and so must their difference and from this it follows that there must be a point for which the denominator vanishes (smooth periodic functions have analytic maxima and minima). These critical points lead to singularities in the density of states, the Van Hove singularities. At very highly symmetrical points in the Brillouin zone, we can have critical points due to the gradient of both conduction and valence energies vanishing, at other critical points only the gradient of the difference vanishes. Critical points are defined by the band structure, and in turn, they help determine the absorption coefficient. Reversing the process, studying the absorption coefficient gives information on the band structure.

10.7 Exciton Absorption (A)

In semiconductors, one may detect absorption for energies just below the energy gap where one might have initially expected transparency. This could be due to absorption by bound electron–hole pairs or excitons. The binding energy of the excitons lowers their absorption below the bandgap energy. It is interesting that one can only think of bound electron–hole pairs if electron and holes move with the same group velocity, in other words the energy gradients of valence and electronic energies need to be the same. That is, excitons form at the critical points of the joint density of states.

One generally talks of two kinds of excitons, the Frenkel excitons and Wannier excitons. The Frenkel excitons are tightly bound and can be described by a variant of tight binding theory. Another way to view Frenkel excitons is as a propagating excited state of a single atom. Thus, we describe it with the Hamiltonian where the states are the localized excited states of each atom. For the Frenkel case let

$$\mathcal{H} = \sum_i \varepsilon |i\rangle\langle i| + \sum_{i,j} V_{ij} |i\rangle\langle j|, \quad (10.96)$$

where with one-dimensional nearest-neighbor hopping

$$V_{ij} = V\delta_i^{j+1} + V\delta_i^{j-1}. \quad (10.97)$$

This can be diagonalized by the substitution:

$$|k\rangle = \sum_j \exp(ijka)|j\rangle, \quad (10.98)$$

which leads to the energy eigenvalues

$$\mathcal{H}|k\rangle = \varepsilon_k |k\rangle, \quad (10.99)$$

where, $\varepsilon_k = \varepsilon + 2V\cos(ka)$. These Frenkel types of excitons are found in the alkali halides.

In semiconductors, the important types of excitons are the Wannier excitons, which have size much larger than typical interatomic dimensions. The Wannier excitons can be analyzed much as a hydrogen atom with reduced mass defined by the electron and hole masses and with the binding Coulomb potential reduced by the appropriate dielectric constant. That is, the energy eigenvalues are

$$E_n = E_g - \frac{\mu e^4}{2\hbar^2 (4\pi\varepsilon)^2 n^2}, \quad n = 1, 2, \ldots, \quad (10.100)$$

where

$$\frac{1}{\mu} = \frac{1}{m_e} + \frac{1}{m_h}. \quad (10.101)$$

Optical absorption in GaAs is shown in Fig. 10.8.

10.8 Imperfections (B, MS, MET)

We will only give a brief discussion here. Reference should be made also to the chapters on semiconductors and defects. Imperfections may produce resonant energy levels in the bands or energy levels that are in the bandgap. Donors and acceptors in semiconductors produce energy levels that may be detected by optical absorption when the thermal energy is much less than their ionization energy. Similarly, deep defects produced in a variety of ways may produce energy levels in the gap, often near the center. Deep defects tend to be very localized in space and therefore to contain a large range of k vectors. Thus, it is possible to have

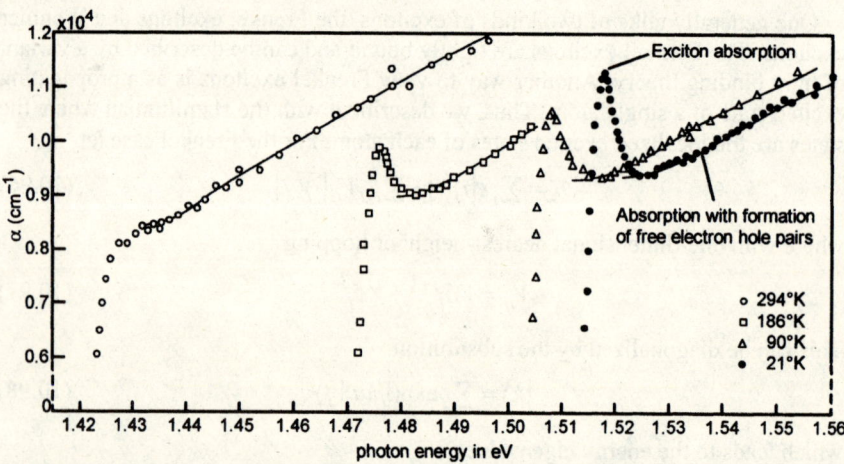

Fig. 10.8. Absorption coefficient near the band edge of GaAs. Note the exciton absorption level below the bandgap E_g [Reprinted with permission from Sturge MD, *Phys Rev* **127**, 771 (1962). Copyright 1962 by the American Physical Society.]

a direct transition from a deep defect to a large range of k values in the conduction band, for example. A shallow level, on the other hand, is well spread out in space and therefore restricted in k value and so direct transitions from it to a band go to quite a restricted range of values. Color centers in alkali halides are examples of other kinds of optically important defects.

Suppose we have some generic defect with energy level in the gap. One could have absorption due to transitions from the valence or conduction band to the level. There could even be absorption between levels due to the defect or different defects. Several types of optical processes are suggested in Fig. 10.9.

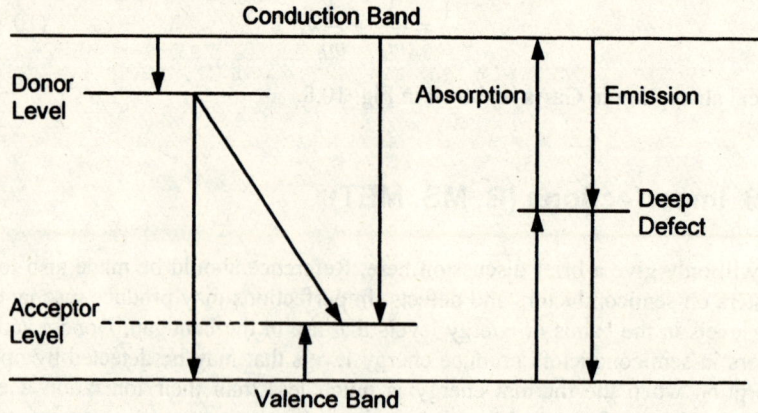

Fig. 10.9. Some typical radiative transitions in semiconductors. Nonradiative (Auger) transitions are also possible

10.9 Optical Properties of Metals (B, EE, MS)

Free-carrier absorption can be viewed as intraband absorption–the electron absorbing the photon remains in the same band.[3] Free-carrier absorption is obviously important for metals, and is often of importance for semiconductors. The electron is accelerated by the photon and gains energy, but since the wave vector of the photon is negligible, something else such as a phonon needs to be involved. For many purposes, the process can be viewed classically by Drude theory with a relaxation time of $\tau \equiv 1/\omega_0$. This relaxation time defines a frictional force constant m^*/τ, where the viscous like frictional force is proportional to the velocity.

We will use classical theory here, but it is worthwhile to make a few comments. It is common to deal with a semiclassical picture of radiation. There we treat the radiation classically, but the underlying electronic systems that absorb and emit the radiation we treat quantum mechanically. Radiation can be treated classically when it is intense enough to have many photons in each mode. Free-electronic systems can be treated classically when their de Broglie wavelengths are small compared to the average interparticle separations.

The de Broglie wavelength can be estimated from the momentum as estimated from equipartition. In practice, this means that for temperatures that are not too low and densities that are not too high, then classical mechanics should be valid. Bound systems are more complicated, but in general, classical mechanics works at higher quantum numbers (higher bound-state energies). In any case, classical and quantum results often overlap in validity well beyond where one might naively expect.

The classical theory can be written, assuming a sinusoidal electric field $E = E_0 \exp(-i\omega t)$ (note these are for free-electrons ($e > 0$) with damping). We also generalize by using an effective mass m^* rather than m:

$$m^* \ddot{x} + \frac{m^*}{\tau} \dot{x} = -eE_0 \exp(-i\omega t). \quad (10.102)$$

Note this is just (9.1) with $\omega_0 = 0$, as appropriate for free charges. Seeking a steady-state solution of the form $x = x_0 \exp(-i\omega t)$, we find

$$x = \frac{-ieE\tau}{m^* \omega (1 - i\omega\tau)}, \quad (10.103)$$

which is (9.2) with $\omega_0 = 0$. Thus, the polarization is given by

$$P = -Nex = (\varepsilon - \varepsilon_L)E, \quad (10.104)$$

where ε_L is the contribution to the dielectric constant of everything except the free carriers (generalizing (9.3)). The frequency-dependent dielectric constant is

$$\varepsilon(\omega) = \varepsilon_L + i\frac{Ne^2\tau}{m^*\omega} \frac{1}{1 - i\omega\tau}, \quad (10.105)$$

---

[3] See also, e.g., Ziman [25, Chap. 8] and Born and Wolf [10.1].

where N is the number of electrons per unit volume. From the real and imaginary parts of ε we find, similar to Sect. 9.2,

$$\varepsilon_r = n^2 - n_i^2 = \frac{\varepsilon_L}{\varepsilon_0} - \frac{\sigma_0 \tau / \varepsilon_0}{1 + \omega^2 \tau^2}, \qquad \sigma_0 = \frac{Ne^2 \tau}{m^*}, \qquad (10.106)$$

and

$$\varepsilon_i = 2nn_i = \frac{\sigma_0}{\varepsilon_0 \omega}\left(\frac{1}{1+\omega^2\tau^2}\right). \qquad (10.107)$$

It is convenient to write this in terms of the plasma frequency

$$\omega_p^2 = \frac{Ne^2}{m\varepsilon_0} = \frac{\sigma_0}{\tau \varepsilon_0} \equiv \frac{\sigma_0 \omega_0}{\varepsilon_0}, \qquad (10.108)$$

and so,

$$\varepsilon_r = \frac{\varepsilon_L}{\varepsilon_0} - \frac{\omega_p^2}{\omega_0^2 + \omega^2}, \qquad (10.109)$$

and

$$\varepsilon_i = \frac{\omega_0}{\omega}\frac{\omega_p^2}{\omega_0^2 + \omega^2}. \qquad (10.110)$$

From here onwards for simplicity we assume $\varepsilon_L \cong \varepsilon_0$. We have three important ω. The plasma frequency ω_p is proportional to the free-carrier concentration, ω_0 measures the electron–phonon coupling and ω is the frequency of light.

We now want to show what these equations predict in three different frequency regions.

(i) $\omega\tau \ll 1$, the low-frequency region. We obtain by (10.109) with $\omega_0 = 1/\tau$

$$n^2 - n_i^2 = 1 - \omega_p^2 \tau^2, \qquad (10.111)$$

which is small, and by (10.110)

$$2nn_i = \omega_p^2 \frac{\tau}{\omega} = \frac{(\omega_p \tau)^2}{\omega \tau}, \qquad (10.112)$$

which is large. Here the imaginary part (of the *dielectric constant*) is much greater than the real part and we have high reflectivity. In this approximation

$$n^2 - n_i^2 = 1 - \omega\tau(2nn_i) \cong 1, \qquad (10.113)$$

but neither n nor n_i are small, so $n \cong n_i$, and $n^2 \cong \sigma_0/2\omega\varepsilon_0$. The reflectivity then becomes

$$R = \frac{(n-1)^2 + n_i^2}{(n+1)^2 + n_i^2} \cong 1 - \frac{2}{n} \cong 1 - 2\sqrt{\frac{2\omega\varepsilon_0}{\sigma_0}}. \qquad (10.114)$$

10.9 Optical Properties of Metals (B, EE, MS)

This is the Hagen–Rubens relation [10.17].

(ii) $1/\tau \ll \omega \ll \omega_p$, the relaxation region. The basic relations become

$$n^2 - n_i^2 = 1 - \frac{\omega_p^2}{\omega^2}, \qquad (10.115)$$

which is large and negative, and

$$2nn_i = \left(\frac{\omega_p}{\omega}\right)^2 \frac{1}{\omega\tau}, \qquad (10.116)$$

which is smaller than $n^2 - n_i^2$. However, this predicts the metal is still strongly reflecting as we now show. Since $\omega\tau \gg 1$ and $\omega_p/\omega \gg 1$, we see

$$(n - n_i)(n + n_i) \cong -\left(\frac{\omega_p}{\omega}\right)^2 \ll 1, \qquad (10.117)$$

or

$$(n_i - n)(n + n_i) = \left(\frac{\omega_p}{\omega}\right)^2 \gg 1. \qquad (10.118)$$

Therefore,

$$n_i \gg n, \qquad (10.119)$$

$$n_i^2 \cong \left(\frac{\omega_p}{\omega}\right)^2, \qquad (10.120)$$

$$n_i \cong \frac{\omega_p}{\omega}, \qquad (10.121)$$

$$2nn_i \cong n_i^2 \frac{1}{\omega\tau}, \qquad (10.122)$$

$$R = \frac{(n-1)^2 + n_i^2}{(n+1)^2 + n_i^2} = \frac{1 + [(n-1)/n_i]^2}{1 + [(n+1)/n_i]^2} \cong 1 + \left(\frac{n-1}{n_i}\right)^2 - \left(\frac{n+1}{n_i}\right)^2, \quad (10.123)$$

and

$$R = 1 - \frac{4n}{n_i^2} = 1 - \frac{2}{\omega_p \tau}. \qquad (10.124)$$

Since $\omega_p \tau \gg 1$, the metal is still strongly reflecting.

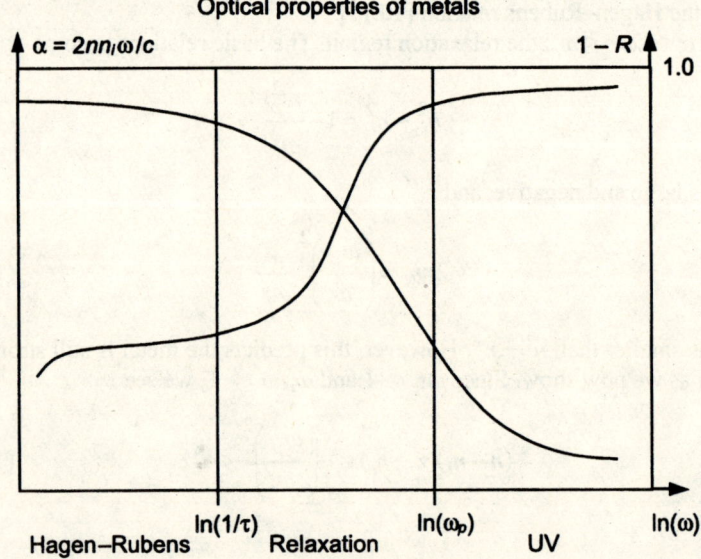

Fig. 10.10. Sketch of absorption and reflection in metals

(iii) $\omega_p \ll \omega$ or $\omega_p/\omega \ll 1$. This is the ultraviolet region where we also assume $\omega \gg \omega_0$.

$$(n^2 - n_i^2) \cong 1, \tag{10.125}$$

so

$$(n - n_i)(n + n_i) = 1. \tag{10.126}$$

$2nn_i = (\omega_p/\omega)^2(1/\omega\tau)$ is very small. Both n and n_i are not very small, therefore n_i is very small. Therefore,

$$n \gg n_i, n \cong 1. \tag{10.127}$$

Therefore,

$$n_i \cong \frac{1}{2}\left(\frac{\omega_p}{\omega}\right)^2 \frac{1}{\omega\tau}. \tag{10.128}$$

So,

$$R \cong \frac{n_i^2}{n_i^2 + 4} \cong \frac{n_i^2}{4} \cong \frac{1}{16}\left(\frac{\omega_p}{\omega}\right)^4 \frac{1}{(\omega\tau)^2} \tag{10.129}$$

is very small. There is little reflectance since this is the ultraviolet transparency region. We summarize our results in Fig. 10.10. See also Seitz [82, p 639], Ziman [25, 1st edn, p 240], and Fox [10.12].

10.9 Optical Properties of Metals (B, EE, MS)

The plasma edge, or the region around the plasma frequency deserves a little more attention. Using Maxwell's equations we have

$$\nabla \times E = -\frac{\partial B}{\partial t}, \quad \nabla \cdot E = \nabla \cdot B = 0 \quad (\rho = 0), \tag{10.130}$$

and

$$\nabla \times B = \mu_0 \frac{\partial^2 D}{\partial t^2} \quad (j = 0), \tag{10.131}$$

and we will include any charge motion in P. Therefore,

$$\nabla^2 E = \mu_0 \frac{\partial^2 D}{\partial t^2}, \quad D = \varepsilon_0 \varepsilon E. \tag{10.132}$$

Note here $\varepsilon \to \varepsilon/\varepsilon_0$. Assume $E = E_0 \exp(-i\omega t)\exp(i\mathbf{k}\cdot\mathbf{r})$. We obtain, as shown below ((10.142), (10.143)), for the wave vector

$$k^2 = \varepsilon(\omega)\mu_0\varepsilon_0\omega^2 \Rightarrow (kc)^2 = \varepsilon(\infty)(\omega^2 - \tilde{\omega}_p^2). \tag{10.133}$$

For a free-electron in an electric field we have already derived the plasma frequency in Sect. 9.4. We give here an alternative simple derivation and bring out a few new features,

$$m\frac{d^2 x}{dt^2} = -eE, \tag{10.134}$$

$$x = x_0 \exp(-i\omega t), \tag{10.135}$$

$$E = E_0 \exp(-i\omega t), \tag{10.136}$$

$$x = \frac{eE}{m\omega^2}. \tag{10.137}$$

Also,

$$P = -Nex = -\frac{Ne^2}{m\omega^2} E, \tag{10.138}$$

$$\varepsilon(\omega) = 1 + \frac{P(\omega)}{\varepsilon_0 E(\omega)} = 1 - \frac{Ne^2}{\varepsilon_0 m\omega^2}, \tag{10.139}$$

$$\omega_p^2 = \frac{Ne^2}{\varepsilon_0 m}, \tag{10.140}$$

$$\varepsilon(\omega) = 1 - \frac{\omega_p^2}{\omega^2}. \tag{10.141}$$

If the positive ion core background has a dielectric constant of $\varepsilon(\infty)$ that is about constant, then (10.141) is modified

$$\varepsilon(\omega) = \varepsilon(\infty)\left[1 - \frac{\tilde{\omega}_p^2}{\omega^2}\right], \qquad (10.142)$$

where

$$\tilde{\omega}_p = \frac{\omega_p}{\sqrt{\varepsilon(\infty)}}. \qquad (10.143)$$

When the frequency is less than the plasma frequency the squared wave vector is negative (10.133) and gives us total reflection. Above the plasma frequency, the wave vector squared is positive and the material is transparent. That is, simple metals should reflect in the visible and be transparent in the ultraviolet, as we have already seen.

It is also good to remember that at the plasma frequency the electrons undergo low-frequency longitudinal oscillations. See Sect. 9.4. Specifically, note that setting $\varepsilon(\omega) = 0$ defines a frequency $\omega = \omega_L$ corresponding to longitudinal plasma oscillations.

$$\varepsilon(\omega) = 1 - \frac{\omega_p^2}{\omega^2}, \text{ so } \varepsilon(\omega) = 0 \text{ implies } \omega = \omega_p. \qquad (10.144)$$

Here we have neglected the dielectric constant of the positive ion cores.

The plasma frequency is also a free longitudinal oscillation. If we have a doped semiconductor with the plasma frequency less than the bandgap over Planck's constant, one can detect the plasma edge, as illustrated in Fig. 10.11. See also Fox op cit, p156. Hence, we can determine the electron concentration.

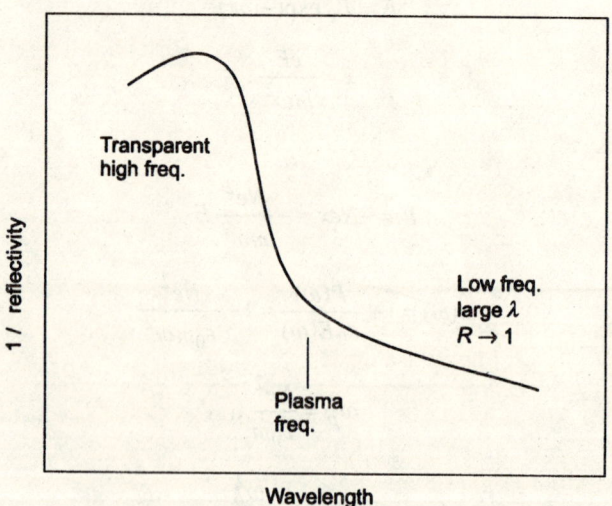

Fig. 10.11. Reflectivity of doped semiconductor, sketch

10.10 Lattice Absorption, Restrahlen, and Polaritons (B)

10.10.1 General Results (A)

Polar solids carry lattice polarization waves and hence can interact with electromagnetic waves (only transverse optical phonons couple to electromagnetic waves by selection rules and conservation laws). The dispersion relations for photons and the phonons of the polarization waves can cross. When these dispersion relations cross, the resulting quanta turn out to be neither photons nor phonons but mixtures called polaritons. One way to view this is shown in Fig. 10.12. We now discuss this process in more detail. We start by considering lattice vibrations in a polar solid. We will later add in a coupling with electromagnetic waves. The displacement of the tth ion in the lth cell for the jth component, satisfies

$$M_t \ddot{v}_{tl}^j = -\sum_{t'h} G_{tt'}^{jj'}(h) v_{t',l+h}^{j'}, \qquad (10.145)$$

where

$$G_{tl,t',l'=l+h}^{jj'} = \frac{\partial^2 U}{\partial v_{tl}^j \partial v_{t'l'}^{j'}}, \qquad (10.146)$$

and U describes the potential of interaction of the ions. If v_{tl} is a constant,

$$\sum_{t'h} G_{tt'}(h) = 0. \qquad (10.147)$$

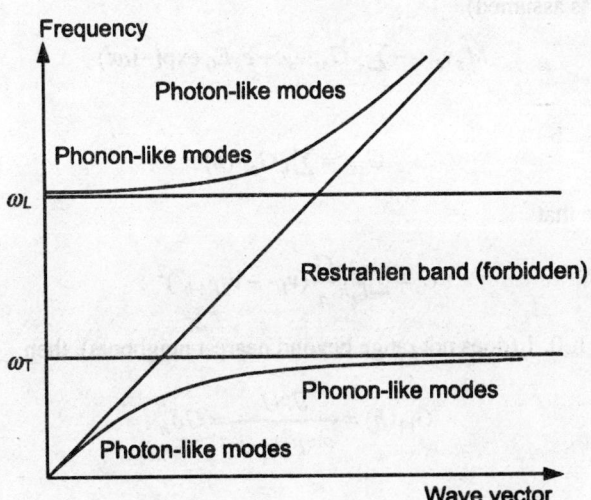

Fig. 10.12. Polaritons as mixtures of photons and transverse phonons. The mathematics of this model is developed in the text

10 Optical Properties of Solids

We will add an electromagnetic wave that couples to the system through the force term.

$$e_l E_0 \exp[i(q \cdot l - \omega t)], \qquad (10.148)$$

where e_l is the charge of the lth ion in the cell. We seek solutions of the form

$$v_{sl}(t) = \exp(iq \cdot l) v_{s,q}(t), \qquad (10.149)$$

(now s labels ions) with $q = K$ (dropping the vector notation of q, h, and l for simplicity from here on) and t is the time. Defining

$$G_{ss'}(K) = \sum_h G_{ss'}(h) \exp(iKh), \qquad (10.150)$$

we have (for one component in field direction)

$$M_s \ddot{v}_{sK} = -\sum_{s'} G_{ss'}(K) v_{s'K} + e_s E_0 \exp(-i\omega t). \qquad (10.151)$$

Note that

$$G_{ss'}(K=0) = \sum_h G_{ss'}(h). \qquad (10.152)$$

Using the above we find

$$\sum_{s'} G_{ss'}(K=0) = 0. \qquad (10.153)$$

Assuming $e_1 = |e|$ and $e_1 = -|e|$ (to build in the polarity of the ions), the following equations can be written (where long wavelengths, $K \cong 0$, and one component of ion location is assumed)

$$M_s \ddot{v}_s = -\sum_{s'} G_{ss'} v_{s'} + e_s E_0 \exp(-i\omega t), \qquad (10.154)$$

where

$$G_{ss'} = \sum_h G_{ss'}(h). \qquad (10.155)$$

If we assume that

$$U = \sum_{l',h} \frac{G}{4} (v_{1l'} - v_{2l'+h'})^2, \qquad (10.156)$$

where $h' = -1, 0, 1$ (does not range beyond nearest neighbors), then

$$G_{11}(h) = \frac{\partial^2 U}{\partial v_{1l} \partial v_{1l+h}} = G \delta_h^0. \qquad (10.157)$$

Similarly,

$$G_{22}(h') = G \delta_{h'}^0, \qquad (10.158)$$

$$G_{11}(h) = -G_{12}(h), \qquad (10.159)$$

and
$$G_{22}(h) = -G_{21}(h). \tag{10.160}$$

Therefore we can write
$$M_1 \ddot{v}_1 = G_{11}(v_2 - v_1) + eE_0 \exp(-i\omega t), \tag{10.161}$$

and
$$M_2 \ddot{v}_2 = G_{22}(v_1 - v_2) - eE_0 \exp(-i\omega t). \tag{10.162}$$

We now apply this to a dielectric where
$$\varepsilon = \varepsilon_0 + P/E, \tag{10.163}$$

and
$$P = \sum_i N_i \alpha_i E_{\text{loc},i}, \tag{10.164}$$

with N_i = the number of ions/vol of type i and α_i is the polarizability. For cubic crystals as derived in the chapter on dielectrics,
$$E_{\text{loc},i} = E + \frac{P}{3\varepsilon_0}. \tag{10.165}$$

Then,
$$\varepsilon = \varepsilon_0 + \frac{1}{E} \sum N_i \alpha_i \left(E + \frac{P}{3\varepsilon_0} \right). \tag{10.166}$$

Let[4]
$$B = \frac{1}{3\varepsilon_0} \sum N_i \alpha_i, \tag{10.167}$$

so
$$\varepsilon = \varepsilon_0 + 3\varepsilon_0 B + B(\varepsilon - \varepsilon_0), \tag{10.168}$$
$$\varepsilon(1 - B) = \varepsilon_0 + 2\varepsilon_0 B, \tag{10.169}$$

and
$$\varepsilon = \varepsilon_0 \frac{1 + 2B}{1 - B}. \tag{10.170}$$

For the diatomic case, define
$$B_{\text{el}} = \frac{1}{3\varepsilon_0} N(\alpha_+ + \alpha_-), \tag{10.171}$$

---

[4] Grosso and Paravicini [55 p342] also introduce B as a parameter and refer to its effects as a "renormalization" due to local field effects.

$$B_{\text{ion}} = \frac{1}{3\varepsilon_0} N\alpha_{\text{ion}}. \qquad (10.172)$$

Then the static dielectric constant is given by

$$\frac{\varepsilon(0)}{\varepsilon_0} = \frac{1 + 2[B_{\text{el}} + B_{\text{ion}}(0)]}{1 - [B_{\text{el}} + B_{\text{ion}}(0)]}, \qquad (10.173)$$

while for high frequency

$$\frac{\varepsilon(\infty)}{\varepsilon_0} = \frac{1 + 2B_{\text{el}}}{1 - B_{\text{el}}}. \qquad (10.174)$$

We return to the equations of motion of the ions in the electric field—which in fact is a local electric field, and it should be so written. After a little manipulation we can write

$$\mu \ddot{v}_1 = \frac{\mu G}{M_1}(v_2 - v_1) + \frac{\mu}{M_1} e E_{\text{loc}}, \qquad (10.175)$$

$$\mu \ddot{v}_2 = \frac{\mu G}{M_2}(v_1 - v_2) - \frac{\mu}{M_2} e E_{\text{loc}}. \qquad (10.176)$$

Using

$$\frac{\mu}{M_1} + \frac{\mu}{M_2} = 1, \qquad (10.177)$$

we can write

$$\mu(\ddot{v}_1 - \ddot{v}_2) + G(v_1 - v_2) = e E_{\text{loc}}. \qquad (10.178)$$

We first discuss this for transverse optical phonons.[5] Here, the polarization is perpendicular to the direction of travel, so

$$E_{\text{loc}} = \frac{P}{3\varepsilon_0} \qquad (10.179)$$

in the absence of an external field. Now the polarization can be written as

$$P = P_{\text{el}} + P_{\text{ion}} = N(\alpha_- + \alpha_+) E_{\text{loc}} + Nev, \quad v = v_1 - v_2, \qquad (10.180)$$

$$P = N(\alpha_+ + \alpha_-)\frac{P}{3\varepsilon_0} + Nev, \qquad (10.181)$$

and

$$P = \frac{Nev}{1 - B_{\text{el}}}, \qquad (10.182)$$

---

[5] A nice picture of transverse and longitudinal waves is given by Cochran [10.7].

so the local field becomes

$$E_{loc} = \frac{1}{3\varepsilon_0} \frac{Nev}{1-B_{el}}. \tag{10.183}$$

The equation of motion can be written

$$\mu\ddot{v} + Gv = \frac{1}{3\varepsilon_0} \frac{Ne^2 v}{1-B_{el}}. \tag{10.184}$$

Seeking sinusoidal solutions of the form $v = v_0 \exp(-i\omega_T t)$ of the same frequency dependence as the local field, then

$$\omega_T^2 = \frac{G}{\mu}\left[1 - \frac{(1/3\varepsilon_0)(Ne^2/G)}{1-B_{el}}\right]. \tag{10.185}$$

We suppose α_{ion} is the static polarizability so that

$$\frac{ev}{E_{loc}} = \alpha_{ion} = \frac{e^2}{G} \tag{10.186}$$

form the equations of motion. So,

$$B_{ion}(0) = \frac{1}{3\varepsilon_0} N\alpha_{ion} = \left(\frac{1}{3\varepsilon_0}\right)\left(\frac{Ne^2}{G}\right), \tag{10.187}$$

or

$$\omega_T^2 = \frac{G}{\mu}\left[1 - \frac{B_{ion}(0)}{1-B_{el}}\right]. \tag{10.188}$$

For the longitudinal case with $q \parallel P$ we have

$$E_{loc} = -\frac{P}{\varepsilon_0} + \frac{1}{3}\frac{P}{\varepsilon_0} = -\frac{2}{3}\frac{P}{\varepsilon_0}. \tag{10.189}$$

So,

$$P = P_{el} + P_{ion} = N(\alpha_+ + \alpha_-)\left(-\frac{2}{3}\frac{P}{\varepsilon_0}\right) + Nev = -2B_{el}P + Nev. \tag{10.190}$$

Then, we obtain the equation of motion,

$$\mu\ddot{v} + Gv = -\frac{2}{3\varepsilon_0}\frac{Ne^2 v}{1+2B_{el}}, \tag{10.191}$$

so

$$\omega_L^2 = \frac{G}{\mu}\left[1 + \frac{(2/3\varepsilon_0)(Ne^2/G)}{1+2B_{el}}\right]. \tag{10.192}$$

By the same reasoning as before, we obtain

$$\omega_L^2 = \frac{G}{\mu}\left[1 + \frac{2B_{ion}(0)}{1+2B_{el}}\right]. \tag{10.193}$$

Thus, we have shown that, in general

$$\omega_L^2 = \frac{G}{\mu}\left[1 + \frac{2[B_{el} + B_{ion}(0)]}{1+2B_{el}}\right], \tag{10.194}$$

and

$$\omega_T^2 = \frac{G}{\mu}\left[1 - \frac{B_{el} + B_{ion}(0)}{1-B_{el}}\right]. \tag{10.195}$$

Therefore, using (10.173), (10.174), (10.194), and (10.195) we find

$$\frac{\varepsilon(\infty)}{\varepsilon(0)} = \frac{\omega_T^2}{\omega_L^2}. \tag{10.196}$$

This is the Lyddane–Sachs–Teller Relation, which was mentioned in Sect. 9.3.2, and also derived in Section 4.3.3 (see 4.79) as an aside in the development of polarons. Compare also Kittel [59, 3rd edn, 1966, p393ff] who gives a table showing experimental confirmation of the LST relation. The original paper is Lyddane et al [10.20]. An equivalent derivation is given by Born and Huang [10.2 p80ff].

For intermediate frequencies $\omega_T < \omega < \omega_L$,

$$\frac{\varepsilon(\omega) - \varepsilon(\infty)}{\varepsilon_0} = \frac{1 + 2[B_{el} + B_i(\omega)]}{1 - [B_{el} + B_i(\omega)]} - \frac{1 + 2B_{el}}{1 - B_{el}}, \tag{10.197}$$

or

$$\frac{\varepsilon(\omega)}{\varepsilon_0} = \frac{\varepsilon(\infty)}{\varepsilon_0} + \frac{3}{[1 - B_{el} - B_i(\omega)](1 - B_{el})} B_i(\omega). \tag{10.198}$$

We need an expression for $B_i(\omega)$. With an external field since only transverse phonons are strongly interacting

$$\mu\ddot{v} + Gv = \frac{1}{3\varepsilon_0}\frac{Ne^2}{1 - B_{el}} v + eE, \tag{10.199}$$

so

$$B_i(0) = \frac{1}{3\varepsilon_0} N\alpha_i(0) = \frac{1}{3\varepsilon_0} N \frac{ev}{E_{\text{loc}}} = \frac{1}{3\varepsilon_0} N \frac{e^2}{G}. \tag{10.200}$$

Seeking a solution of the form $v = v_0 \exp(-i\omega t)$ we get

$$\omega_T^2 v \mu + Gv - \frac{Gv}{1 - B_{\text{el}}} B_i(0) = eE. \tag{10.201}$$

So,

$$\omega_T^2 = \left(\frac{G}{\mu}\right)\left(1 - \frac{B_i(0)}{1 - B_{\text{el}}}\right), \tag{10.202}$$

or

$$\mu(\omega_T^2 - \omega^2)v = eE. \tag{10.203}$$

So,

$$\alpha_i(\omega) = \frac{ev}{E_{\text{loc}}} = \frac{e}{E_{\text{loc}}} \frac{eE}{\mu(\omega_T^2 - \omega^2)}. \tag{10.204}$$

Using the local field relations, we have

$$\begin{aligned}E_{\text{loc}} &= E + \frac{P}{3\varepsilon_0} = E + \frac{1}{3\varepsilon_0} \frac{Nev}{1 - B_{\text{el}}} \\ &= E + \frac{1}{3\varepsilon_0} \frac{Ne}{1 - B_{\text{el}}} \frac{eE}{\mu} \frac{1}{(\omega_T^2 - \omega^2)},\end{aligned} \tag{10.205}$$

so,

$$\frac{E}{E_{\text{loc}}} = \left(\frac{1}{1+F}\right)\left(\frac{e^2}{\mu} \frac{1}{\omega_T^2 - \omega^2}\right), \tag{10.206}$$

where,

$$F = \frac{G}{\mu} \frac{B_i(0)}{(1 - B_{\text{el}})(\omega_T^2 - \omega^2)}. \tag{10.207}$$

Or,

$$B_i(\omega) = \frac{1}{3\varepsilon_0} N\alpha_i(\omega) = (1 - B_{\text{el}}) \frac{F}{1+F}, \tag{10.208}$$

or

$$\frac{\varepsilon(\omega)}{\varepsilon_0} = \frac{\varepsilon(\infty)}{\varepsilon_0} + \frac{3(1-B_{el})F/(1+F)}{(1-B_{el})[(1-B_{el})-(1-B_{el})F/(1+F)]}, \quad (10.209)$$

or

$$\frac{\varepsilon(\omega)}{\varepsilon_0} = \frac{\varepsilon(\infty)}{\varepsilon_0} + \frac{3}{1-B_{el}}\frac{G}{\mu}\frac{B_i(0)}{(1-B_{el})(\omega_T^2 - \omega^2)}. \quad (10.210)$$

Defining

$$c = 3\frac{G}{\mu}\frac{B_i(0)}{(1-B_{el})^2}, \quad (10.211)$$

after some algebra we also find

$$\omega_T^2 + \frac{c\varepsilon_0}{\varepsilon(\infty)} = \omega_L^2. \quad (10.212)$$

10.10.2 Summary of the Properties of ε(q, ω) (B)

Since $n = \varepsilon^{1/2}$ with $\sigma = 0$ (see (10.8)), if $\varepsilon < 0$, one gets high reflectivity (by (10.15) with n_c pure imaginary). Note if

$$\omega_T^2 < \omega^2 < \omega_T^2 + \frac{c\varepsilon_0}{\varepsilon(\infty)}, \quad (10.213)$$

then $\varepsilon(\omega) < 0$, since by (10.210), (10.211), and (10.212) we can also write

$$\varepsilon(\omega) = \varepsilon(\infty)\frac{\omega_L^2 - \omega^2}{\omega_T^2 - \omega^2},$$

and one has high reflectivity ($R \rightarrow 1$). Thus, one expects a whole band of forbidden nonpropagating electromagnetic waves. ω_T is called the Restrahl frequency and the forbidden gap extends from ω_T to ω_L. We only get Restrahl absorption in semiconductors that show ionic character; it will not happen in Ge and Si. We give some typical values in Table 10.2. See also Born and Huang [2, p118].

Table 10.2. Selected lattice frequencies and dielectric constants

Crystal	ω_T (cm^{-1})	ω_L (cm^{-1})	$\varepsilon(0)$ (cgs)	$\varepsilon(\infty)$ (cgs)
InSb	185	197	17.88	15.68
GaAs	269	292	12.9	10.9
NaCl	164	264	5.9	2.25
KBr	113	165	4.9	2.33
LiF	306	659	8.8	1.92
AgBr	79	138	13.1	4.6

From Anderson HL (ed), *A Physicists Desk Reference* 2nd edn, American Institute of Physics, Article 20: Frederikse HPR, Table 20.02.B.1 p.312, 1989, with permission of Springer-Verlag. Original data from Mitra SS, *Handbook on Semiconductors*, Vol 1, Paul W (ed), North-Holland, Amsterdam, 1982, and from *Handbook of Optical Constants of Solids*, Palik ED (ed), Academic Press, Orlando, FL, 1985.

10.10.3 Summary of Absorption Processes: General Equations (B)

Much of what we have discussed can be summed up in Fig. 10.13. Summary expressions for the dielectric constants are given in (10.67) and (10.68). See also Yu and Cardona [10.27, p. 251], and Cohen [10.8] as well as Cohen and Chelikowsky [10.9, p31].

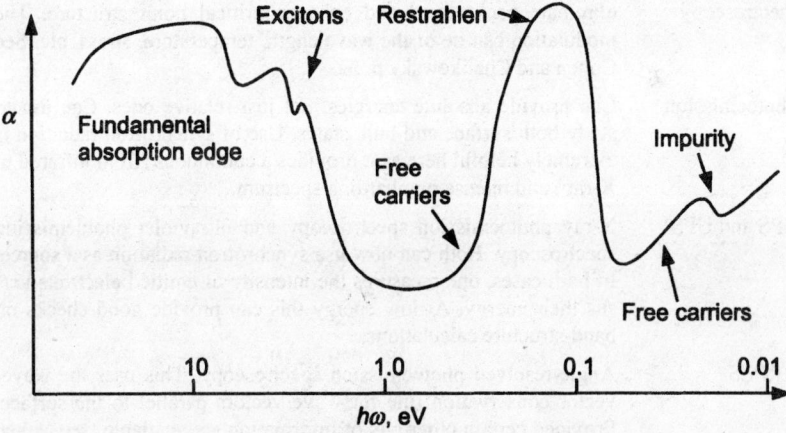

Fig. 10.13. Sketch of absorption coefficient of a typical semiconductor such as GaAs. Adapted from Elliott and Gibson [10.11, p. 208]

10.11 Optical Emission, Optical Scattering and Photoemission (B)

10.11.1 Emission (B)

We will only tread lightly on these topics, but they are important to mention. For example, photoemission (the ejection of electrons from the solid due to photons) can often give information that is not readily available otherwise, and it may be easier to measure than absorption. Photoemission can be used to study electron structure. Two important kinds are XPS – X-ray photoemission from solids, and UPS ultraviolet photoemission. Both can be compared directly with the valence-band density of states. See Table 10.3. A related discussion is given in Sect. 12.2.

Table 10.3. Some optical experiments on solids

High-energy reflectivity	The low-energy range below about 10 eV is good for investigating transitions between valence and conduction bands. The use of synchrotron radiation allows one to consider much higher energies that can be used to probe transitions between the conduction-band and core states. Since core levels tend to be well defined, such measurements provide direct data about conduction band states including critical point structure. The penetration depth is large compared to the depth of surface irregularities and thus this measurement is not particularly sensitive to surface properties. Only relative energy values are measured.
Modulation spectroscopy	This involves measuring derivatives of the dielectric function to eliminate background and enhance critical point structure. The modulation can be of the wavelength, temperature, stress, etc. See Cohen and Chelikowsky p. 52.
Photoemission	Can provide absolute energies, not just relative ones. Can use to study both surface and bulk states. Use of synchrotron radiation is extremely helpful here as it provides a continuous (from infrared to X-ray) and intense bombarding spectrum.
XPS and UPS	X-ray photoemission spectroscopy and ultraviolet photoemission spectroscopy. Both can now use synchrotron radiation as a source. In both cases, one measures the intensity of emitted electrons versus their energy. At low energy this can provide good checks on band-structure calculations.
ARPES	Angle-resolved photoemission spectroscopy. This uses the wave-vector conservation rule for wave vectors parallel to the surface. Provided certain other bits of information are available (see Cohen and Chelikowsky, p.68), information about the band structure can be obtained (see also Sect. 3.2.2).

Reference: Cohen and Chelikowsky [10.8]. See also Brown [10.3].

Also, the topic of emission is important because it involves applications—fluorescent lighting and television are obviously important and based on emission not on absorption. There are perhaps four principal aspects of optical emission. First, there are many types of transitions allowed. A second aspect is the excitation mechanism that positions the electron for emission. Third are the mechanisms that delay emission and give rise to luminescence. Finally, there are those combinations of mechanisms that produce laser action. Luminescence is often defined as light emission that is not due just to the temperature of the emitting body (that is, it is not black-body emission). There are several different kinds of luminescence depending on the source of the energy. For example, one uses the term photoluminescent if the energy comes from IR, visible, or UV light. Although there seems to be no universal agreement on the terms phosphorescence and fluorescence, phosphorescence is used for delayed light emission and fluorescence sometimes just means the light emitted due to excitation. Metals have high absorption at most optical frequencies, and so when we deal with photoemission, we normally deal with semiconductors and insulators.

10.11.2 Einstein *A* and *B* Coefficients (B, EE, MS)

We give now a brief discussion of emission as it pertains to the lasers and masers. The MASER (microwave amplification by stimulated emission of radiation) was developed by C. H. Townes in 1951, also independently by N. G. Basov and A. M. Prokhorov at about the same time). The first working LASER (light amplification by stimulated emission of radiation) was achieved by T. H. Maiman in 1960 using a ruby crystal. Ruby is sapphire (Al_2O_3) with a small amount of chromium impurities.

The Einstein *A* and *B* coefficients are easiest to discuss in terms of discrete levels, and exhibit a main idea of lasers. See Fig. 10.14. For a complete discussion of how lasers produce intense, coherent, and monochromatic beams of light see the references on applied physics [32–35]. Let the spontaneous emission and the induced transition rates be defined as follows:

Spontaneous emission $n \to m$ A_{nm}
Induced emission $n \to m$ B_{nm}
Induced absorption $m \to n$ B_{mn}

From the Planck distribution we have for the density of photons

$$\rho(v) = \frac{8\pi h^3 v^2}{c^3} \frac{1}{\exp(hv/kT)-1}. \tag{10.214}$$

Thus, generalizing to band-to-band transitions, we can write the generation rate as

$$G_{mn} = B_{mn} N_m f_m N_n (1 - f_n) \rho(v_{mn}), \tag{10.215}$$

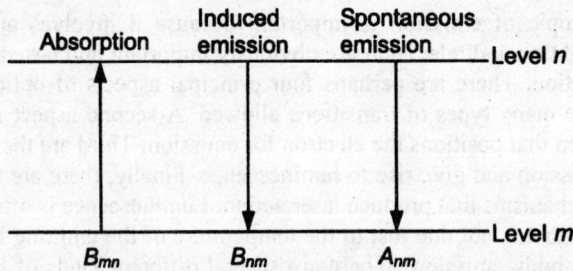

Fig. 10.14. The Einstein A and B coefficients

where N represents the number and f is the Fermi function. Also, we can write the recombination rate as

$$R_{nm} = B_{nm}N_n f_n N_m(1-f_m)\rho(v_{mn}) + A_{nm}N_n f_n N_m(1-f_m). \quad (10.216)$$

In steady state, $G_{mn} = R_{nm}$. From the Fermi function we can show

$$\frac{f_m(1-f_n)}{f_n(1-f_m)} = \exp\left(\frac{E_n - E_m}{kT}\right). \quad (10.217)$$

Thus, since $B_{nm} = B_{mn}$, we have from (10.215) and (10.216)

$$B_{nm}\rho(v)\left[\exp\left(\frac{E_n - E_m}{kT}\right) - 1\right] = A_{nm}, \quad (10.218)$$

and

$$E_n - E_m = hv_{mn}, \quad (10.219)$$

we find for the ratio between the A and B coefficients,

$$\frac{A}{B} = \frac{8\pi n^3 v^2}{c^3}. \quad (10.220)$$

10.11.3 Raman and Brillouin Scattering (B, MS)

The laser has facilitated many optical experiments such as, for example, Raman scattering. We now discuss briefly Raman and Brillouin scattering. One refers to the inelastic scattering of light by phonons as Raman scattering if optical phonons are involved, and Brillouin scattering if acoustic phonons are. If phonons are emitted one speaks of the Stokes line and if absorbed as the anti-Stokes line. Note that these processes are two-photon processes (there is one photon in and one out). Raman and Brillouin scattering are made possible by the strain dependence of the electronic polarization. The relevant conservation equations can be written:

$$\omega_k = \omega_{k'} \pm \omega_K, \quad (10.221)$$

$$k = k' \pm K, \quad (10.222)$$

where ω and k refer to photons and ω_K and K to phonons. Since the value of the wave vector of photons is very small, the phonon wave vector can be at most twice that of the photon, and hence is very small compared to the Brillouin zone width. Hence, the energy of the optical phonons is very nearly constant at the optical phonon energy of zero wave vectors.

Brillouin scattering from longitudinal acoustic waves can be viewed as scattering from a density grating that moves at the speed of sound. Raman scattering can be used to determine the frequency of the zone-center phonon modes. Since the processes depend on phonons, a temperature dependence of the relative intensity of the Stokes and anti-Stokes lines can be predicted.

A simple idea as to the temperature dependence of the Stokes and the anti-Stokes lines is as follows [23, p. 323]. (For a more complete analysis see [10.2, p272]. See also Fox op. cit. p222.)

Stokes: $\qquad Intensity \propto \left|\langle n_K + 1 | a_k^\dagger | n_K \rangle\right|^2 \propto n_K + 1$, $\qquad$ (10.223)

Anti-Stokes: $\qquad Intensity \propto \left|\langle n_K - 1 | a_k | n_K \rangle\right|^2 \propto n_K$ $\qquad$ (10.224)

$$\frac{I(\omega + \omega_K)}{I(\omega - \omega_K)} = \frac{n_K}{n_K + 1} = \exp(-\hbar\beta\Omega) \, . \qquad (10.225)$$

A diagram of Raman/Brillouin scattering involving absorption of a phonon (anti-Stokes) is shown in Fig. 10.15. As we have shown above, the intensity of the anti-Stokes line goes to zero at absolute zero, simply because there are no phonons available to absorb.

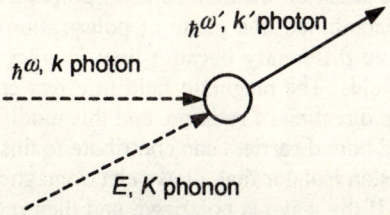

Fig. 10.15. Raman and Brillouin scattering. The diagram shows absorption. Acoustic phonons are involved for Brillouin scattering, and optical phonons for Raman

An expression for the frequency shift of both of these processes is now given. For absorption

$$k + K = k', \qquad (10.226)$$

and

$$\omega_k + \omega_K = \omega_{k'} \, . \qquad (10.227)$$

Assuming the wavelength of the phonon is much greater than the wavelength of light, we have $k \cong k'$. If we let θ be the angle between k and k', then it is easy to see that

$$K = 2k \sin\left(\frac{\theta}{2}\right). \tag{10.228}$$

The shift in frequency of the scattered light is ω_K. For Brillouin scattering, with $V \cong \omega_K/K$ being the phonon velocity and n being the index of refraction, one finds

$$\omega_K = \left(\frac{2n\omega_k V}{c}\right) \sin\left(\frac{\theta}{2}\right), \tag{10.229}$$

and thus n can be determined. When phonons are absorbed, the photons are shifted up in frequency by ω_K, and when phonons are emitted, they are shifted down in frequency by this amount.

10.12 Magneto-Optic Effects: The Faraday Effect (B, EE, MS)

The rotation of the plane of polarization of plane-polarized light, which is propagating along an external magnetic field, is called the Faraday effect.[6] Substances for which this occurs naturally without an applied field are said to be optically active. One way of understanding this effect is to resolve the plane-polarized light into counterrotating circularly polarized components. Each component will have (see below) a different index of refraction and so propagates at a different speed, thus when they are recombined, the plane of polarization has been rotated. The two components behave differently because they interact with electrons via the two rotating electric fields. The magnetic field in effect causes a different radial force depending on the direction of rotation, and this modifies the effective spring constant. Both free and bound carriers can contribute to this effect. A major use of the Faraday effect is as an isolator that allows electromagnetic waves to propagate only in one direction. If the wave is polarized, and then rotated by 45 degrees by the Faraday rotator, any wave reflected back through the rotator will be rotated another 45 degrees in the same direction and hence be at 90 degrees to the polarizer and so cannot travel that way.

A simple classical picture of the effect works fairly well. We assume an electron bound by an isotropic Hooke's law spring in an electric and a magnetic field. By Newton's second law ($e > 0$):

$$m\ddot{r} = -kr - e(E + \dot{r} \times B). \tag{10.230}$$

---

[6] A comprehensive treatment has been given by Caldwell [10.5].

10.12 Magneto-Optic Effects: The Faraday Effect (B, EE, MS)

Defining $\omega_0^2 = k/m$ (a different use of ω_0 from that in (10.108)!), letting $\mathbf{B} = B\mathbf{k}$, and assuming the electric field is in the (x,y)-plane, if we write out the x and y components of the above equation we have

$$\ddot{x} + \frac{e}{m}\dot{y}B + \omega_0^2 x = -\frac{e}{m}E_x, \qquad (10.231)$$

$$\ddot{y} - \frac{e}{m}\dot{x}B + \omega_0^2 y = -\frac{e}{m}E_y. \qquad (10.232)$$

We define $w_\pm = x \pm iy$ and $E_\pm = E_x \pm iE_y$. Note that the real and imaginary parts of E_+ correspond to "right-hand waves" (thumb along z) and the real and imaginary parts of E_- correspond to "left-hand waves".

We assume for the two circularly polarized components,

$$E_\pm = E_0 \exp[\pm i(\omega t - k_\pm z)], \qquad (10.233)$$

which when added together gives a plane-polarized beam along x at $z = 0$. We seek steady-state solutions for which

$$w_\pm = \exp[\pm i(\omega t - k_\pm z)]. \qquad (10.234)$$

Substituting we find

$$w_\pm = \frac{(-e/m)E_\pm}{(\omega_0^2 - \omega^2) \pm (e/m)B\omega}. \qquad (10.235)$$

The polarization P is given by

$$\mathbf{P} = -Ne\mathbf{r}, \qquad (10.236)$$

where N is the number of electrons/volume:

$$P_\pm = \frac{(Ne^2/m)E_\pm}{(\omega_0^2 - \omega^2) \pm (e/m)B\omega}. \qquad (10.237)$$

It is convenient to write this in terms of two special frequencies. The cyclotron frequency is

$$\omega_c = \frac{eB}{m}, \qquad (10.238)$$

and the plasma frequency is

$$\omega_p = \sqrt{\frac{Ne^2}{m\varepsilon_0}}. \qquad (10.239)$$

Thus (10.237) can be written

$$P_\pm = \frac{\varepsilon_0 \omega_p^2 E_\pm}{(\omega_0^2 - \omega^2) \pm \omega_c \omega}. \qquad (10.240)$$

As usual we write

$$D_\pm = \varepsilon_0 E_\pm + P_\pm, \qquad (10.241)$$

or
$$D_\pm = \varepsilon_\pm E_\pm. \qquad (10.242)$$

Using (10.240), (10.241), (10.242), and
$$n_\pm^2 = \frac{\varepsilon_\pm}{\varepsilon_0}, \qquad (10.243)$$

we find
$$n_\pm^2 = 1 + \frac{\omega_p^2}{(\omega_0^2 - \omega^2) \pm \omega_c \omega}. \qquad (10.244)$$

The total angle that the polarization turns through is
$$\Theta = \frac{1}{2}(\Theta_+ - \Theta_-), \qquad (10.245)$$

where in a distance l (and with period of rotation T)
$$\Theta_\pm = 2\pi \frac{l}{v_\pm T} = \frac{\omega l}{v_\pm} = \frac{\omega l}{c} n_\pm. \qquad (10.246)$$

Thus,
$$\Theta = \frac{1}{2} \frac{\omega l}{c}(n_+ - n_-). \qquad (10.247)$$

If
$$\omega_p^2 \ll (\omega_0^2 - \omega^2) \pm \omega_c \omega, \qquad (10.248)$$

then
$$n_\pm \cong 1 + \frac{1}{2} \frac{\omega_p^2}{(\omega_0^2 - \omega^2) \pm \omega_c \omega}. \qquad (10.249)$$

So, combining (10.247) and (10.249)
$$\Theta = -\frac{\omega_c \omega_p^2 \omega^2 l}{2c} \frac{1}{(\omega_0^2 - \omega^2)^2 - \omega_c^2 \omega^2}. \qquad (10.250)$$

For free carriers $\omega_0 = 0$, we find if $\omega_c \ll \omega$,
$$\Theta = -\frac{l\omega_p^2 \omega_c}{2c\omega^2}. \qquad (10.251)$$

Note a positive B (along z) with propagation along z will give a negative Verdet constant (the proportionality between the angle and the product of the field and path length) and a clockwise Θ when it is viewed along (i.e. in the direction of) $-z$.

Problems

10.1 In a short paragraph explain what photoconductivity is, and describe any photoconductivity experiment.

10.2 Describe, very briefly, the following magneto-optical effects: (a) Zeeman effect, (b) inverse Zeeman effect, (c) Voigt effect, (d) Cotton–Mouton effect, (e) Faraday effect, (f) Kerr magneto-optic effect.

Describe briefly the following electro-optic effects: (g) Stark effect, (h) inverse Stark effect, (i) electric double refraction, (j) Kerr electro-optic effect.

Descriptions of these effects can be found in any good optics text.

10.3 Given a plane wave $E = E_0 \exp[i(k \cdot r - \omega t)]$ normally incident on a surface, detail the assumptions, conditions and steps to show $n_c E_0 = E_1 - E_2$, (cf. (10.26)).

10.4 (a) From $[x, p_x] = i\hbar$, show that

$$[H, \hat{e} \cdot r] = -\frac{i\hbar}{m} \hat{e} \cdot p,$$

(b) For $\hat{e} = \hat{i}$, show the oscillator strength f_{ij} obeys the sum rule $\sum_j f_{ij} = 1$.

10.5 For intermediate frequencies $\omega_T < \omega < \omega_L$, given (by (10.198))

$$\frac{\varepsilon(\omega)}{\varepsilon_0} - \frac{\varepsilon(\infty)}{\varepsilon_0} = \frac{3 B_{\text{ion}}(\omega)}{[1 - B_{\text{el}} - B_{\text{ion}}(\omega)](1 - B_{\text{el}})},$$

and the equation of motion (by (10.199))

$$\mu \ddot{v} + G v = \frac{1}{3\varepsilon_0} \frac{N e^2}{1 - B_{\text{el}}} v + eE,$$

derive the equation

$$\omega_T^2 + \frac{c \varepsilon_0}{\varepsilon(\infty)} = \omega_L^2,$$

where c is a defined as constant within the derivation. In this process, show intermediate derivations for the following equations defining constants as necessary:

$$\mu(\omega_T^2 - \omega^2) v = eE,$$

$$\frac{E}{E_{\text{loc}}} = \frac{1}{1 + F},$$

$$\varepsilon(\omega) = \varepsilon(\infty) + \frac{c \varepsilon_0}{\omega_T^2 - \omega^2}.$$

10.6 This problem fills in the details of Sect. 10.11.2.
   (a) Describe the factors that make up the generation rate
   $$G_{mn} = B_{mn} N_m f_m N_n (1 - f_n) \rho(v_{mn}).$$
   (b) Show from the Fermi function that
   $$\frac{f_n(1 - f_m)}{f_m(1 - f_n)} = \exp\left(\frac{E_m - E_n}{kT}\right).$$
   (c) Starting from $G_{mn} = R_{nm}$, show that
   $$\frac{A}{B} = \frac{8\pi n^3 v^2}{c^3}.$$

11 Defects in Solids

11.1 Summary About Important Defects (B)

A defect in a solid is any deviation from periodicity in the solid. All solids have defects, but for some applications, they can be neglected, while for others, the defects can be very important. By now, simple defects are well understood, but for more complex defects, a considerable amount of fundamental work remains to be accomplished for a thorough understanding.

Some discussion of defects has already been made. In Chap. 2, the effects of defects on the phonon spectrum of a one-dimensional lattice were discussed, whereas in Chap. 3 the effects of defects on the electronic states in a one-dimensional lattice were considered. In the semiconductor chapter, donor and acceptor states were used, but some details were postponed until this chapter.

There is only one way to be perfect, but there are numerous ways to be imperfect. Thus, we should not be surprised that there are many kinds of defects. The mere fact that no crystal is infinite is enough to introduce *surface* defects, which could be electronic or vibrational. Electronic surface states are classified as Tamm states (if they are due to a different potential in the last unit cell at the surface edge with atoms far apart) or Shockley states (the cells remain perfectly repetitive right up to the edge, but with atoms close enough so as to have band crossing[1]). Whether or not Tamm and Shockley states should be distinguished has been the subject of debate that we do not wish to enter into here. In any case, the atoms on the surface are not in the same environment as interior atoms, and so, their contribution to the properties of the solid must be different. The surface also acts to scatter both electrons and phonons. The properties of surfaces are of considerable practical importance. All input and output to solids goes through the surfaces. Thermionic and cold field emission from surfaces is discussed in Sects. 11.7 and 11.8. Surface reconstruction is discussed in Chap. 12. Another important application of surface physics is to better understand corrosion.

Besides surfaces, we briefly review other ways crystals can have defects, starting with *point defects* (see Crawford and Slifkin [11.7]). When a crystal is grown, it is not likely to be pure. Foreign impurity atoms will be present, leading to *substitutional* or *interstitial* defects (see Fig. 11.1). Interstitial atoms can originate from atoms of the crystal as well as foreign atoms. These may be caused by thermal effects (see below) or may be introduced artificially by *radiation damage*. Radiation damage (or thermal effects) may also cause vacancies. Also,

---

[1] See, e.g., Davison and Steslicka [11.8].

when a crystal is composed of more than one element, these elements may not be exactly in their proper chemical proportions. The stoichiometric derivations can result in vacancies as well as *antisite defects* (an atom of type A occupying a site normally occupied by an atom of type B in an AB compound material).

Vacancies are always present in any real crystal. Two sorts of point defects involving vacancies are so common that they are given names. These are the Schottky and Frenkel defects, shown for an ionic crystal in Fig. 11.2. Defects such as *Schottky* and *Frenkel* defects are always present in any real crystal at a finite temperature in equilibrium. The argument is simple. Suppose we assume that the free energy $F = U - TS$ has a minimum in equilibrium. The defects will increase U, but they cause disorder, so they also cause an increase in the entropy S. At high enough temperatures, the increase in U can be more than compensated by the decrease in $-TS$. Thus, the stable situation is the situation with defects.

Mass transport is largely possible because of defects. Vacancies can be quite important in controlling diffusion (discussed later in Sect. 11.5). Ionic conductivity studies are important in studying the motion of lattice defects in ionic crystals. *Color centers* are another type of point defect (or complex of point defects). We will discuss them in a little more detail later (Sect. 11.4). Color centers are formed by defects and their surrounding potential, which trap electrons (or holes).

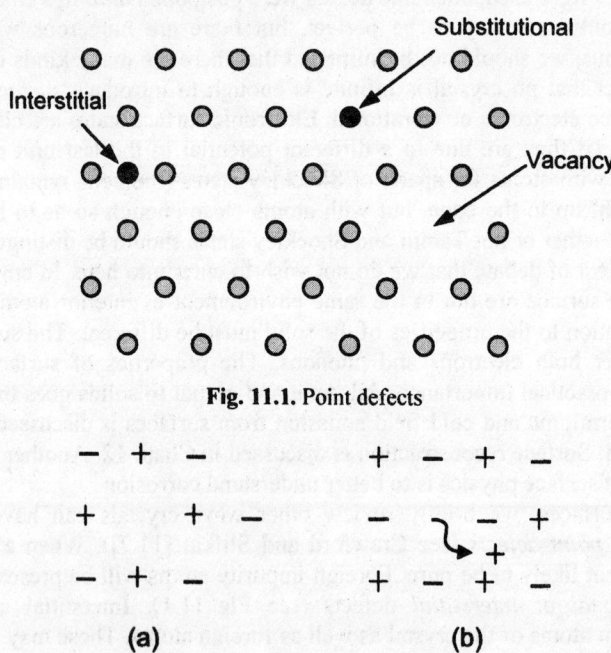

Fig. 11.1. Point defects

Fig. 11.2. (a) Schottky and (b) Frenkel defects

Table 11.1. Summary of common crystal lattice defects

Point defects	Comments
Foreign atoms	Substitutional or interstitial
Vacancies	Schottky defect is vacancy with atom transferred to surface
Antisite	Example: A on a B site in an AB compound
Frenkel	Vacancy with foreign atom transferred to interstice
Color centers	Several types – F is vacancy with trapped electron (ionic crystals – see Sect. 11.4
Donors and acceptors	Main example are shallow defects in semiconductors – see Sects. 11.2 and 11.3
Deep levels in semiconductors	See Sects. 11.2 and 11.3

Line defects	Comments
Dislocations	Edge and screw – see Sect. 11.6 – General dislocation is a combination of these two

Surface defects	Comments
External	
Tamm and Shockley electronic states	See Sect. 11.1
Reconstruction	See Sect. 12.2
Internal	
Stacking fault	Example: a result of an error in growth[2]
Grain boundaries	Tilt between adjacent crystallites – can include low angle (with angle, in radians, being the ratio of the Burgers vector (magnitude) to the dislocation spacing) to large angle (which includes twin boundaries)
Heteroboundary	Between different crystals

Volume defects	Comments
Many examples	Three-dimensional precipitates and complexes of defects

See, e.g., Henderson [11.16].

Vacancies, substitutional atoms, and interstitial atoms are all point defects. Surfaces are planar defects. There is another class of defects called *line defects*. *Dislocations* are important examples of line defects, and they will be discussed

---

[2] A fcc lattice along (1,1,1) is composed of planes ABCABC etc. If an A plane is missing then we have ABCBCABC, etc. This introduces a local change of symmetry. See, e.g., Kittel [23, p. 18].

later (Sect. 11.6). They are important for determining how easily crystals deform and may also relate to crystal growth.

Finally, there are defects that occur over a whole volume. It is usually hard to grow a *single crystal*. In a single crystal, the lattice planes are all arranged as expected–in a perfectly regular manner. When we are presented with a chunk of material, it is usually in a polycrystal form. That is, many little crystals are stuck together in a somewhat random way. The boundary between crystals is also a two-dimensional defect called a *grain boundary*. We have summarized these ideas in Table 11.1.

11.2 Shallow and Deep Impurity Levels in Semiconductors (EE)

We start by considering a simple chemical model of shallow donor and acceptor defects. We will give a better definition later, but for now, by "shallow", we will mean energy levels near the bottom of the conduction band for donor level and near the top of the valence band for acceptors.

Consider Si^{14} as the prototype semiconductor. In the usual one-electron shell notation, its electron structure is denoted by

$$1s^2 2s^2 2p^6 3s^2 3p^2.$$

There are four valence electrons in the $3s^2 3p^2$ shell, which requires eight to be filled. We think of neighboring Si atoms sharing electrons to fill the shells. This sharing lowers energy and binds the electrons. We speak of covalent bonds. Schematically, in two dimensions, we picture this occurring as in Fig. 11.3. Each line represents a shared electron. By sharing, each Si in the outer shell has eight electrons. This is of course like the discussion we gave in Chap. 1 of the bonding of C to form diamond.

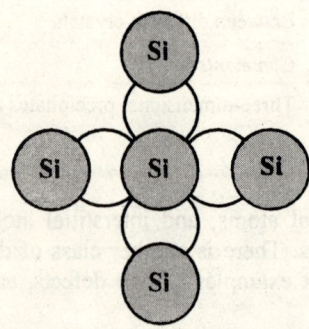

Fig. 11.3. Chemical model of covalent bond in Si

Now, suppose we have an atom, say As, which substitutionally replaces a Si. The sp shell of As has five electrons ($4s^24p^3$) and only four are needed to "fill the shell". Thus, As acts as a donor with an additional loosely bound electron (with a large orbit encompassing many atoms), which can be easily excited into the conduction band at room temperature.

An acceptor like In (with three electrons in its outer sp shell ($5s^25p$)) needs four electrons to complete its covalent bonds. Thus, In can accept an electron from the valence band, leaving behind a hole. The combined effects of effective mass and dielectric constant cause the carrier to be bound much less tightly than in an analogous hydrogen atom. The result is that donors introduce energy levels just below the conduction-band minimum and acceptors introduce levels just above the top of the valence band. We discuss this in more detail below.

In brief, it turns out that the ground-state donor energy level is given by (atomic units, see the appendix)

$$E_n = -\frac{m^*/m}{2n^2\varepsilon^2}, \tag{11.1}$$

where m^*/m is the effective mass ratio typically about 0.25 for Si and ε is the dielectric constant (about 11.7 in Si). E_n in (11.1) is measured from the bottom of the conduction band. Except for the use of the dielectric constant and the effective mass, this is the same result as obtained from the theory of the energy levels of hydrogen. A similar, remarkably simple result holds for acceptor states. These results arise from pioneering work by Kohn and Luttinger as discussed in [11.17], and we develop the basics below.

11.3 Effective Mass Theory, Shallow Defects, and Superlattices (A)

11.3.1 Envelope Functions (A)

The basic model we will use here is called the envelope approximation.[3] It will allow us to justify our treatments of effective mass theory and of shallow defects in semiconductors. With a few more comments, we will then be able to relate it to a simple approach to superlattices, which will be discussed in more detail in Chap. 12.

Let

$$\mathcal{H}_0 = -\frac{\hbar^2}{2m}\nabla^2 + V(r), \tag{11.2}$$

---

[3] Besides [11.17], see also Luttinger and Kohn [11.22] and Madelung [11.23].

where $V(r)$ is the periodic potential. Let $\mathcal{H} = \mathcal{H}_0 + U$ where $U = V_D(r)$ is the extra defect potential. Now, $\mathcal{H}_0 \psi_n(k, r) = E_n \psi_n(k, r)$ and $\mathcal{H}\psi = E\psi$.

We expand the wave function in Bloch functions

$$\psi = \sum_{n,k} a_n(k) \psi_n(k, r), \qquad (11.3)$$

where n is the band index. Also, since $E_n(k)$ is a periodic function in k-space, we can expand it in a Fourier series with the sum restricted to lattice points

$$E_n(k) = \sum_m F_{nm} e^{i k \cdot R_m}. \qquad (11.4)$$

We define an operator $E_n(-i\nabla)$ by substituting $-i\nabla$ for k:

$$\begin{aligned} E_n(-i\nabla) \psi_n(k, r) &= \sum_m F_{nm} e^{R_m \cdot \nabla} \psi_n(k, r) \\ &= \sum_m F_{nm} [1 + R_m \cdot \nabla + \tfrac{1}{2}(R_m \cdot \nabla)^2 + \ldots] \psi_n(k, r) \quad (11.5) \\ &= \sum_m F_{nm} \psi_n(k, r + R_m), \end{aligned}$$

by the properties of Taylor's series. Then using Bloch's theorem

$$E_n(-i\nabla) \psi_n(k, r) = \sum_m F_{nm} e^{i k \cdot R_m} \psi_n(k, r), \qquad (11.6)$$

and by (11.4)

$$E_n(-i\nabla) \psi_n(k, r) = E_n(k) \psi_n(k, r). \qquad (11.7)$$

Substituting (11.3) into $\mathcal{H}\psi = E\psi$, we have (using the fact that ψ_n is an eigenfunction of $\mathcal{H}_0$ with eigenvalue $E_n(k)$)

$$\sum_{n,k} E_n(k) a_n(k) \psi_n(k, r) + \sum_{n,k} V_D a_n(k) \psi_n(k, r) = \sum_{n,k} E a_n(k) \psi_n(k, r). \qquad (11.8)$$

If we use (11.4) and (11.6), this becomes

$$\sum_{n,k} a_n(k) [E_n(-i\nabla) + V_D] \psi_n(k, r) = E\psi. \qquad (11.9)$$

11.3.2 First Approximation (A)

We neglect band-to-band interactions and hence, neglect the summation over n. Dropping n entirely from (11.9), we have

$$\psi = \sum_k a(k) \psi(k, r), \qquad (11.10)$$

and

$$[E(-i\nabla) + V_D] \psi(k, r) = E\psi. \qquad (11.11)$$

11.3.3 Second Approximation (A)

We assume a large extension in real space that means that only a small range of k values are important – say the ones near a parabolic (assumed for simplicity) minimum at $k = 0$ (Madelung op. cit. Chap. 9).

We assume, then,

$$\psi(k,r) = e^{ik \cdot r} u(k,r) \cong e^{ik \cdot r} u(0,r) = e^{ik \cdot r} \psi(0,r) \tag{11.12}$$

so using (11.10) and (11.12),

$$\psi = F(r)\psi(0,r), \tag{11.13}$$

where

$$F(r) = \sum_k a(k) e^{ik \cdot r}. \tag{11.14}$$

So, we have by (11.11)

$$[E(-i\nabla) + V_D]F(r)\psi(0,r) = EF(r)\psi(0,r). \tag{11.15}$$

Using the definition of $E(-i\nabla)$ as in (11.5) we have with n suppressed,

$$\sum_m F_m F(r + R_m)\psi(0, r + R_m) = (E - V_D)F(r)\psi(0,r). \tag{11.16}$$

But, $\psi(0, r + R_m) = \psi(0, r)$, so it can be cancelled. Thus retracing our steps, we have

$$[E(-i\nabla) + V_D]F(r) = EF(r). \tag{11.17}$$

This simply means that a rapidly varying function has been replaced by a slowly varying function $F(r)$ called the "envelope" function. This immediately leads to the concept of shallow donors. Consider the bottom of a parabolic conductor band near $k = 0$ and expand about $k = 0$;

$$E(-i\nabla) = E_c + \left.\frac{dE}{dk}\right|_{k=0}(-i\nabla) + \frac{1}{2}\frac{d^2E}{dk^2}(-i\nabla)^2. \tag{11.18}$$

Also,

$$\left.\frac{dE}{dk}\right|_{k=0} = 0, \tag{11.19}$$

and

$$\left.\frac{d^2E}{dk^2}\right|_{k=0} = \frac{\hbar^2}{m^*}, \tag{11.20}$$

where m^* is the effective mass. Thus, we find

$$E(-i\nabla) = E_c - \frac{\hbar^2}{2m^*}\nabla^2. \tag{11.21}$$

And, if $V_D = e^2/4\pi\varepsilon r$, our resulting equation is

$$\left(-\frac{\hbar^2}{2m^*}\nabla^2 - \frac{e^2}{4\pi\varepsilon r}\right)F(r) = (E - E_c)F(r). \quad (11.22)$$

Except for the use of ε and m^*, these solutions are just hydrogenic wave functions and energies, and so our use of the hydrogenic solution (11.1) is justified.

Now let us discuss briefly electron and hole motion in a perfect crystal. If $U = 0$, we simply write

$$\left(-\frac{\hbar^2}{2m_e^*}\nabla^2 + E_c\right)F = EF. \quad (11.23)$$

On the other hand, suppose U is still 0, but consider a valence band with a maximum at $k = 0$. We then can expand about that point with the following result:

$$E(-i\nabla) = E_v + \frac{1}{2}\frac{d^2E}{dk^2}(-i\nabla)^2. \quad (11.24)$$

Using the hole mass, which has the opposite sign for the electron mass ($m_h = -m_e$), we can write

$$E(-i\nabla) = E_v + \frac{\hbar^2}{2m_h^*}(\nabla^2), \quad (11.25)$$

so the relevant Schrödinger equation becomes

$$\left[\frac{-\hbar^2}{2m_h^*}(\nabla^2) - E_v\right]F = -EF. \quad (11.26)$$

Looking at (11.23) and (11.26), we see how discontinuities in band energies can result in effective changes in the potential for the carriers, and we see why the hole energies are inverted from the electron energies.

Now let us consider superlattices with a set of layers so there is both a lattice periodicity in each layer and a periodicity on a larger scale due to layers (see Sect. 12.6). The layers A and B could for example be laid down as ABABAB...

There are several more considerations, however, before we can apply these results to superlattices. First, we have to consider that if we are to move from a region of one band structure (layer) to another (layer), the effective mass changes since adjacent layers are different. With the possibility of change in effective mass, the Hamiltonian is often written as

$$H = -\frac{\hbar^2}{2}\frac{\partial}{\partial z}\left(\frac{1}{m^*(z)}\frac{\partial}{\partial z}\right) + V(z), \quad (11.27)$$

rather than in the more conventional way. This allows the Hamiltonian to remain Hermitian, even with varying m^*, and it leads to a probability current density of

$$j_z(z) = \frac{\hbar}{2i}\left[\frac{\psi^*}{m^*}\cdot\frac{\partial\psi}{\partial z} - \frac{\psi}{m^*}\cdot\frac{\partial\psi^*}{\partial z}\right], \qquad (11.28)$$

from which we apply the requirement of continuity on ψ and $\partial\psi/(m^*\partial z)$ rather than ψ and $\partial\psi/\partial z$.

We have assumed the thickness of each layer is sufficient that the band structure of the material can be established in this thickness. Basically, we will need both layers to be several monolayers thick. Also, we assume in each layer that the electron wave function is an envelope function (different for different monolayers) times a Bloch function (see (11.13)). Finally, we assume that in each layer $U = U_0$ (a constant appropriate to the layer) and the carrier motion perpendicular to the layers is free-electron-like so,

$$F(r) = \varphi(z)e^{i(k_x x + k_y y)}, \qquad (11.29)$$

$$\left[-\frac{\hbar^2}{2m^*}(k_x^2 + k_y^2) - \frac{\hbar^2 d^2}{2m^* dz^2} + U_0\right]\varphi(z)e^{i(k_x x + k_y y)} = EF, \qquad (11.30)$$

which means

$$\left[-\frac{\hbar^2}{2m^*}\frac{d^2}{dz^2} + U_0\right]\varphi(z) = E_z\varphi(z), \qquad (11.31)$$

where for each layer

$$E = E_z + \frac{\hbar^2}{2m^*}(k_x^2 + k_y^2). \qquad (11.32)$$

There are many, many complications to the above. We have assumed, e.g., that $m^*_{x,y} = m^*_z$ which may not be so in all cases. The book by Bastard [11.1] can be consulted. See also, Mitin et al [11.25].

In semiconductors, shallow levels are often defined as being near a band edge and deep levels as being near the center of the forbidden energy gap. In more recent years, a different definition has been applied based on the nature of the causing agent. Shallow levels are now defined as defect levels produced by the long-range Coulomb potential of the defect and deep levels[4] are defined as being produced by the central cell potential of the defect, which is short ranged. Since

---

[4] See, e.g., Li and Patterson [11.20, 11.21] and references cited therein.

the potential is short range, a modification of the Slater-Koster model, already discussed in Chap. 2, is a convenient starting point for discussing deep defects. Some reasons for the significance of shallow and deep defects are given in Table 11.2. Deep defects are commonly formed by substitutional, interstitial, and antisite atoms and by vacancies.

Table 11.2. Definition and significance of deep and shallow levels

Shallow levels are defect levels produced by the long-range Coulomb potential of defects. Deep Levels are defect levels produced by the central cell potential of defect.

	Deep level	Shallow level
Energy	May or may not be near band edge. Spectrum is not hydrogen-like.	Near band edge. Spectrum is hydrogen-like.
Typical properties	Recombination centers. Compensators. Electron–hole generators.	Suppliers of carriers.

11.4 Color Centers (B)

The study of color centers arose out of the curiosity as to what caused the yellow coloration of rock salt (NaCl) and other coloration in similar crystals. This yellow color was particularly noted in salt just removed from a mine. Becquerel found that NaCl could be colored by placing the crystal near a discharge tube. From a fundamental point of view, NaCl should have an infrared absorption due to vibrations of its ions and an ultraviolet absorption due to excitation of the electrons. A perfect NaCl crystal should not absorb visible light, and should be uncolored. The coloration of NaCl must be due, then, to defects in the crystal. The main absorption band in NaCl occurs at about 4650 Å (the "F"-band). This blue absorption is responsible for the yellow color that the NaCl crystal can have. A further clue to the nature of the absorption is provided by the fact that exposure of a colored crystal to white light can result in the bleaching out of the color. Further experiments show that during the bleaching, the crystal becomes photoconductive, which means that electrons have been promoted to the conduction band. It has also been found that NaCl could be colored by heating it in the presence of Na vapor. Some of the Na atoms become part of the NaCl crystal, resulting in a deficiency of Cl and, hence, Cl^- vacancies. Since photoconductive experiments show that F-band defects can release electrons, and since Cl^- vacancies can trap electrons, it seems very suggestive that the defects responsible for the F-band (called "F-centers") are electrons trapped at Cl^- vacancies. (Note: the "F" comes from the German *farbe*, meaning "color".) This is the explanation accepted today. Of course, since some Cl^- vacancies are always present in a NaCl crystal in thermodynamic equilibrium, any sort of radiation that causes electrons to be knocked into the Cl⁻ vacancies will form F-centers. Thus, we have an explanation of Becquerel's early results as mentioned above.

More generally, color centers are formed when point defects in crystals trap electrons with the resultant electronic energy levels at optical frequencies. Color centers usually form "deep" traps for electrons, rather than "shallow" traps, as donor impurities in semiconductors do, and, their theoretical analysis is complex. Except for relatively simple centers such as F-centers, the analysis is still relatively rudimentary.

Typical experiments that yield information about color centers involve optical absorption, paramagnetic resonance and photoconductivity. The absorption experiments give information about the transition energies and other properties of the transition. Paramagnetic resonance gives wave function information about the trapped electron, while photoconductivity yields information on the quantum efficiency (number of free electrons produced per incident photon) of the color centers.

Mostly by interpretation of experiment, but partly by theoretical analysis, several different color centers have been identified. Some of these are listed below. The notation is

[missing ion | trapped electron | added ion],

where our notation is p ≡ proton, e ≡ electron, − ≡ halide ion, + ≡ alkaline ion, and M^{++} ≡ doubly charged positive ion. The usual place to find color centers is in ionic crystals.

$$[-|e|] = \text{F-center}$$
$$[-|2e|] = \text{F'-center}$$
$$[\mp-|2e|] = \text{M-center (?)}$$
$$[|e|p] = U_2\text{-center}$$
$$[+|e|M^{++}] = Z_1\text{-center (?)}.$$

In Figs 11.4 and 11.5 we give models for two of the less well-known color centers. In these two figures, ions enclosed by boxes indicate missing ions, a dot means an added electron, and a circle includes a substitutionally added ion. We include several references to color centers. See, e.g., Fowler [11.12] or Schulman and Compton [11.28].

Color centers turn out to be surprisingly difficult to treat theoretically with precision. But success has been obtained using modern techniques on, e.g., F centers in LiCl. See, e.g., Louie p. 94, in Chelikowsky and Louie [11.4]. In recent years tunable solid-state lasers have been made using color centers at low temperatures.

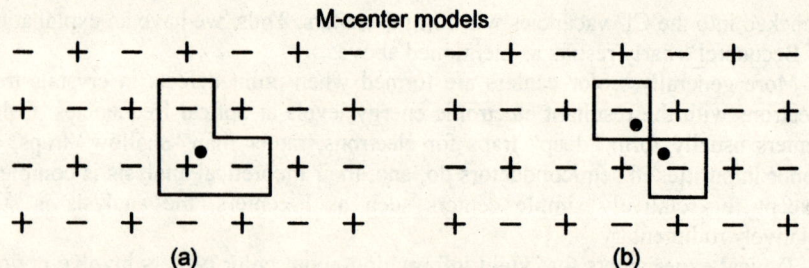

Fig. 11.4. Models of the M-center: (a) Seitz, (b) Van Doorn and Haven. [Reprinted with permission from Rhyner CR and Cameron JR, *Phys Rev* **169**(3), 710 (1968). Copyright 1968 by the American Physical Society.]

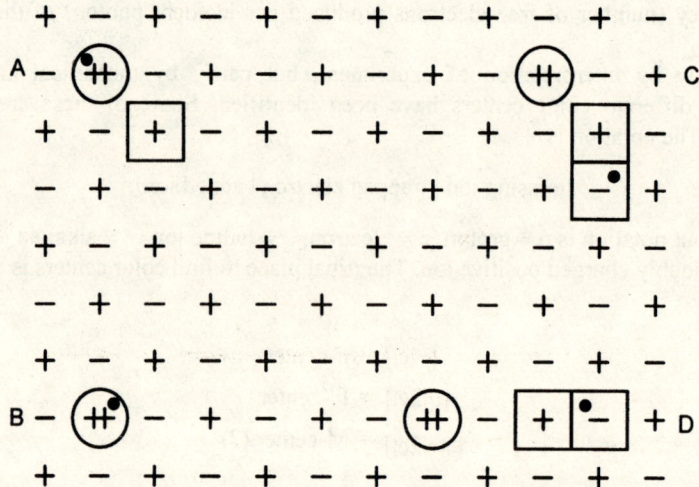

Fig. 11.5. Four proposed models for Z_1-centers [Reprinted with permission from Paus H and Lüty F, *Phys Rev Lett* **20**(2), 57 (1968). Copyright 1968 by the American Physical Society.]

11.5 Diffusion (MET, MS)

Point defects may diffuse through the lattice, while vacancies may provide a mechanism to facilitate diffusion. Diffusion and defects are intimately related, so we give a brief discussion of diffusion. If C is the concentration of the diffusing quantity, Fick's Law says the flux of diffusing quantities is given by

$$J = -D\frac{\partial C}{\partial x}, \tag{11.33}$$

where D is, by definition, the diffusion constant. Combining this with the equation of continuity

$$\frac{\partial J}{\partial x} + \frac{\partial C}{\partial \tau} = 0, \qquad (11.34)$$

leads to the diffusion equation

$$\frac{\partial C}{\partial \tau} = D\frac{\partial^2 C}{\partial x^2}. \qquad (11.35)$$

For solution of this equation, we refer to several well-known treatises as referred to in Borg and Dienes [11.2]. Typically, the diffusion constant is a function of temperature via

$$D = D_0 \exp(-E_0/kT), \qquad (11.36)$$

where E_0 is the activation energy that depends on the process. Interstitial defects moving from one site to an adjacent one typically have much less E_0 than say, vacancy motion. Obviously, the thermal variation of defect diffusion rates has wide application.

11.6 Edge and Screw Dislocation (MET, MS)

Any general dislocation is a combination of two basic types: the *edge* and the *screw* dislocations. The edge dislocation is perhaps the easiest to describe. If we imagine the pages in a book as being crystal planes, then we can visualize an edge dislocation as a book with half a page (representing a plane of atoms) missing. The edge dislocation is formed by the missing half-plane of atoms. The idea is depicted in Fig. 11.6. The motion of edge dislocations greatly reduces the shear strength of crystals. Originally, the shear strength of a crystal was expected to be much greater than it was actually found to be for real crystals. However, all large crystals have dislocations, and the movement of a dislocation can greatly aid the shearing of a crystal. The idea involves similar reasoning as to why it is easier to move a rug by moving a wrinkle through it rather than moving the whole rug. The force required to move the wrinkle is much less.

Crystals can be strengthened by introducing impurity atoms (or anything else), which will block the motion of dislocations. Dislocations themselves can interfere with each other's motions and bending crystals can generate dislocations, which then causes *work hardening*. Long, but thin, crystals called *whiskers* have been grown with few dislocations (perhaps one screw dislocation to aid growth – see below). Whiskers can have the full theoretical strength of ideal, perfect crystals.

The other type of dislocation is called a screw dislocation. *Screw* dislocations can be visualized by cutting a book along A (see Fig. 11.7), then moving the upper half of the book a distance of one page and taping the book into a spiral staircase. Another view of the dislocation is shown in Fig. 11.8 where successive atomic planes are joined together to form one surface similar to the way a kind of Riemann surface can be defined. Screw dislocations greatly aid crystal growth. During the growth, a wandering atom finds two surfaces to "stick" to at the growth edge (or jog) (see Fig. 11.8) rather than only one flat plane. Actual crystals have shown little spirals on their surface corresponding to this type of growth.

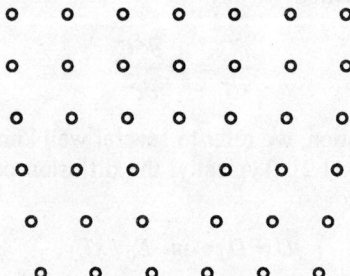

Fig. 11.6. An edge dislocation

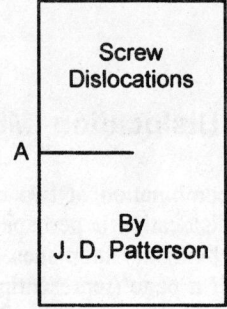

Fig. 11.7. A book can be used to visualize screw dislocations

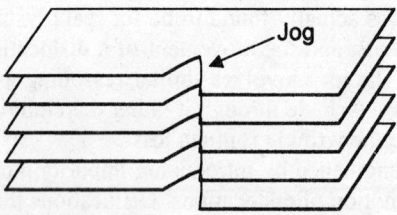

Fig. 11.8. A screw dislocation

We have already mentioned that any general dislocation is a combination of the edge and screw. It is well at this point to make the idea more precise by the use of the *Burgers vector*, which is depicted in Fig. 11.9. We take an atom-to-atom path around a dislocation line. The path is drawn in such a way that it would close on itself as if there were no dislocations. The additional vector needed to close the

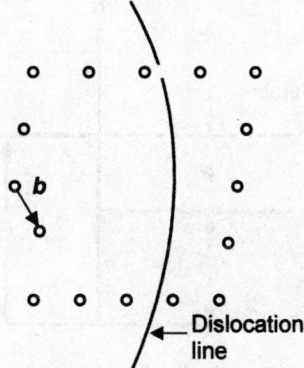

Fig. 11.9. Diagram used for the definition of the Burgers vector *b*

path is the Burgers vector. For a pure edge dislocation, the Burgers vector is perpendicular to the dislocation line; for a pure screw dislocation, it is parallel. In general, the Burgers vector can make any angle with the dislocation line, which is allowed by crystal symmetry. The book by Cottrell [11.6] is a good source of further details about dislocations. See also deWit [11.9].

11.7 Thermionic Emission (B)

We now discuss two very classic and important properties of the surfaces of metals – in this Section thermionic emission and in the next Section cold-field emission.

So far, we have mentioned the role of Fermi–Dirac statistics in calculating the specific heat, Pauli paramagnetism, and Landau diamagnetism. In this Section we will apply Fermi–Dirac statistics to the emission of electrons by heated metals. It will turn out that the fact that electrons obey Fermi–Dirac statistics is relatively secondary in this situation.

It is also possible to have cold (no heating) emission of electrons. Cold emission of electrons is obtained by applying an electric field and allowing the electrons to tunnel out of the metal. This was one of the earliest triumphs of quantum mechanics in explaining hitherto unexplained phenomena. It will be explained in the next section.[5]

For the purpose of the calculation in this section, the surface of the metal will be pictured as in Fig. 11.10. In Fig. 11.10, E_F is the Fermi energy, ϕ is the work function, and E_0 is the barrier height of the potential. The barrier can at least be partially understood by an image charge calculation.

---

[5] A comprehensive review of many types of surface phenomena is contained in Gundry and Tompkins [11.14].

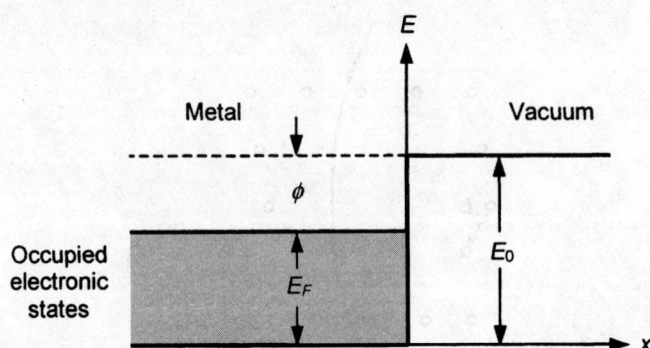

Fig. 11.10. Model of the surface of a metal used to explain thermionic emission

We wish to calculate the current density as a function of temperature for the heated metal. If $n(p)\,d^3p$ is the number of electrons per unit volume in p to $p+d^3p$ and if v_x is the x component of velocity of the electrons with momentum p, we can write the rate at which electrons with momentum from p to $p+d^3p$ hit a unit area in the (x,y)-plane as

$$v_x n(p)\mathrm{d}^3 p = n(p)(p_x/m)\mathrm{d}p_x\mathrm{d}p_y\mathrm{d}p_z . \qquad (11.37)$$

Now

$$n(p)\mathrm{d}^3 p = n(k)\mathrm{d}^3 k = 2f(E)\frac{\mathrm{d}^3 k}{(2\pi)^3} \qquad (11.38)$$

$$= 2f(E)\frac{\mathrm{d}^3 p}{(2\pi\hbar)^3}$$
$$= \frac{2f(E)}{h^3}\mathrm{d}^3 p, \qquad (11.39)$$

so that

$$n(p) = 2f(E)/h^3 . \qquad (11.40)$$

In (11.40), $f(E)$ is the Fermi function and the factor 2 takes the spin degeneracy of the electronic states into account. Finally, we need to consider that only electrons whose x component of momentum p_x satisfies

$$p_x^2/2m > \phi + E_F \qquad (11.41)$$

will escape from the metal.

If we assume the probability of reflection at the surface of the metal is R and is constant (or represents an average value), the emission current density j is e (the electronic charge, here $e > 0$) times the rate at which electrons of sufficient energy strike unit area of the surface times $T_r \equiv 1 - R$. Thus, the emission current density is given by

$$j = -\frac{2e}{h^3}T_r \int_{-\infty}^{\infty}\left(\int_{-\infty}^{\infty}\left\{\int_{\phi+E_F}^{\infty}\frac{d(p_x^2/2m)}{\exp[(E-E_F)/kT]+1}\right\}dp_z\right)dp_y. \quad (11.42)$$

Since $E = (1/2m)(p_x^2 + p_y^2 + p_z^2)$, we can write this expression as

$$j = -kT\frac{2e}{h^3}T_r \int_{-\infty}^{\infty}\left[\int_{-\infty}^{\infty}\left(\int_0^{\infty}\frac{dE'}{e^{E'/kT}\exp\{[\phi+(p_x^2+p_y^2)/2m]/kT\}+1}\right)dp_z\right]dp_y,$$

where $E' = (p_x^2/2m) - \phi - E_F$. But

$$\int_0^{\infty}\frac{dn}{ae^n+1} = \ln(1+a^{-1}),$$

so that

$$j = -(T_r)\frac{2kTe}{h^3}\int_{-\infty}^{\infty}\left[\int_{-\infty}^{\infty}\ln(1+e^{-G})dp_z\right]dp_y, \quad (11.43)$$

where

$$G = \frac{\phi + (1/2m)(p_x^2 + p_y^2)}{kT}. \quad (11.44)$$

At common operating temperatures, $G \gg 1$, so since $\ln(1 + \varepsilon) \approx \varepsilon$ (for small ε) we can write (this approximation amounts to replacing Fermi–Dirac statistics by Boltzmann statistics for all electrons that get out of the metal)

$$j = -\frac{T_r 2kTe}{h^3}\int_{-\infty}^{\infty}\left\{\int_{-\infty}^{\infty}\exp(-\phi/kT)\cdot\exp\left[\frac{-1}{2mkT}(p_x^2+p_y^2)\right]dp_z\right\}dp_y.$$

Thus, so far as the temperature dependence goes, we can write

$$j = AT^2 e^{-\phi/kT}, \quad (11.45)$$

where A is a quantity that can be determined from the above expressions. In actual practice there is little point to making this evaluation. Our A depends on having an idealized surface, which is never realized. Typical work functions ϕ, as determined from thermionic emission data, are of the order of 5 eV, see Table 11.3.

Equation (11.45) is often referred to as the *Richardson–Dushmann* equation. It agrees with experimental results at least qualitatively. Account must be taken, however, of adsorbates that can lower the effective work function.[6]

---

[6] See Zanquill [11.33].

Table 11.3. Work functions

Element	φ (eV)
Ag	
100	4.64
110	4.52
111	4.74
poly	4.26
Co poly	5
Cu poly	4.65
Fe poly	4.5
K poly	2.3
Na poly	2.75
Ni poly	5.15
W poly	4.55

From Anderson HL (ed), *A Physicists Desk Reference* 2nd edn, Article 21: Hagstum HD, Surface Physics, p. 330, American Institute of Physics, (1989) by permission of Springer-Verlag. Original data from Michaelson HB, *J Appl Phys* **48**, 4729 (1977).

11.8 Cold-Field Emission (B)

To have a detectable cold-field emission it is necessary to apply a strong electric field. The strong electric field can be obtained by using a sharp point, for example. We shall assume that we have applied an electric field E_1 in the $-x$ direction to the metal so that the electron's potential energy (with $-e$ the charge of the electron) produced by the electric field is $V = E_0 - eE_1 x$. The form of the potential function near the surface of the metal will be assumed to be as in Fig. 11.11.

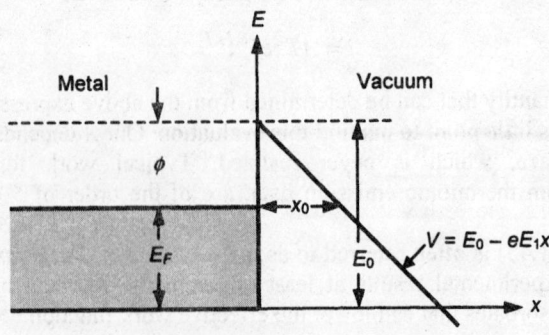

Fig. 11.11. Potential energy for tunneling from a metal in the presence of an applied electric field

11.8 Cold-Field Emission (B)

To calculate the current density, which is emitted by the metal when the electric field is applied, it is necessary to have the transmission coefficient for tunneling through the barrier. This transmission coefficient can perhaps be adequately evaluated by use of the WKB approximation. For a high and broad barrier, the WKB approximation gives for the transmission coefficient

$$T = \exp\left(-2\int_{x=0}^{x_0} K(x)dx\right), \tag{11.46}$$

where

$$K(x) = \sqrt{2m/\hbar^2}\sqrt{V(x) - E}, \tag{11.47}$$

x_0 is the second classical turning point, and E is, of course, the energy.

The upper limit of the integral is determined (for an electron of energy E) from

$$E = E_0 - eE_1 x_0 \quad \text{or} \quad x_0 = \frac{E_0 - E}{eE_1}.$$

Therefore

$$\int_{x=0}^{x_0} K(x)dx = \frac{2}{3}\sqrt{2m(E_0 - E)^3/(\hbar eE_1)^2}.$$

Since $(E_0 - E_F) = \phi$, the transmission coefficient for electrons with the Fermi energy is given by

$$T = \exp\left(-\frac{4}{3}\sqrt{\frac{2m}{\hbar^2}\frac{\phi^3}{(eE_1)^2}}\right). \tag{11.48}$$

Further analysis shows that the current density for field-emitted electrons is given approximately by $J \propto E_1^2 T$ so,

$$J = aE_1^2 e^{-b/E_1}, \tag{11.49}$$

where a and b are different constants for different materials. Equation (11.49), where b is commonly proportional to $\phi^{3/2}$, is often referred to as the *Fowler–Nordheim* equation. The ideas of Fowler–Nordheim tunneling are also used for the tunneling of electrons in a metal-oxide-semiconductor (MOS) structure. See also Sarid [11.27].

There is another type of electron emission that is present when an electric field is applied. When an electric field is applied, the height of the potential barrier is slightly lowered. Thus more electrons can be classically emitted (without tunneling) by thermionic emission than previously. This additional emission due to the lowering of the barrier is called *Schottky emission*. If we imagine the barrier is caused by image charge attraction, it is fairly easy to see why the maximum barrier height should decrease with field strength. Simple analysis predicts the barrier lowering to be proportional to the square root of the magnitude of the electric field. The idea is shown in Fig. 11.12. See Problem 11.7.

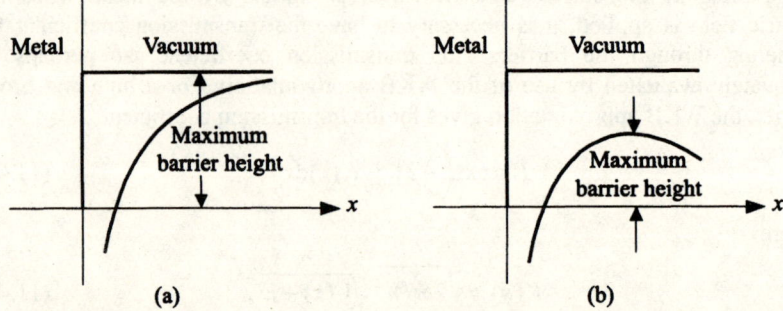

Fig. 11.12. The effect of an electric field on the surface barrier of a metal: (**a**) with no field, and (**b**) with a field

11.9 Microgravity (MS)

It is believed that crystals grown in microgravity will often be more perfect. S. Lehoczky of Marshall Space Flight Center has been experimenting for years with growing mercury cadmium telluride in microgravity (on the Space Shuttle) with the idea of producing more perfect crystals that would yield better infrared detectors. See, e.g., Lehoczky et al [11.19].

First, we should talk about what microgravity is and what it is not. It is not the absence of gravity, or even a region where gravity is very small. Unless one goes very far from massive bodies, this is impossible. Even at a Shuttle orbit of 300 km above the Earth, the force of gravity is about 90% the value experienced on the Earth.

Newton himself understood the principle. If one mounts a cannon on a large mountain on an otherwise flat Earth and fires the cannonball horizontally, it will land some distance away from the base of the mountain. Adding more powder will cause the ball to go further. Finally, a point will be reached when the ball falls exactly the same amount that the earth curves. The ball will then be in free-fall and in orbit. The effects of gravity for objects inside the ball will be very small. In an orbiting satellite, there will be exactly one surface where the effects of gravity are negligible. At other places inside, one has "microgravity".

There are many ways to produce microgravity; all you have to do is arrange to be in free-fall. Drop towers and drop tubes offer two ways of accomplishing this. The first commercial use of microgravity was probably the drop tower used in 1785 in England to make spherical lead balls. Marshall Space Flight Center had both a drop tower and a drop tube 100 meters high–this alone allowed free-fall, or microgravity, for about 5 s. In a drop tower, the entire experimental package is dropped. For crystal growth experiments, this means the furnace as well as the instrumentation and the specimen are all placed in a special canister and dropped.

In a drop tube, there is an enclosure in which, for example, only the molten sample would be dropped. Special aerodynamic design, vacuum, or other means is used to reduce air drag and, hence, obtain real free-fall. For slightly longer times (20 s or so), the KC 135 aircraft can be put into a parabolic path to produce microgravity. Extending this idea, rockets have been used to produce microgravity for periods of about 400 s.

Problems

11.1 Give a simple derivation of Ivey's law. Ivey's law states that fa^2 = constant where f is the frequency of absorption in the F-band and a is the lattice spacing in the colored crystal. Use as a model for the F-center an electron in a box and assume that the absorption is due to a transition between the ground and first excited energy states of the electron in the box.

11.2 The F-center absorption energy in NaCl is about 2.7 eV. For a particle in a box of side a_{NaCl} = 5.63 × 10^{-10} m, find the excitation energy of an electron from the ground to the first excited state.

11.3 A low-angle grain boundary is found with a tilt angle of about 20 s on a (100) surface of Ge. What is the prediction for the linear dislocation density of etch pits predicted?

11.4 Find the allowed energies of a hydrogen atom in two dimensions. The answer you should get is [12.54]

$$E_n = -\frac{R}{(n-\frac{1}{2})^2},$$

where n is a nonzero integer. R is the Rydberg constant that can be written as

$$R = -\frac{me^4}{2(4\pi\varepsilon_0 K\hbar)^2},$$

where $K = \varepsilon/\varepsilon_0$ with ε the appropriate dielectric constant. Since the Bohr radius is

$$a_B = \frac{4\pi\varepsilon_0 K\hbar^2}{me^2},$$

one can also write

$$R = -\frac{\hbar^2}{2ma_B^2}.$$

Note that the result is the same as the three-dimensional hydrogen atom if one replaces n by $n - \frac{1}{2}$.

11.5 Quantum wells will be discussed in Chap. 12. Find the allowed energies of a donor atom, represented by a hydrogen atom with electron mass m and in a region of dielectric constant as above. Suppose the quantum well is of width w and with infinite sides with potential energy $V(z)$. Also suppose $w \ll a_B$. In this case the wave function for a donor in a quantum well is

$$\left[-\frac{\hbar^2}{2m}\nabla^2 - \frac{e^2}{4\pi\varepsilon_0 K\sqrt{x^2+y^2}} + V(z)\right]\psi = E\psi,$$

where $V(z) = 0$ for $0 < z < w$ and is infinite otherwise. The answer is

$$E = E_{p,n} = \frac{\hbar^2\pi^2 p^2}{2mw^2} - \frac{R}{(n-\frac{1}{2})^2},$$

p, n are nonzero integers and R is the Rydberg constant [12.54].

11.6 a) Show that a solution of the one-dimensional diffusion equation is

$$C(x,t) = \frac{A}{\sqrt{t}}\exp\left(-\frac{x^2}{4Dt}\right).$$

b) If $\int_{-\infty}^{\infty} C(x,t)dx = Q$, show that

$$A = \frac{Q}{2\sqrt{\pi D}}.$$

11.7 This problem illustrates the Schottky effect. See Fig. 11.11 and Fig. 11.12. Suppose the attraction outside the metal is caused by an image charge.

a) Show that in the absence of an electric field we can write the potential energy as

$$V(x) = E_0 - \frac{e^2}{16\pi\varepsilon_0 x},$$

so that with an external field

$$V(x) = E_0 - eE_1 x - \frac{e^2}{16\pi\varepsilon_0 x}.$$

b) Thus show that with the electric field E_1, the barrier height is reduced from E_0 to $E_0 - \Delta$, where

$$\Delta = \frac{1}{2}\sqrt{\frac{e^3 E_1}{\pi\varepsilon_0}}.$$

12 Current Topics in Solid Condensed–Matter Physics

This chapter is concerned with some of the newer areas of solid condensed-matter physics and so contains a variety of topics in nanophysics, surfaces, interfaces, amorphous materials, and soft condensed matter.

There was a time when the living room radio stood on the floor and people gathered around in the evening and "watched" the radio. Radios have become smaller and smaller and thus, increasingly cheaper. Eventually, of course, there will be a limit in smallness of size to electronic devices. Fundamental physics places constraints on how small the device can be and still operate in a "conventional way". Recently people have realized that a limit for one kind of device is an opportunity for another. This leads to the topic of new ways of using materials, particularly semiconductors, for new devices.

Of course, the subject of electronic technology, particularly semiconductor technology, is too vast to consider here. One main concern is the fact that quantum mechanics places basic limits on the size of devices. This arises because quantum mechanics associates a wavelength with the electrons that carry current and electrical signals. Quantum effects become important when electron wavelength becomes comparable to component size. In particular, the phenomenon of tunneling, which is often assumed to be of no importance for most ordinary microelectronic devices becomes important in this limit. We will discuss some of the basic physics needed to understand these devices, in which tunneling and related phenomena are important. Here we get into the area of bandgap engineering to attain structures that have desired properties not attainable with homostructures. Generally, these structures are *nanostructures*. A nanostructure is a condensed-matter structure having at least one minimum dimension between about 1 nm to 10 nm.

We will start by discussing surfaces and then consider how to form nanostructures on surfaces by molecular beam epitaxy. Nanostructures may be two dimensional (quantum wells), one dimensional (quantum wires), or "zero" dimensional (quantum dots). We will discuss all of these and also talk about heterostructures, superlattices, quantum conductance, Coulomb blockade, and single-electron devices.

Another reduced-dimensionality effect is the quantum Hall effect, which arises when electrons in a magnetic field are confined two dimensionally. As we will see, the ideas and phenomena involved are quite novel.

Carbon, carbon nanotubes, and fullerene nanotechnology may lead to entirely new kinds of devices and they are also included in this chapter, as the nanotubes are certainly nanostructures.

Amorphous, noncrystalline disordered solids have become important and we discuss them as examples of new materials if not reduced dimensionality.

Finally, the new area of soft condensed-matter physics is touched on. This area includes liquid crystals, polymers, and other materials that may be "soft" to the touch. The unifying idea here is the ease with which the materials deform due to external forces.

12.1 Surface Reconstruction (MET, MS)

As already mentioned, the input and output of a device go through the surface, so physical understanding of surfaces is critical. Of course, the nature of the surface also affects crystal growth, chemical reactions, thermionic emission, semiconducting properties, etc.

One generally thinks of the surface of a material as being the top two or three layers. The rest can be called the bulk or substrate. The distortion near the surface can be both perpendicular (stretching or contracting) as well as parallel. Below we concentrate on that which is parallel.

If we project the bulk with its periodicity on the surface and if no reconstruction occurs we say the surface is 1×1. More likely the lack of bonding forces on the surface side will cause the surface atoms to find new locations of minimum energy. Then the projection of the bulk on the surface is different in symmetry from the surface. For the special case where the projection defines primitive surface vectors a and b, while the actual surface has primitive vectors $a_S = Na$ and $b_S = Mb$ then one says one has an $N \times M$ reconstruction. If there also is a rotation R of β associated with a_S and b_S primitive cell compared to the a, b primitive cell we write the reconstruction as

$$\left(\frac{|a_S|}{|a|} \times \frac{|b_S|}{|b|}\right) R\beta .$$

Note that the vectors a and b depend on whether the original (unreconstructed or unrelaxed) surface is (1, 1, 1) or (1, 0, 0), or in general (h, k, l). For a complete description the surface involved would also have to be included. The reciprocal lattice vectors A, B associated with the surface are defined in the usual way as

$$A \cdot a_S = B \cdot b_S = 2\pi , \qquad (12.1a)$$

and

$$A \cdot b_S = B \cdot a_S = 0 , \qquad (12.1b)$$

where the 2π now inserted in an alternative convention for reciprocal lattice vectors. One uses these to discuss two-dimensional diffraction.

Low-energy electron diffraction (LEED, see Sects. 1.2.7 and 12.2) is commonly used to examine the structure of surfaces. This is because electrons, unlike photons, have charge and thus, do not penetrate too far into materials. There are theoretical techniques, including those using the pseudopotential, which are available. See Chen and Ho [12.12].

Since surfaces are so important for solid-state properties we briefly review techniques for their characterization in the next section.

12.2 Some Surface Characterization Techniques (MET, MS, EE)

AFM: Atomic Force Microscopy—This instrument detects images of surfaces on an atomic scale by sensing atomic forces between the sample and a cantilevered tip (in one kind of mode, there are various modes of operation). Unlike STMs (see below), this instrument can be used for nonconductors as well as conductors.

AES: Auger Electron Spectroscopy—uses an alternative (to X-ray emission) decay scheme for an excited core hole. The core hole is often produced by the impact of energetic electrons. An electron from a higher level makes a transition to fill the hole, and another bound electron escapes with the left-over energy. The Auger process leaves two final-state holes. The energy of the escaping electron is related to the characteristic energies of the atom from which it came, and therefore chemical analysis is possible.

EDX: Energy Dispersive X-ray Spectroscopy—electrons are incident at a grazing angle and the energy of the grazing X-rays that are produced, are detected and analyzed. This technique has sensitivities comparable to Auger electron spectroscopy.

Ellipsometry—study of the reflection of plane-polarized light from the surface of materials to determine the properties of these materials by measuring the ellipticity of the reflected light.

EELS: Electron Energy Loss Spectroscopy—electrons scattered from surface atoms may lose amounts of energy dependent on surface excitations. This can be used to examine surface vibrational modes. It is also used to detect surface plasmons.

EXAFS: Extended X-ray Absorption Spectroscopy—photoelectrons caused to be emitted by X-rays are backscattered from surrounding atoms. They interfere with the emitted photoelectrons and give information about the geometry of the atoms that surround the original absorbing atom. When this technique is surface specific, as for detecting Auger electrons, it is called *SEXAFS*.

FIM: Field Ion Microscopy—this can be used to detect individual atoms. Ions of the surrounding imaging gas are produced by field ionization at a tip and are detected on a fluorescent screen placed at a distance, to which ions are repelled.

LEED: Low-Energy Electron Diffraction—due to their charge, electrons do not penetrate deeply into a surface. *LEED* is the coherent reflection or diffraction of electrons typically with energy less than hundreds of electron volts from the surface layers of a solid. Since it is from the surface, the diffraction is two-dimensional and can be used to examine surface reconstruction.

RHEED: Reflection High-Energy Electron Diffraction–high-energy electrons can also be diffracted from the surface, provided they are at grazing incidence and so do not greatly penetrate.

SEM: Scanning Electron Microscopy–a focused electron beam is scanned across a surface. The emitted secondary electrons are used as a signal that, in a synchronous manner, is displayed on the surface of an oscilloscope. An electron spectrometer can be used to only display electrons whose energies correspond to an Auger peak, in which case the instrument is called a scanning Auger microscope (SAM).

SIMS: Secondary Ion Mass Spectrometry–a destructive but sensitive surface technique. Kiloelectron-volt ions bombard a surface and knock off or sputter ions, which are analyzed by a mass spectrometer and thus can be chemically analyzed.

TEM: Transmission Electron Microscopy–this is like SEM except that the electrons transmitted through a thin specimen are examined. Both elastically and inelastically scattered electrons can be examined, and high contrast is possible.

STM: Scanning Tunneling Microscopy–A sample (metal or semiconductor) has a sharp metal tip placed within 10 Å or less of its surface. A small voltage of order 1 V is established between the two. Since the wave functions of the atoms on the surface of the sample and the tip overlap, in equilibrium the Fermi energies of the sample and tip equalize and under the voltage difference a tunneling current of order nanoamperes will flow between the two. Since the current flow is due to tunneling, it depends exponentially on the distance from the sample to the tip. The exponential dependence makes the tunneling sensitive to sub-angstrom changes in distance, and hence it is possible to use this technique to detect and image individual atoms. The current depends on the local density of states (LDOS) at the surface of the sample and hence is used for LDOS mapping. The position of the tip is controlled by piezoelectric transducers. The apparatus is operated in either the constant-distance or constant-current mode.

UPS: Ultraviolet Photoelectron (or Photoemission) Spectroscopy–the binding energy of a core electron is measured by measuring the energy of the core electron ejected by the ultraviolet photon. For energies not too high, the energy distribution of emitted electrons is dominated by the joint density of initial and final states. An angle-resolved mode is often used since the parallel (to the surface) component of the k vector as well as the energy is conserved. This allows experimental determination of the energy of the initial occupied state for which k parallel is thus known (see Sect. 3.2.2). See also Table 10.3.

XPS: X-ray Photoelectron (or Photoemission) Spectroscopy–the binding energy of a core electron is measured by measuring the energy of the core electron ejected by the X-ray photon–also called *ESCA*. See also Table 10.3

There are of course many other characterization techniques that we could discuss. There are many kinds of scanning probe microscopes, for example. There are many kinds of characterization techniques that are not primarily related to surface properties. Some ideas have already been discussed. Elastic and inelastic X-ray and neutron scattering come immediately to mind. Electrical conductivity and other electrical measurements can often yield much information, as can the many kinds of magnetic resonance techniques. Optical techniques can yield important information (see, e.g., Perkowitz [12.49], as well as Chap. 10 on optical properties in this book). Raman scattering spectroscopy is often important in the infrared. Spectroscopic data involves information about intensity versus frequency. In Raman scattering, the incident photon is inelastically scattered by phonons. Commercial instruments are available, as they are for FTIR (Fourier transform infrared spectroscopy), which use a Michelson interferometer to increase the signal-to-noise ratio and get the Fourier transform of the intensity versus frequency. A FFT (fast Fourier transform) algorithm is then used to get the intensity versus frequency in real time. Perkowitz also mentions photoluminescence spectroscopy, where in general after photon excitation an electron returns to its initial state. Commercial instruments are also available. This technique gives fingerprints of excited states. Considerable additional information about characterization can be found in Bullis et al [12.5]. For a general treatment see Prutton [12.52] and Marder (preface ref. 6, pp73-82).

12.3 Molecular Beam Epitaxy (MET, MS)

Molecular beam epitaxy (MBE) was developed in the 1970s and is by now a common technology for use in making low-dimensional solid-state structure. MBE is an ultrathin film vacuum technique in which several atomic and/or molecular beams collide with and stick to the substrate. Epitaxy means that at the interface of two materials, there is a common crystal orientation and registry of atoms. The substrates are heated to temperature T and mounted suitably. Each effusion cell, from which the molecular beams originate, are held at appropriate temperatures to maintain a suitable flux. The effusion cells also have shutters so that the growth of layers due to the molecular beams can be controlled (see Fig. 12.1). MBE produces high-purity layers in ultrahigh vacuum. Abrupt transitions on an atomic scale can be grown at a rate of a few (tens of) nanometers per second. See, e.g., Joyce [12.31]. Other techniques for producing layered structures include chemical vapor deposition and electrochemical deposition.

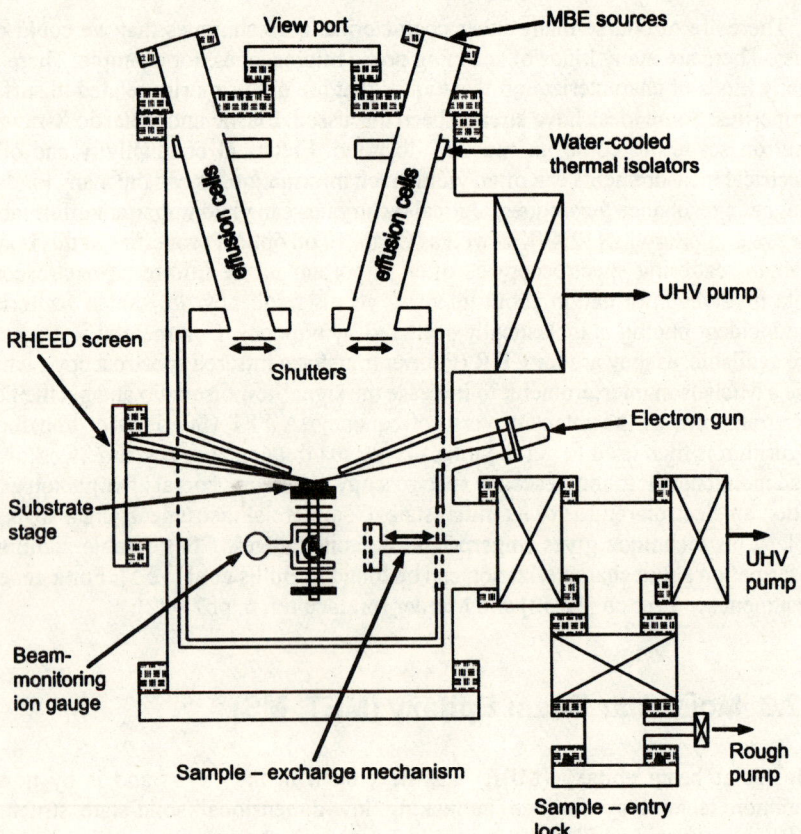

Fig. 12.1. Schematic diagram of an ultrahigh vacuum, molecular beam growth system (adapted from Joyce BA, *Rep Prog Physics* **48**, 1637 (1985), by permission of the Institute of Physics). Reflection high-energy electron diffraction (RHEED) is used for monitoring the growth

12.4 Heterostructures and Quantum Wells

By use of MBE or other related techniques, heterostructures and quantum wells can be formed. Heterostructures are layers of semiconductors with the same crystal structure, grown coherently, but with different bandgaps. Their properties depend heavily on their type. Two types are shown in Fig. 12.2: normal (example GaAlAs-GaAs) and broken (example GaSb-InAs). There are also other types. See Butcher et al [12.6 p. 15]. ΔE_c is the conduction-band offset.

Two-dimensional quantum wells are formed by sandwiching a small-bandgap material between two large-bandgap materials. Energy barriers are formed that quantize the motion in one direction. These can be used to form resonant tunneling devices (e.g. by depositing small-bandgap – large-bandgap – small-bandgap – large – small, etc. See applications of superlattices in Sect. 12.6.1). A quantum well can show increased tunneling currents due to resonance at allowed energy levels in the well. The current versus voltage can even show a decrease with voltage for certain values of voltage. See Fig. 12.11. Diodes and transistors have been constructed with these devices.

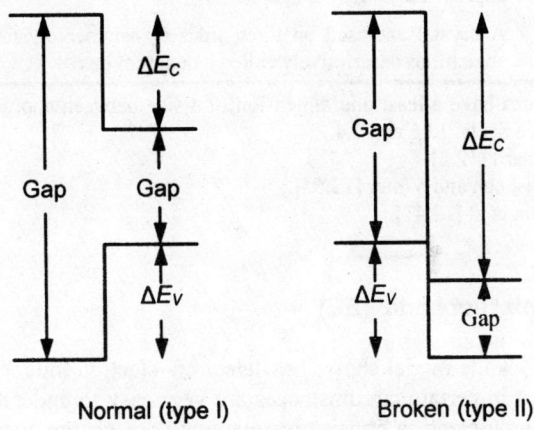

Fig. 12.2. Normal and broken heterostructures

12.5 Quantum Structures and Single-Electron Devices (EE)

Dimension is an important aspect of small electronic devices. Dimensionality can be controlled by sandwiching. If the center of the sandwich is bordered by planar materials for which the electronic states are higher (wider bandgaps), then three-dimensional motion can be reduced to two, producing quantum wells. Similarly one can make linear one-dimensional "quantum wires" and nearly zero-dimensional or "quantum dot" materials. That is, a quantum wire is made by laying down a line of narrow-gap semiconductors surrounded by a wide-gap one with the carriers confined in two dimensions, while a quantum dot involves only a small volume of narrow-gap material surrounded by wide-gap material and the carriers are confined in all three dimensions. With the quantum-dot structure, one may confine or exchange one electron at a time and develop single-electron transistors that would be fast, low power, and have essentially error-free signals. These three types of quantum structures are summarized in Table 12.1.

Table 12.1. Summary of three types of quantum structures

Nanostructures	Comments
Quantum wells	Superlattices can be regarded as quantum well layers – alternating layers of different crystals (when the wells are not too far apart)
Quantum wires	A crystal enclosed on two sides by another crystalline material, with appropriate wider bandgaps
Quantum dots	A crystal enclosed on three sides by another crystalline material – sometimes descriptively called a quantum box

Note: Nanostructures have a least one dimension of a size between approximately one to ten nanometers. See Sects. 12.6 and 7.4.
References: 1. Bastard [12.2]
           2. Weisbuch and Vinter [12.65]
           3. Mitin et al [12.47]

12.5.1 Coulomb Blockade[1] (EE)

The Coulomb blockade model shows how electron–electron interactions can give rise to effects that in certain circumstances are very easy to understand. It relates to the ideas of single-electron transistors, quantum dots, charge quantization leading to an energy gap in the density of states for tunneling, and is sometimes even qualitatively likened to a dripping faucet. For purposes of illustration, we consider a simple model of an artificial atom represented by the metal particle shown in Fig. 12.3.

Experimentally, the conductance (current per voltage bias) from source to drain shows large changes with gate voltage. We wish to analyze this with the Coulomb blockade model. Let C be the total capacitance between the metal particle and the rest of the system, which we will assume is approximately the capacitance between the metal particle and the gate. Let V_g be the gate voltage, relative to source, and assume the source, particle, and drain voltages are close (but sufficiently different to have the possibility of drawing current from source to drain). If there is a charge Q on the metal particle, then its electrostatic energy is

$$U = QV_g + \frac{Q^2}{2C}. \tag{12.2}$$

Setting $\partial U/\partial Q = 0$, we find U has a minimum at

$$Q = Q_0 = -CV_g. \tag{12.3}$$

If N is an integer, let $Q_0 = -(N + \eta)e$, where $e > 0$, so

$$CV_g = (N + \eta)e. \tag{12.4}$$

---

[1] See Kastner [12.32]. See also Kelly [12.33, pp. 300-305].

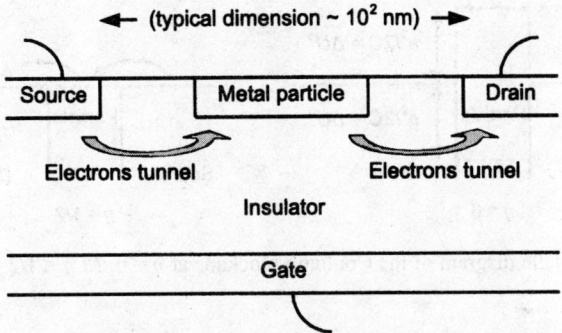

Fig. 12.3. Model of a single-electron transistor

Note that while Q_0 can be any value, the actual physical situation will be determined by the integer number of electrons on the artificial atom (metal particle) that makes U the smallest. This will only be at a mathematical minimum if V_g is an integral multiple of e/C.

For $-1/2 < \eta < 1/2$, and $V_g = (N + \eta)e/C$, the minimum energy is obtained for N electrons on the metal particle. The Coulomb blockade arises because of the energy required to transfer an electron to (or from) the metal particle (you can't transfer less than an electron). We can easily calculate this as follows. Let us consider η between zero and one half. Combining (12.2) and (12.4),

$$U = \frac{1}{C}\left[Q(N+\eta)e + \frac{Q^2}{2}\right]. \tag{12.5}$$

Let the initial charge on the particle be $Q_i = -Ne$ and the final charge be $Q_f = -(N \pm 1)e$. Then for the energy difference,

$$\Delta U^\pm = U_f^\pm - U_i,$$

we find

$$\Delta U^+ = \frac{e^2}{C}\left(\frac{1}{2} - \eta\right), \quad \Delta U^- = \frac{e^2}{C}\left(\frac{1}{2} + \eta\right).$$

We see that for $\eta < 1/2$ there is an energy gap for tunneling: the Coulomb blockade. For $\eta = 1/2$ the energies for the metal particle having N and $N + 1$ electrons are the same and the gap disappears. Since the source and the drain have approximately the same Fermi energy, one can understand this result from Fig. 12.4. Note ΔU^+ is the energy to add an electron and ΔU^- is the energy to take away an electron (or to add a hole). Thus the gap in the allowed states of the particle is e^2/C. Just above $\eta = 1/2$, the number of electrons on the artificial atoms increases by 1 (to $N + 1$) and the process repeats as V_g is increased. It is indeed reminiscent of a dripping faucet.

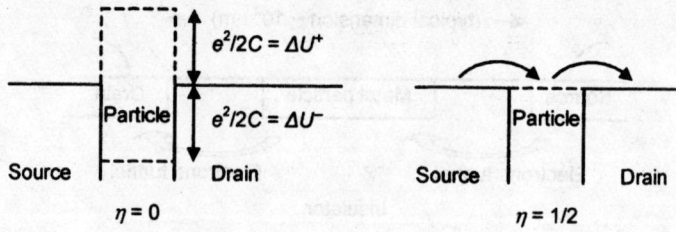

Fig. 12.4. Schematic diagram of the Coulomb blockade at $\eta = 0$. At $\eta = 1/2$ the energy gap ΔE disappears

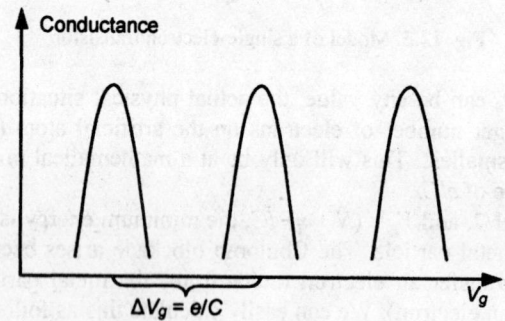

Fig. 12.5. Periodic conductance peaks

The total voltage change from one turn on to the next turn on occurs, e.g., when η goes from 1/2 to 3/2 or

$$\Delta V_g = \frac{e}{C}\left[N+\frac{3}{2}-\left(N+\frac{1}{2}\right)\right] = \frac{e}{C}.$$

A sketch of the conductance versus gate voltage in Fig. 12.5 shows periodic peaks. In order to conduct, an electron must go from source to particle, and then from particle to drain (or a hole from drain to particle, etc.).

Low temperatures are required to see this effect, as one must have

$$kT < \frac{e^2}{2C},$$

so that thermal effects do not wash out the gap. This condition requires small temperatures and small capacitances, such as encountered in nanodevices. In addition the metal particle–artificial atom has discrete energy levels that may be observed in tunneling experiments by fixing V_g and varying the drain-to-source voltage. See Kasner op cit.

12.5.2 Tunneling and the Landauer Equation (EE)

Metal-Barrier-Metal Tunneling (EE)

We start by considering tunneling through a barrier as suggested in Fig. 12.6. We assume each (identical) metal is in local equilibrium with a chemical potential μ. Due to an applied external potential difference φ, we assume the chemical potential is shifted down by $-e\varphi/2$ ($e > 0$) for metal 1 and up by $e\varphi/2$ for metal 2 (see Fig. 12.7).

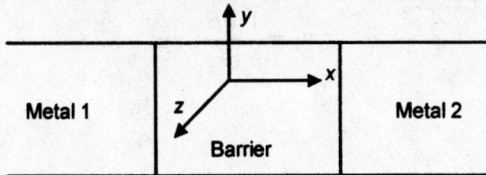

Fig. 12.6. Schematic diagram for barrier tunneling

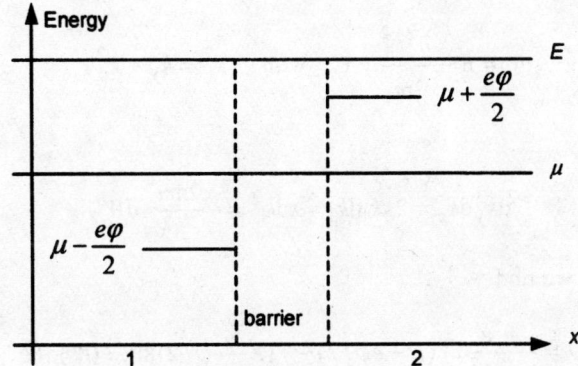

Fig. 12.7. Tunneling sketch

We consider an electron of energy E and assume it tunnels through the barrier without changing energy. We write its energy as (with W defined by the equation and assuming for simplicity the same effective mass in all directions)

$$E = W + E_{\parallel} + C = \frac{\hbar^2 k_x^2}{2m^*} + \frac{\hbar^2}{2m^*}(k_y^2 + k_z^2) + C,$$

where C is a constant that determines the bottom of the conduction band and m^*, assumed constant, is the effective mass. We assume, for this case, that the transmission coefficient T across the barrier depends only on W, $T = T(W)$. We insert a factor of 2 for the spin and consider electron flow in the $\pm x$ directions. With $\varphi = 0$, let the chemical potential in each metal be μ and the Fermi function

$$f(E,\mu) = \frac{1}{\exp[(E-\mu)/kT]+1}.$$

Notice $\mu \to \mu - e\varphi/2$ is the same as $E \to E + e\varphi/2$. Then the current density J is (considering current flowing each way, $\pm x$)

$$dJ = -2ev_x[f(E+e\varphi/2,\mu)(1-f(E-e\varphi/2,\mu)) \\ - f(E-e\varphi/2,\mu)(1-f(E+e\varphi/2,\mu))]T(W)\frac{d^3k}{(2\pi)^3}.$$

Since

$$v_x = \frac{1}{\hbar}\frac{\partial E}{\partial k_x},$$

then

$$v_x dk_x = \frac{1}{\hbar}dE.$$

Also $d^3k = dk_x dk_y dk_z$ and since

$$W = E - \frac{\hbar^2 k_\parallel^2}{2m^*} - C \quad \text{with} \quad (k_\parallel^2 = k_y^2 + k_z^2),$$

we have

$$dk_y dk_z = 2\pi k_\parallel dk_\parallel = \pi dk_\parallel^2 = -\frac{2\pi m^*}{\hbar^2}dW,$$

so substituting we find

$$dJ = \frac{m^* e}{2\pi^2 \hbar^3}[f(E+e\varphi/2) - f(E-e\varphi/2)]dE\, T(W)dW.$$

When the form of the barrier is known and is suitably simple, the transmission coefficient is often evaluated by the WKB approximation. J can then be calculated by integrating over appropriate limits (W from 0 to $E - C$ and E from C to infinity). This is the standard simple way of looking at tunneling conductance. A different situation is presented below.

Landauer Equation and Quantum Conductance (EE)

In mesoscopic (intermediate between atomic and macroscopic sizes) channels at small sizes, it may be necessary to have a different picture of transport because of quantum effects. In mesoscopic channels at low voltage and low temperatures and few inelastic collisions, Landauer has derived that the electronic conductance is $2e^2/h$ times the number of conductance channels corresponding to all (quantized) transverse energies from zero to the Fermi energy. Transverse energy is defined as the total energy minus the kinetic energy for velocities in the direction of the

12.5 Quantum Structures and Single-Electron Devices (EE)

channel. We derive this result below (see, e.g., Imry I and Landauer R, *More Things in Heaven and Earth*, Bederson B (ed), Springer-Verlag, 1999, p515ff.)

We here write the electron energy as

$$E = \frac{\hbar^2 k_x^2}{2m} + E_{ny,nz},$$

where $E_{ny,nz}$ represents the quantized energy corresponding to the y and z directions. We have replaced the barrier by a device of conductance length L in the x direction and with small size in the y and z directions. We assume this small size is of order of the electron wavelength and thus $E_{ny,nz}$ is clearly quantized. We also regard the two metals as leads to the device and we continue to assume we can treat each lead as essentially in thermal equilibrium.

We assume $T_{ny,nz}(E)$ is the transmission coefficient of the device. Note we have allowed for the possibility that T depends on the quantized motion in the y and z directions. Thus the current is

$$I = -\frac{2e}{L}\sum_{ny,nz} L\int \frac{dk_x}{2\pi} v_x T_{ny,nz}(E)[f(E + e\varphi/2,\mu) - f(E - e\varphi/2,\mu)].$$

Note that $dk_x/2\pi$ is the number of states per unit length, so we multiply by L. then we end up with (effectively) the number of electrons, but we want the number per unit length so we divide by L. If φ and the temperature are small then

$$[f(E + e\varphi/2,\mu) - f(E - e\varphi/2,\mu)] = \left.\frac{\partial f}{\partial E}\right)_{\varphi=0}(e\varphi)$$
$$= -\delta(E - \mu)e\varphi.$$

Then using $v_x dk_x = (1/\hbar)dE$ as before, we have

$$I = \frac{2e}{h}\sum_{ny,nz} T_{ny,nz}(\mu)e\varphi.$$

We thus obtain for the conductance

$$G = \frac{I}{\varphi} = \frac{2e^2}{h}\sum_{ny,nz} T_{ny,nz}(\mu).$$

Note that the sum is only over states with total energy μ so $E_{ny,nz} \leq \mu$.

The quantity e^2/h is called the quantum of conductance G_0 so

$$G = 2G_0 \sum_{ny,nz} T_{ny,nz}(\mu),$$
$$(E_{ny,nz} \leq \mu)$$

which is the Landauer equation. This equation has been verified by experiment. Recently, a similar effect has been seen for thermal conduction by phonons. Here the unit of thermal conduction is $(\pi k_B)^2 T/3h$ (see Schab [12.53]).

12.6 Superlattices, Bloch Oscillators, Stark–Wannier Ladders

A superlattice is a set of essentially epitaxial layers (with thickness in nanometers) laid down in a periodic way so as to introduce two periodicities: the lattice periodicity, and the layer periodicity. One can introduce this additional periodicity by doping variations or by compositional variations. A particularly interesting type of superlattice is the strained layer. This is a superlattice in which the lattice constants do not exactly match. It has been found that one can do this without introducing defects provided the layers are sufficiently thin. The resulting strain can be used to productively modify the energy levels.

Minibands can appear in a superlattice. These are caused by quantum wells with discrete levels that are split into minibands due to tunneling between the wells. Some applications of superlattices will be discussed later. For a more quantitative discussion of superlattices, see the sections on Envelope Functions, Effective Mass Theory, Shallow Defects, and Superlattices in Sect. 11.3, and also Mendez and Bastard [12.46].

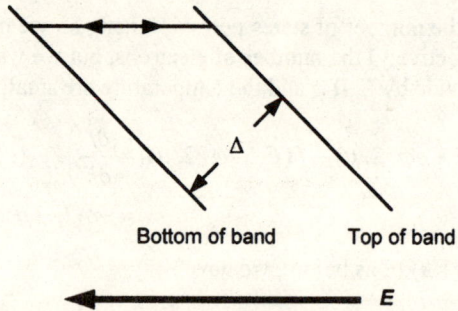

Fig. 12.8. Miniband "tilted" by electric field, and Bloch oscillations

Bloch oscillations can occur in minibands. Consider a portion of a miniband when it is "tilted" by an electric field as shown in Fig. 12.8. An electron in the band will lower its potential energy in the electric field while gaining in kinetic energy, and thus, follow a constant energy path from the bottom of the band to the top, as illustrated above. For very narrow minibands, there is a good chance it will reach the top before phonon emission. In such cases, it could be Bragg reflected. Several reflections between the top and bottom could be possible. These are the Bragg reflections.

We can be slightly more quantitative about Bloch oscillations. The equation of motion of an electron in a lattice is

$$\hbar \frac{dk}{dt} = -eE, \quad e > 0. \tag{12.6}$$

12.6 Superlattices, Bloch Oscillators, Stark–Wannier Ladders

The width of the Brillouin zone associated with the superlattice is

$$K = \frac{2\pi}{p}, \tag{12.7}$$

where p is the length of the fundamental repeat distance for the superlattice and K is thus a reciprocal lattice vector of the superlattice. Integrating (12.6) from one side of the zone to the other, we find

$$\hbar K = -eEt. \tag{12.8}$$

The Bloch frequency for an oscillation corresponding to the time required to cross the Brillouin zone boundary is given by

$$\omega_B = \frac{2\pi}{t} = \frac{peE}{\hbar}. \tag{12.9}$$

In a tight binding approximation, the energy band structure is given by

$$E_k = A - B\cos(kp), \quad -\frac{\pi}{p} \leq k \leq \frac{\pi}{p}. \tag{12.10}$$

The group velocity can then be calculated by

$$v_g = \frac{1}{\hbar}\frac{dE_k}{dk}. \tag{12.11}$$

In time zero to t_1, the electron moves

$$x_1 = \int_0^{t_1} v_g dt = \int_0^{-eEt_1/\hbar} v_g \frac{dt}{dk} dk. \tag{12.12}$$

Combining (12.12), (12.11), (12.10), (12.9), and (12.6), we find

$$x_1 = \frac{B}{eE}[\cos(\omega_B t_1) - 1]. \tag{12.13}$$

The electron oscillates in real space with the Bloch frequency ω_B, as expected. In a normal material (nonsuperlattice), the band width is much larger than the miniband width Δ, so that phonon emission before Bloch oscillations set in is overwhelmingly probable. Note that the time required to cross the (superlattice) Brillouin zone is also the time required to go from $k = 0$ to π/p (assuming bottom of band is at 0 and top at $2\pi/p$) then be Bragg reflected to $-\pi/p$ and hence go from $k = -\pi/p$ to 0. So the Bloch oscillation is a complete oscillation of the band to the top and back.

Consider a superlattice of quantum wells producing a narrow miniband. On applying an electric field, the whole drawing "tilts" producing a set of discrete energy levels known as a Stark–Wannier Ladder (see Fig. 12.9). The presence of the (sufficiently strong) electric field may cause the extended wave functions of the miniband to become localized wave functions. If p is the thickness of the period of

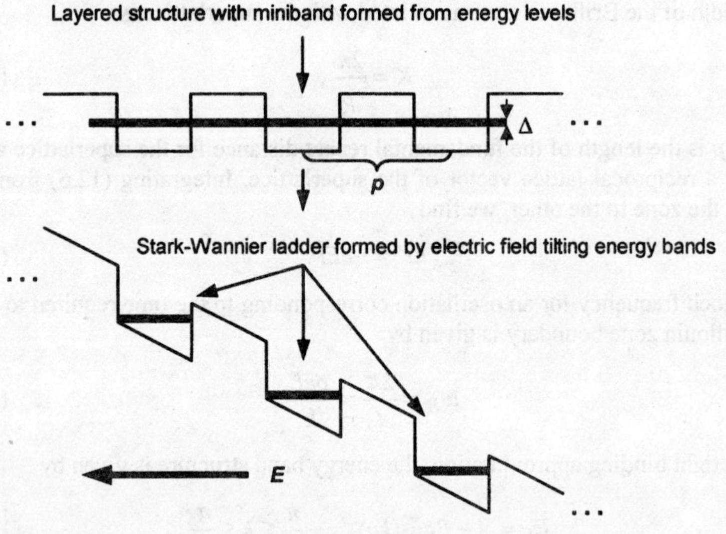

Layered structure with miniband formed from energy levels
...
Δ
...
p
Stark-Wannier ladder formed by electric field tilting energy bands
...
E
...

Fig. 12.9. An applied electric field to a superlattice may create a Stark–Wannier ladder when electrons in the discrete levels have no states to easily tunnel to. Only one miniband is shown and the tilt is exaggerated

the superlattice and Δ is the width of the miniband, the Stark–Wannier ladder occurs where $|eEp| \geq \Delta$. Actual realistic calculation gives a set of sharp resonances rather than discrete levels, and the Stark–Wannier ladder has been verified experimentally. Stark–Wannier ladders were predicted by Wannier [12.64]. See also Lyssenko et al [12.45], and [55, p31ff].

GaAlAs
GaAlAs
Bandgaps
GaAs
GaAs
Energy
bands

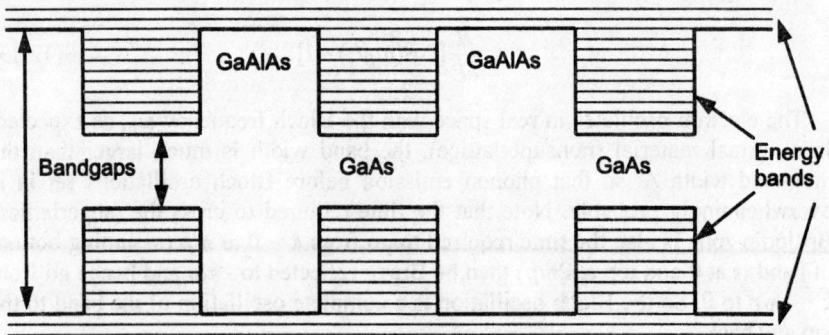

Fig. 12.10. GaAs-GaAlAs superlattice

12.6.1 Applications of Superlattices and Related Nanostructures (EE)

High Mobility (EE)

See Fig. 12.10. Suppose the GaAlAs is heavily donor doped. The donated electrons will fall into the GaAs wells where they would be *separated* from the impurities (donor ions) that furnished them and could scatter them. Thus, high mobility would be created. So, this structure would create high-conductivity semiconductors. Superlattices were proposed by Esaki and Tsu [12.18]. They have since become a very large part of basic and applied research in solid-state physics.

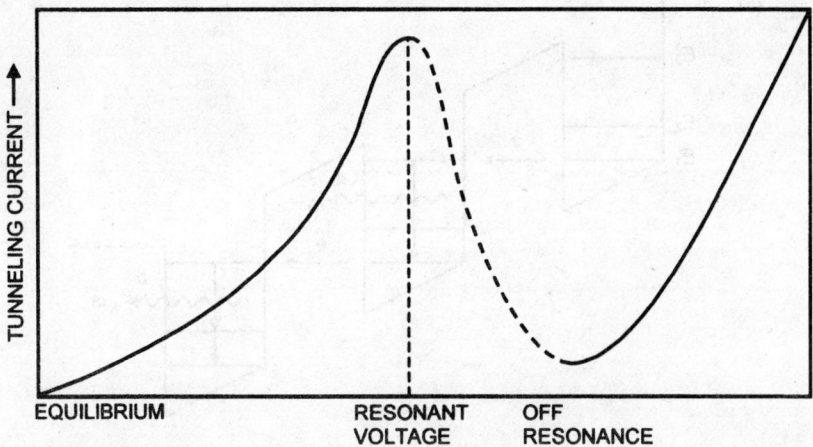

Fig. 12.11. *V-I* curve showing the peak and valley indicating resonant tunneling for a double barrier structure with metals (Fermi energy E_F) on each side

Resonant Tunneling Devices (EE)

A quantum well is formed by layers of wide-bandgap, narrow-bandgap, and wide-bandgap semiconductors. Quantum barriers can be formed from narrow-gap, wide gap, narrow-gap semiconductors. A resonant tunneling device can be formed by surrounding a well with two barriers. Outside the barrier, electrons populate states up to the Fermi energy. If a voltage is applied across the device, the (quasi) Fermi energy on the input side can be moved until it equals the energy of one of the discrete energies within the well.

Typically, the current increases with increasing voltage until a match is obtained, and as the voltage is further increased, the current decreases. The decrease in current with increasing voltage is called negative differential resistance, which can be applied in making high-frequency devices (See Fig. 12.11). See, e.g., Beltram and Capasso in Butcher et al [12.6 Chap. 15]. See also Capasso and Datta [12.8].

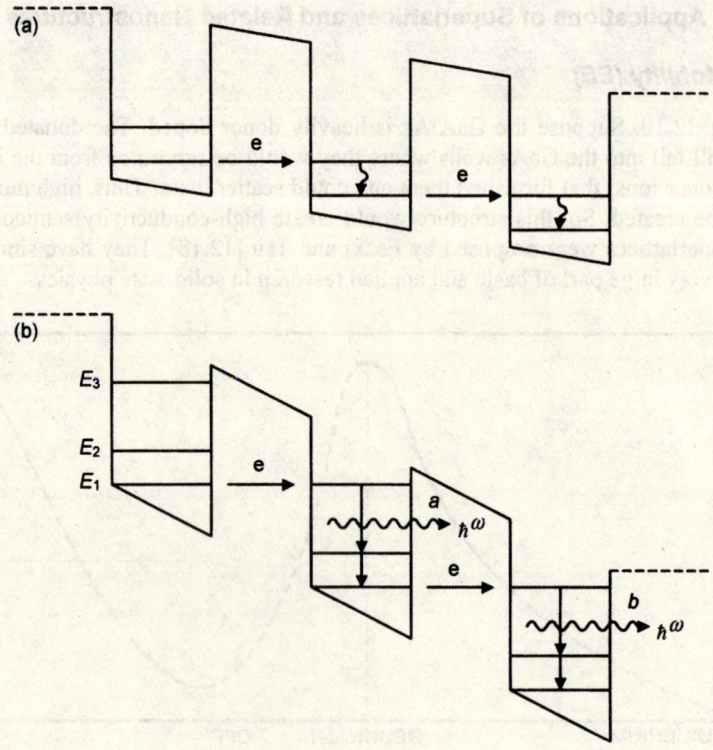

Fig. 12.12. (a) Resonant tunneling through a superlattice with a discrete Stark-Wannier "ladder" of states. (b) Resonant tunneling laser (emission a may trigger emission b, etc.). Note that in (a) we are considering non radiative transitions while (b) has indicated radiative transitions a and b. Adapted from Capasso F, *Science* **235**, 175 (1987).

Lasers (EE)

We start with a superlattice (or at least a multiple quantum well structure) of alternating wide-gap, narrow-gap materials. The quantum wells form where we have narrow-gap semiconductors and the electrons settle into discrete ground states in the quantum wells. Now, apply an electric field so that the ground state of one level is in resonance with the excited state of the next level. One then gets resonant tunneling between these two states. In effect, one can obtain a population inversion leading to lasing action (see Fig. 12.12). For further details, discussion of relevance of minibands, etc., see Capasso et al [12.9]. Lasers using quantum wells are used in compact disk players.

Infrared Detectors (EE)

This can be made similarly to the way the laser is made, except one deals with excitations to the conduction band and subsequent collection by the electric field. See Fig. 12.13 where the idea is sketched. One assumes the excitation energy is in the infrared.

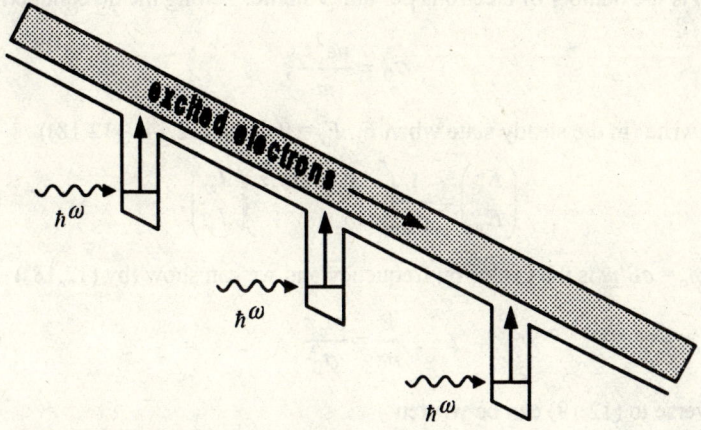

Fig. 12.13. Infrared photodetector made with quantum wells. As shown, the electrons in the wells are excited into the conduction band states and then can be collected and detected. Adapted from Capasso and Datta [12.8, p. 81]

12.7 Classical and Quantum Hall Effect (A)

12.7.1 Classical Hall Effect – CHE (A)

The Hall effect has been important for many reasons. For example, in semiconductors it can be used for determining the sign and the concentration of charge carriers. The fractional quantum Hall effect, in terms of basic physics ideas, may be the most important discovery in solid-state physics in the last quarter of a century. To start, we first reconsider the classical quantum Hall effect for electrons only.

Let electrons move in the (x,y)-plane with a magnetic field in the z direction and an electric field also in the (x,y)-plane. In MKS units and standard notation $(e > 0)$

$$F_x = -eE_x - eV_yB - \frac{mV_x}{\tau}, \qquad (12.14)$$

$$F_y = -eE_y + eV_xB - \frac{mV_y}{\tau}, \qquad (12.15)$$

where the term involving the relaxation time τ is due to scattering. The current density is given by

$$J_x = -neV_x, \qquad (12.16)$$

$$J_y = -neV_y, \qquad (12.17)$$

where n is the number of electrons per unit volume. Letting the dc conductivity be

$$\sigma_0 = \frac{ne^2\tau}{m}, \qquad (12.18)$$

we can write (in the steady state when $F_x, F_y = 0$ using (12.14)–(12.18))

$$\begin{pmatrix} E_x \\ E_y \end{pmatrix} = \frac{1}{\sigma_0}\begin{pmatrix} 1 & \omega_c\tau \\ -\omega_c\tau & 1 \end{pmatrix}\begin{pmatrix} J_x \\ J_y \end{pmatrix}, \qquad (12.19)$$

where $\omega_c = eB/m$ is the cyclotron frequency and we can show (by (12.18))

$$\frac{B}{ne} = \frac{\omega_c\tau}{\sigma_0}. \qquad (12.20)$$

The inverse to (12.19) can be written

$$\begin{pmatrix} J_x \\ J_y \end{pmatrix} = \frac{\sigma_0}{1+(\omega_c\tau)^2}\begin{pmatrix} 1 & -\omega_c\tau \\ \omega_c\tau & 1 \end{pmatrix}\begin{pmatrix} E_x \\ E_y \end{pmatrix}. \qquad (12.21)$$

We will use the geometry as shown in Fig. 12.14. We rederive the Hall coefficient. Setting $J_y = 0$, then

$$E_y = |V \times B| = -\frac{\omega_c\tau}{\sigma_0}J_x = -\frac{B}{ne}J_x, \qquad (12.22)$$

where $V = V_x = J_x/ne$ from (12.16). The Hall coefficient is defined as

$$R_H = \frac{E_y}{J_x B_z} = -\frac{1}{ne} \qquad (12.23)$$

as usual. The Hall voltage over the length w would then be

$$V_H = -E_y w = \frac{BJ_x w}{ne}. \qquad (12.24)$$

The current through the segment of area tw is

$$I_x = J_x tw, \qquad (12.25)$$

so

$$V_H = \frac{BI_x}{nte}. \qquad (12.26)$$

12.7 Classical and Quantum Hall Effect (A)

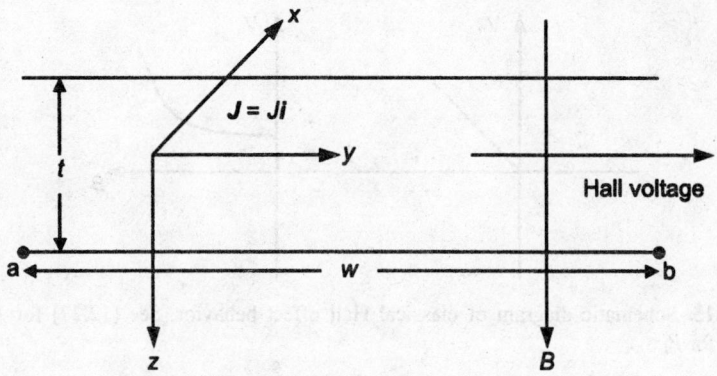

Fig. 12.14. Schematic diagram of classical Hall effect

Define n_a as the number of electrons per unit area (projected into the (x,y)-plane) so the Hall voltage can be written

$$V_H = \frac{I_x B}{n_a e}. \tag{12.27}$$

The Hall conductance $1/R_{xy}$ is

$$\frac{1}{R_{xy}} = \frac{I_x}{V_H} = \frac{n_a e}{B}. \tag{12.28}$$

Longitudinally over a length L, the voltage change is

$$V_L = E_x L = \frac{J_x L}{\sigma_0} = \frac{I_x L}{tw\sigma_0}, \tag{12.29}$$

which we find to be independent of B. This is the usual Drude result. However, this result is based on the assumption that all electrons are moving with the same velocity. If we allow the electrons to have a distribution of velocities by doing a proper Boltzmann equation calculation, we find there is a magnetoresistance effect. The result is (Blakemore [12.3]).

$$\sigma = \frac{\sigma_0}{1 + (\sigma_0 R_H)^2 \frac{|\mathbf{J} \times \mathbf{B}|^2}{|\mathbf{J}|^2}}. \tag{12.30}$$

In addition, when band-structure effects are taken into account one finds there also may be a magnetoresistance even when $\mathbf{J} \times \mathbf{B} = 0$. Classically then we predict behavior for the Hall effect (with I_x = constant) as shown schematically in Fig. 12.15.

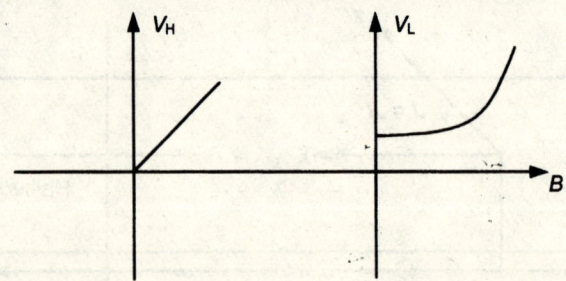

Fig. 12.15. Schematic diagram of classical Hall effect behavior. See (12.27) for V_H and (12.29) for V_L

12.7.2 The Quantum Mechanics of Electrons in a Magnetic Field: The Landau Gauge (A)

We start by solving the problem of electrons moving in two dimensions (x, y) in a magnetic field in the z direction (see, e.g., [12.41, 12.51, 12.56, 12.59]). The essential ideas of the quantum Hall effect can be made clear by ignoring electron spin, and so we do. The limit to two dimensions is necessary for the quantum Hall effect as we will discuss later. The discussion of Landau diamagnetism (Sect. 3.2.2) may be helpful here as a review of the quantum mechanics of electrons in magnetic fields.

For $\boldsymbol{B} = B\boldsymbol{k}$, one choice of A is:

$$A = -\frac{1}{2} \boldsymbol{r} \times \boldsymbol{B} , \qquad (12.31)$$

which is a cylindrically symmetric gauge. Instead, we use the Landau gauge where $A_x = -yB$, $A_y = 0$, and $A_z = 0$. This yields a simpler solution for the Hall situation that we consider.

The free-electron Hamiltonian can then be written

$$\mathcal{H} = \frac{1}{2m}[p - qA]^2 , \qquad (12.32)$$

where $q = -e < 0$. In two dimensions this becomes (compare Sect. 3.2.2)

$$\mathcal{H} = \frac{1}{2m}\left[\left(\frac{\hbar}{i}\frac{\partial}{\partial x} - eyB\right)^2 - \hbar^2 \frac{\partial^2}{\partial y^2}\right]. \qquad (12.33)$$

Introducing the "magnetic length"

$$l_\mu = \sqrt{\frac{\hbar}{eB}} , \qquad (12.34)$$

we can then write the Schrödinger equation as

$$-\frac{\hbar^2}{2m}\left[\frac{\partial^2}{\partial y^2}-\left(\frac{1}{i}\frac{\partial}{\partial x}-\frac{y}{l_\mu^2}\right)^2\right]\psi = E\psi. \quad (12.35)$$

We seek a solution of the form

$$\psi = Ae^{ikx}\varphi(y),$$

and thus

$$\left[-\frac{\hbar^2}{2m}\frac{\partial^2}{\partial y^2}+\frac{\hbar^2}{2ml_\mu^4}(y-l_\mu^2 k)^2\right]\varphi = E\varphi. \quad (12.36)$$

Since also

$$l_\mu = \sqrt{\frac{\hbar}{m\omega_c}},$$

and from (12.34) and from the preceding equation for l_μ, we have

$$\frac{\hbar^2}{2ml_\mu^4} = \frac{1}{2}m\omega_c^2.$$

This may be recognized as a harmonic oscillator equation with the quantized energies

$$E_n = \hbar\omega_c\left(n+\frac{1}{2}\right), \quad n = 0,1,2\ldots,$$

and the eigenfunctions are

$$\varphi_n = \left(\frac{m\omega_c}{\pi\hbar}\right)^{1/4}\frac{1}{\sqrt{2^n n!}}H_n\left(\frac{y}{l_\mu}-l_\mu k\right)\exp\left[-\frac{1}{2}\left(\frac{y-l_\mu^2 k}{l_\mu}\right)^2\right], \quad (12.37)$$

where the $H_n(x)$ are the Hermite polynomials

$$H_0(x)=1, \quad H_1(x)=2x, \quad H_2(x)=4x^2-2, \quad \text{etc.}$$

For the Hall effect we now solve for the case in which there is also an electric field in the y direction (the Hall field). This adds a potential of

$$U = eEy. \quad (12.38)$$

The drift velocity in crossed E and B fields is

$$V = \frac{E}{B},$$

so by (12.38), the above, and $\omega_c = eB/m$

$$U = m\omega_c Vy. \tag{12.39}$$

Thus we can write from (12.38):

$$\left[-\frac{\hbar^2}{2m}\frac{\partial^2}{\partial y^2} + \frac{1}{2}m\omega_c^2(y - l_\mu^2 k)^2 + m\omega_c Vy\right]\varphi = E\varphi. \tag{12.40}$$

Now since V is very small, we can neglect terms involving the square of V. Then if we define the origin so $y = y' - aV$, with $a = 1/\omega_c$, the Schrödinger equation simplifies to the same form as (12.36):

$$\left[-\frac{\hbar^2}{2m}\frac{\partial^2}{\partial y^2} + \frac{1}{2}m\omega_c^2(y' - l_\mu^2 k)^2\right]\varphi = [E - m\omega_c l_\mu^2 kV]\varphi. \tag{12.41}$$

Thus using (12.37) in new notation,

$$\varphi_n \propto H_n\left(\frac{y + V/\omega_c}{l_\mu} - l_\mu k\right)\exp\left[-\frac{1}{2}\left(\frac{y + V/\omega_c - l_\mu^2 k}{l_\mu}\right)^2\right], \tag{12.42}$$

and

$$E_n = \hbar\omega_c\left(n + \frac{1}{2}\right) + m\omega_c l_\mu^2 kV. \tag{12.43}$$

Now let us discuss some qualitative results related to these states.

12.7.3 Quantum Hall Effect: General Comments (A)

We first present the basic experimental results of the quantum Hall effect and then indicate how it can be explained. We have already described the Hall geometry. The Hall resistance is V_H/I_x, where I_x may be held constant. The longitudinal resistance is V_L/I_x. One finds plateaus at values of $(h/e^2)/v$ with e^2/h being called the quantum of conductance and v is an integer for the integer quantum Hall effect and a fraction for the fractional quantum Hall effect.

As shown in Fig. 12.16, V_L/I_x appears to be zero when the Hall resistance is on a plateau. The figures only schematically illustrate the effect for $v = 2$, 1, and 1/3. There are many other plateaus, which we have omitted.

The quantum Hall effect requires two dimensions, low temperatures, electrons, and a large external magnetic field. Two dimensions are necessary so the gaps in between the Landau levels ($E_g = \hbar\omega_c$) are not obliterated by the continuous energy introduced by motion in the third dimension. (The IQHE involves filled or empty Landau levels. Gaps for the FQHE, which involve partially filled Landau levels, are introduced by electron–electron interactions.) Low temperatures are necessary so as not to wipe out the quantization of levels by thermal-broadening effects.

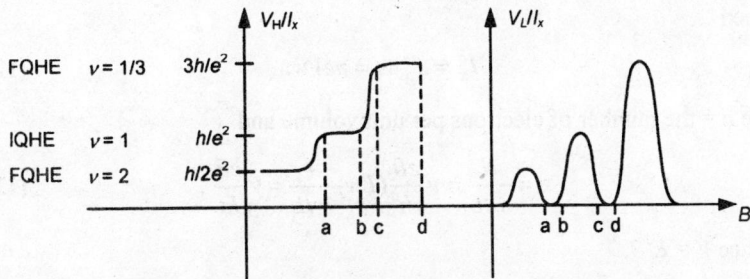

Fig. 12.16. Schematic diagram of quantum Hall effect behavior

There are two convenient ways to produce the two dimensional electron systems (2DES). One way is with MOSFETs. In a MOSFET a positive gate voltage can create a 2DES in an inversion region at the Si and SiO_2 interface. One can also use GaAs and AlGaAs heterostructures with donor doping in the AlGaAs so the electrons go to the GaAs region that has lower potential. This separates the electrons from the donor impurities and hence the electrons can have high mobility due to low scattering of them.

The IQHE was discovered by Klaus von Klitzing in 1980 and for this he was awarded the Nobel prize in 1985. About two years later, Stormer and Tsui discovered the FQHE and they along with Laughlin (for theory) were awarded the 1998 Nobel prize for this effect.

Qualitatively, the IQHE can be fairly simply explained. As each Landau level is filled there is a gap to the next Landau level. The gap is filled by localized nonconducting states, and as the Fermi level moves through this gap, no change in current is observed. The Landau levels themselves are conducting. For the IQHE the electron–electron interactions effects are really not important, but the disorder that causes the localized states in the gap is crucial.

The fractional quantum Hall effect occurs for partially filled Landau levels and electron–electron interaction effects are crucial. They produce an excitation gap reminiscent of the gap produced in the Mott insulating transition. Potential fluctuations cause localized states and plateau formation.

The Integer Quantum Hall Effect – IQHE–Simple Picture (A)

We give an elementary picture of the IQHE. We start with four results.
a. The Landau degeneracy per spin is eB/h. (This follows because the number of states per unit area in k-space (ΔA) and in real space is $(\Delta A)/(2\pi)^2$. Then from (6.29), $(\Delta A) = (2\pi)^2(eB/h)$. Thus, the number of states per unit area in real space is $n_B \equiv eB/h$).
b. The drift velocity perpendicular to E and B field is $V = E/B$.
c. Flux quanta have the value $\Phi_0 = h/e$ (see (8.47)).
d. The number of filled Landau levels $v = N/N_\Phi$, where N is the number of electrons and N_Φ is the number of flux quanta. This follows from $v = N/(eBLw/h) = N/(\Phi/\Phi_0)$.

Then

$$I_x = J_x wt = neVwt, \qquad (12.44)$$

where n = the number of electrons per unit volume and

$$n = \frac{N}{wtL} = v\frac{eB}{h}(Lw)\frac{1}{wtL} = v\frac{eB}{ht}. \qquad (12.45)$$

So since $V = E/B$,

$$I_x = v\frac{eB}{ht}e\frac{E}{B}wt = ve^2\frac{Ew}{h} = ve^2\frac{V_H}{h}, \qquad (12.46)$$

or

$$\frac{I}{V_H} = \frac{1}{R_{xy}} = \text{the Hall conductance} = \frac{ve^2}{h}. \qquad (12.47)$$

If B changes, as long as the Fermi level stays in the gap, the Landau levels are filled or empty and the current over the voltage remains on a plateau of fixed n. (It can be shown that the total current carried by a full Landau level remains constant even as the number of electrons that fill it varies with the Landau degeneracy.)

Incidentally, when $1/R_{xy} = ve^2/h$ then $1/R_{xx} = I/V_L \to \infty$ or $R_{xx} \to 0$. This is because the electrons in conducting states have no available energy states into which they can scatter.

Fractional Quantum Hall Effect – FQHE (A)

One needs to think about the FQHE both by thinking about the Laughlin wave functions and by thinking of their physical interpretation. For example, for the $v = 1/3$ case with $m = 3$ (see general comments, next section), the wave function is (see [12.41]):

$$\psi(z_1,...,z_N) = \prod_{j<k}^{N}(z_j - z_k)^m \exp\left(-\frac{1}{4l_\mu^2}\sum_{j=1}^{N}|z_j|^2\right), \qquad (12.48)$$

where $z_j = x_j + iy_j$ locates the jth electron in 2D. Positive and negative excitations at $z = z_0$ are given by (see also [12.59])

$$\psi^+ = \exp\left(-\frac{1}{4l_\mu^2}\sum_{j=1}^{N}|z_j|^2\right)\prod_{j}^{N}(z_j - z_0)\prod_{j<k}^{N}(z_j - z_k)^m, \qquad (12.49)$$

$$\psi^- = \exp\left(-\frac{1}{4l_\mu^2}\sum_{j=1}^{N}|z_j|^2\right)\prod_{j}^{N}\left(2l_\mu^2\frac{\partial}{\partial z_j} - z_0^*\right)\prod_{j<k}^{N}(z_j - z_k)^m. \qquad (12.50)$$

For $m = 3$, these excitations have effective charges of magnitude $e/3$. The ground state of the FQHE is considered to be like an incompressible fluid as the density is determined by the magnetic field and is fairly rigidly locked. The papers by Laughlin should be consulted for full details.

These wave functions have led to the idea of composite particles (CPs). Rather than considering electrons in 2D in a large magnetic field, it turns out to be possible to consider an equivalent system of electrons plus attached field vortices (see Fig. 16, p. 885 in [12.51]). The attached vortices account for most of the magnetic field and the new particles can be viewed as weakly interacting because the vortices minimize the electron–electron interactions.

General Comments (A)

It turns out that the composite particles may behave as either bosons or fermions according to the number of attached flux quanta. Electrons plus an odd number of surrounding flux quanta are Bose CPs and electrons with an even number of attached quanta are Fermi CPs. The $v = 1/3$, $m = 3$ case involves electrons with three attached quanta and hence these CPs are bosons that can undergo a Bose–Einstein-like condensation, produce an energy gap, and have a FQHE with plateaus. For the $v = 1/2$ case, there are two attached quanta, the systems behaves as a collection of fermions, there is no Bose–Einstein condensation and no FQHE.

In general, when the magnetic field increases, electrons can "absorb" some field and become "anyons." These can be shown to obey fractional statistics and seem to be intermediate between fermions and bosons. This topic takes us too far afield and references should be consulted.[2]

There are different ways to construct CPs to describe the same physical situation, but normally one tries to use the simplest. Also, there are still problems connected with the understanding of some values of v. A complete description would take us further than we intend to go, but the chapter references listed at the end of this book can be a good starting point for further investigation.

The quantum Hall effects are very rich in physical effects. So far, they are not so rich in applications. However, the experiments do determine e^2/h to three parts in ten million or better, and hence they provide an excellent resistance standard. Also, since the speed of light is a defined quantity, the QHE also determines the fine structure constant $e^2/\hbar c$ to high accuracy. It is interesting that the quantum Hall effect determines e^2/h, while we found earlier that e/h could be determined by the Josephson effect. Thus the two can be used to determine both e and h individually.

---

[2] See Lee [12.43].

12.8 Carbon – Nanotubes and Fullerene Nanotechnology (EE)

Carbon is very versatile and important both to living tissues and to inanimate materials. Carbon of course forms diamond and graphite. In recent years the ability of carbon to aggregate into fullerenes and nanotubes has been much discussed.

Fullerenes are stable, cage-like molecules of carbon with often a nearly spherical appearance. A C_{60} molecule is also called a Buckyball. Both are named after Buckminster Fuller because of their resemblance to the geodesic domes he designed. Buckyballs were discovered in 1985 as a byproduct of laser-vaporized graphite. Some of the fullerides (salts such as $K-C_{60}$) can be superconductors (see, e.g., Hebard [12.25]).

Carbon nanotubes are one or more cylindrical and seamless shells of graphitic sheets. Their ends are capped by half of a fullerene molecule. They were discovered in 1993 by Sumio Iijima and mass produced in 1995 by Rick Smalley. For more details see, e.g., Dresselhaus et al [12.17]. While carbon nanotubes are now easy to produce, they are not easy to produce in a controlled fashion.

To form them, start with a single sheet of graphite called graphene whose band structure leads to a semimetal (where the conduction band edge is very close to the valence band edge). A picture of the dispersion relations show a two-dimensional E vs. k relationship where two cones touch at their tips with the same conic axis and in an end-to-end fashion. See Fig. 12.17. Where the cones touch is the Fermi energy, or as it is called, the Fermi point.

Nanotubes can be semiconductors or metals. It depends on the boundary conditions on the wave function as determined by how the sheet is rolled up. Both the circumference and twist are important. This, in turn, affects whether a bandgap is introduced where the Fermi point in graphene was. The semiconducting bandgap can be varied by the circumference. Multiwalled nanotubes are more complex.

Semiconductor nanotubes can be made to act as transistors by using a gate voltage. A negative bias (to the gate) induces holes and makes them conduct. Positive bias makes the conductance shut off. They have even been made to act as simple logic devices. See McEven PL, "Single-Wall Carbon Nanotubes," *Physics World*, pp. 32-36 (June 2000). One interesting feature about nanotubes is that they provide a way around the fundamental size limits of Si devices. This is because they can be made very small and are not plagued with surface states (they have no surface formed by termination of a 3D structure and as cylinders they have no edges).

Carbon nanotubes are a fascinating example of one-dimensional transport in hopefully easy to make structures. They are quantum wires with ballistic electrons – and they show many interesting quantum effects.

An additional feature of interest is that carbon nanotubes show significant mechanical strength. Their strength arises from the carbon bond.[3]

---

[3] Carbon is becoming an increasingly interesting material with the suggestion that under certain circumstances it can even be magnetic. See Coey M and Sanvito S, Physics World, Nov 2004, p33ff.

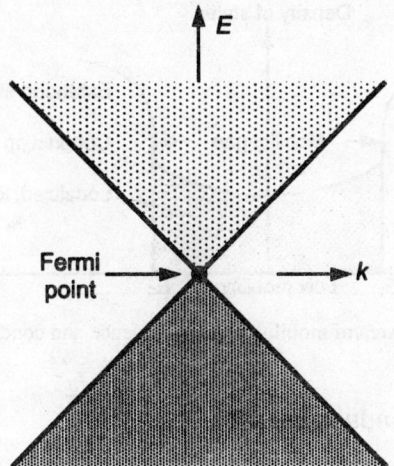

Fig. 12.17. Dispersion relation for graphene

12.9 Amorphous Semiconductors and the Mobility Edge (EE)

By amorphous, we will mean noncrystalline. Here, rather than an energy gap one has a mobility gap separating localized and nonlocalized states. The localization of electron states is an important concept. The electron–electron interaction itself may give rise to localization as shown by Mott [12.48], as we have discussed earlier in the book. In effect, the electron–electron interaction can split the originally partially filled band into a filled band and an empty band separated by a bandgap. We are more interested here in the Anderson localization transition caused by random local field fluctuations due to disorder. In amorphous semiconductors, this can lead to "mobility edges" rather than band edges (see Fig. 12.18).

The dc conductivity of an amorphous semiconductor is of the form

$$\sigma = \sigma_0 \exp\left(-\frac{\Delta E}{kT}\right), \qquad (12.51)$$

for charge transport by *extended state carriers*, where ΔE is of the order of the mobility gap and σ_0 is a conductor. For *hopping of localized carriers*

$$\sigma = \sigma_0 \exp\left(-\frac{T_0}{T}\right)^{1/4}, \qquad (12.52)$$

where σ_0 and T_0 are constants. Memory and switching devices have been made with amorphous chalcogenide semiconductors. The meaning of (12.52) is amplified in the next section.

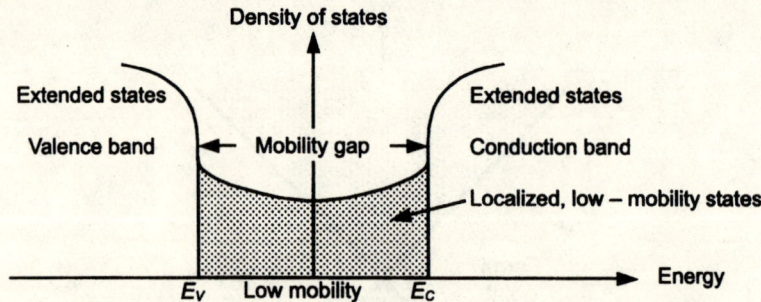

Fig. 12.18. Area of mobility between valence and conduction bands

12.9.1 Hopping Conductivity (EE)

So far, we have discussed band conductivity. Here electrons move along at constant energy, in the steady state the energy they gain from the field is dissipated by collisions. One can even have band conductivity in impurity bands when the impurity wave functions overlap sufficiently to form a band. One usually thinks of impurity states as being localized, and for localized states there is no dc conductivity at absolute zero. However, at nonzero temperatures, an electron in a localized state may make a transition to an empty localized state, getting any necessary energy from a phonon, for example. We say the electron hops from state to state. In general, then, an electron hop is a transition of the electron involving both its position and energy.

The topic of hopping conductivity is very complicated and a thorough treatment would take us too far afield. The books by Shoklovskii and Efros [12.55], and Mott [12.48], together with copious references cited therein, can be consulted. In what is given below, we are primarily concerned with hopping conductivity in lightly doped semiconductors.

Suppose the electron jumps to a state a distance R. We assume very low temperatures with the relevant states localized near the Fermi energy. We assume states just below the Fermi energy hop to states just above gaining the energy E_a (from a phonon). Letting $N(E)$ be the number of states per unit volume, we estimate:

$$\frac{1}{E_a} \approx \frac{4}{3}\pi R^3 N(E_F), \qquad (12.53)$$

thus we estimate (see Mott [12.48]) the hopping probability and hence the conductivity is proportional to

$$\exp(-2\alpha R - E_a/kT), \qquad (12.54)$$

where α is a constant denoting the rate of exponential decrease of the wave function of the localized state $\exp(-\alpha r)$.

Substituting (12.53) into (12.54) and maximizing the expression with regard to the hopping range R gives:

$$\sigma = \sigma_0 \exp[(-T_0/T)^{1/4}], \quad (12.55)$$

where

$$T_0 = 1.5\alpha^3 \beta / N(E_F), \quad (12.56)$$

and β is a constant, whose value follows from the derivation, but in fact needs to be more precisely evaluated in a more rigorous presentation.

Maximizing also yields

$$R = \text{constant}(1/T)^{1/4}, \quad (12.57)$$

so the theory is said to be for variable-range hopping (VRH); the lower the temperature, the longer the hopping range and the less energy is involved.

Equation (12.55), known as Mott's law, is by no means a universal expression for the hopping conductivity. This law may only be true near the Mott transition, and even then that is not certain. Electron–electron interactions may cause a Coulomb gap (Coulombic correlations may lead the density of states to vanish at the Fermi level), and lead to a different exponent (from one quarter–actually to 1/2 for low-temperature VRH).

12.10 Amorphous Magnets (MET, MS)

Magnetic effects are typically caused by short-range interactions, and so they are preserved in the amorphous state although the Curie temperature is typically lowered. A rapid quench of a liquid metallic alloy can produce an amorphous alloy. When the alloy is also magnetic, this can produce an amorphous magnet. Such amorphous magnets, if isotropic, may have low anisotropy and hence low coercivities. An example is $Fe_{80}B_{30}$, where the boron is used to lower the melting point, which makes quenching easier. Transition metal amorphous alloys such as $Fe_{75}P_{15}C_{10}$ may also have very small coercive forces in the amorphous state.

On the other hand, amorphous NdFe may have a high coercivity if the quench is slow so as to yield a multicrystalline material. Rare earth alloys (with transition metals) such as $TbFe_2$ in the amorphous state may also have giant coercive fields (~ 3 kOe). For further details, see [12.20, 12.26, 12.36].

We should mention that bulk amorphous steel has been made. It has approximately twice the strength of conventional steel. See Lu et al [12.44].

Nanomagnetism is also of great importance, but is not discussed here. However, see the relevant chapter references at the end of this book.

12.11 Soft Condensed Matter (MET, MS)

12.11.1 General Comments

Soft condensed-matter physics occupies an intermediate place between solids and fluids. We can crudely say that soft materials will not hurt your toe if you kick them.

Generally speaking, hard materials are what solid-state physics discusses and the focus of this book was crystalline solids. Another way of contrasting soft and hard materials is that soft ones are typically not describable by harmonic excitations about the ground-state equilibrium positions. Soft materials are also often complex, as well as flexible. Soft materials have a shape but respond more easily to forces than crystalline solids.

Soft condensed-matter physics concerns itself with *liquid crystals* and *polymers*, which we will discuss, and fluids as well as other materials that feel soft. Also included under the umbrella of soft condensed matter are *colloids*, *emulsions*, and *membranes*. As a reminder, colloids are solutes in a solution where the solute clings together to form 'particles,' and emulsions are two-phase systems with the dissolved phase being minute drops of a liquid. A membrane is a thin, flexible sheet that is often a covering tissue. Membranes are two-dimensional structures built from molecules with a hydrophilic head and a hydrophobic tail. They are important in biology.

For a more extensive coverage the books by Chaikin and Lubensky [12.11], Isihara [12.27], and Jones [12.30] can be consulted.

We will discuss liquid crystals in the next Section and then we have a Section on polymers, including rubbers.

12.11.2 Liquid Crystals (MET, MS)

Liquid crystals involve phases that are intermediate between liquids and crystals. Because of their intermediate character some call them *mesomorphic* phases. Liquid crystals consist of highly anisotropic weakly coupled (often rod-like) molecules. They are liquid-like but also have some anisotropy. The anisotropic properties of some liquid crystals can be changed by an electric field, which affects their optical properties, and thus watch displays and screens for computer monitors have been developed. J. L. Fergason [12.19] has been one of the pioneers in this as well as other applications.

There are two main classes of liquid crystals: nematic and smectic. In nematic liquid crystals the molecules are partly aligned but their position is essentially random. In smectic liquid crystals, the molecules are in planes that can slide over each other. Nematic and smectic liquid crystals are sketched in Fig. 12.19. An associated form of the nematic phase is the cholesteric. Cholesterics have a director (which is a unit vector along the average axis of orientation of the rod-like molecules) that has a helical twist.

Liquid crystals still tend to be somewhat foreign to many physicists because they involve organic molecules, polymers, and associated structures. For more details see deGennes PG and Prost [12.15] and Isihara [12.27 Chap. 12].

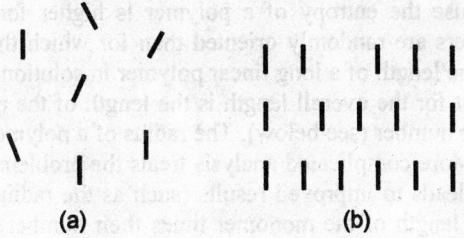

Fig. 12.19. Liquid crystals. (a) Nematic (long-range orientational order but no long-range positional order), and (b) Smectic (long-range orientational order and in one dimension long-range positional order)

12.11.3 Polymers and Rubbers (MET, MS)

Polymers are a classic example of soft condensed matter. In this section, we will discuss polymers[4] and treat rubber as a particular example.

A monomer is a simple molecule that can join with itself or similar molecules (many times) to form a giant molecule that is referred to as a polymer. (From the Greek, polys – many and meros – parts). A polymer may be either naturally occurring or synthetic. The number of repeating units in the polymer is called the degree of polymerization (which is typically of order 10^3 to 10^5). Most organic substances associated with living matter are polymers, thus examples of polymers are myriad. Plastics, rubbers, fibers, and adhesives are common examples. Bakelite was the first thermosetting plastic found. Rayon, Nylon, and Dacron (polyester) are examples of synthetic fibers. There are crystalline polymer fibers such as cellulose (wood is made of cellulose) that diffract X-rays and by contrast there are amorphous polymers (rubber can be thought of as made of amorphous polymers) that don't show diffraction peaks.

There are many subfields of polymers of which rubber is one of the most important. A rubber consists of many long chains of polymers connected together somewhat randomly. The chains themselves are linear and flexible. The random linking bonds give shape. Rubbers are like liquids in that they have a well-defined volume, but not a well-defined shape. They are like a solid in that they maintain their shape in the absence of forces. The most notable property of rubbers is that

---

[4] As an aside we mention the connection of polymers with fuel cells, which have been much in the news. In 1839 William R. Grove showed the electrochemical union of hydrogen and oxygen generates electricity—the idea of the fuel cell. Hydrogen can be extracted from say methanol, and stored in, for example, metal hydrides. Fuel cells can run as long as hydrogen and oxygen are available. The only waste is water from the fuel-cell reaction. In 1960 synthetic polymers were introduced as electrolytes.

they have a very long and reversible elasticity. Vulcanizing soft rubber, by adding sulfur and heat treatment makes it harder and increases its strength. The sulfur is involved in linking the chains.

A rubber can be made by repetition of the isoprene group (C_5H_8, see Fig. 12.20).[5] Because the entropy of a polymer is higher for configurations in which the monomers are randomly oriented than for which they are all aligned, one can estimate the length of a long linear polymer in solution by a random-walk analysis. The result for the overall length is the length of the monomer times the square root of their number (see below). The radius of a polymer in a ball is given by a similar law. More complicated analysis treats the problem as a self-avoiding random walk and leads to improved results (such as the radius of the ball being approximately the length of the monomer times their number to the 3/5 power). Another important feature of polymers is their viscosity and diffusion. The concept of reptation (which we will not discuss here, see Doi and Edwards [12.16]), which means snaking, has proved to be very important. It helps explain how one polymer can diffuse through the mass of the others in a melt. One thinks of the Brownian motion of a molecule along its length as aiding in disentangling the polymer.

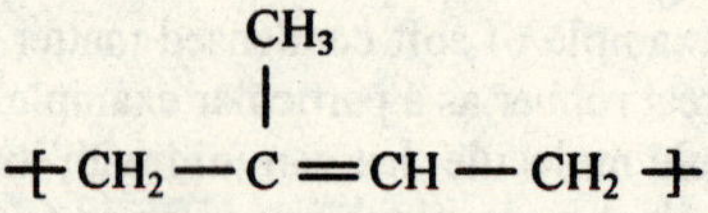

$$+CH_2-\underset{\underset{CH_3}{|}}{C}=CH-CH_2+$$

Fig. 12.20. Chemical structure of isoprene (the basic unit for natural rubber)

We first give a one-dimensional model to illustrate how the length of a polymer can be estimated from a random-walk analysis. We will then discuss a model for estimating the elastic constant of a rubber.

We suppose N monomers of length a linked together along the x-axis. We suppose the ith monomer to be in the $+x$ direction with probability of $1/2$ and in the $-x$ direction with the same probability. The rms length R of the polymer is calculated below.

Let $x_i = a$ for the monomer in the $+x$ direction and $-a$ for the $-x$ direction. Then the total length is $x = \sum x_i$ and the average squared length is

$$\left\langle x^2 \right\rangle = \left\langle \sum x_i \right\rangle^2 = \sum \left\langle x_i^2 \right\rangle, \quad (12.58)$$

since the cross terms drop out, so

$$\left\langle x^2 \right\rangle = Na^2, \quad (12.59)$$

or

$$R = a\sqrt{N}. \quad (12.60)$$

---

[5] See, e.g., Brown et al [12.4]. See also Strobl [12.57].

We have already noted that a similar scaling law applies to the radius of a N-monomer polymer coiled in a ball in three dimensions.

In a similar way, we can estimate the tension in the polymer. This model or generalizations of it to two or three dimensions (See, e.g., Callen [12.7]) seem to give the basic idea. Let n^+ and n^- represent the links in the + and − directions. The length x is

$$x = (n^+ - n^-)a, \qquad (12.61)$$

and the total number of monomers is

$$N = (n^+ + n^-). \qquad (12.62)$$

Thus

$$n^+ = \frac{1}{2}\left(N + \frac{x}{a}\right), \quad n^- = \frac{1}{2}\left(N - \frac{x}{a}\right). \qquad (12.63)$$

The number of ways we can arrange N monomers with n^+ in the $+x$ direction and n^- in the − direction is

$$W = \frac{N!}{n^+! n^-!}. \qquad (12.64)$$

Using $S = k\ln(W)$ and using Stirling's approximation, we can find the entropy S. Then since $dU = TdS + Fdx$, where T is the temperature, U the internal energy and F the tension, we find

$$F = -T\frac{\partial S}{\partial x} + \frac{\partial U}{\partial x}, \qquad (12.65)$$

so we find (assuming we use a model in which $\partial U/\partial x$ can be neglected)

$$F = \frac{kT}{2a}\ln\left(\frac{1 + x/Na}{1 - x/Na}\right) = \frac{kTx}{Na^2} \quad \text{(if } x \ll Na\text{)}. \qquad (12.66)$$

The tension F comes out to be proportional to both the temperature and the extension x (it becomes stiffer as the temperature is raised!). Another way to look at this is that the polymer contracts on warming. In 3D, we think of the polymer curling up at high temperatures and the entropy increasing.

Problems

12.1 If the periodicity $p = 50$ Å and $E = 5 \times 10^4$ V/cm, calculate the fundamental frequency for Bloch oscillations. Compare the results to relaxation times τ typical for electrons, i.e. compute $\omega_B \tau$.

12.2 Find the minimum radius of a spherical quantum dot whose electron binding energy is at least 1 eV.

12.3 Discuss how the Kronig–Penny model can be used to help understand the motion of electrons in superlattices. Discuss both transverse and in-plane motion. See, e.g., Mitin et al [12.47 pp. 99-106].

12.4 Consider a quantum well parallel to the (x,y)-plane of width w in the z direction. For simplicity assume the depth of the quantum well is infinite. Assume also for simplicity that the effective mass is a constant m for motion in all directions, See, e.g., Shik [12.54, Chaps. 2 and 4].

a) Show the energy of an electron can be written

$$E = \frac{\hbar^2 \pi^2 n^2}{2mw^2} + \frac{\hbar^2}{2m}(k_x^2 + k_y^2),$$

where $p_x = \hbar k_x$ and $p_y = \hbar k_y$ and n is an integer.

b) Show the density of states can be written

$$D(E) = \frac{m}{\pi \hbar^2} \sum_n \theta(E - E_n),$$

where $D(E)$ represents the number of states per unit area per unit energy in the (x,y)-plane and

$$E_n = \frac{\hbar^2 \pi^2}{2mw^2} n^2.$$

$\theta(x)$ is the step function $\theta(x) = 0$ for $x < 0$ and $= 1$ for $x > 0$.

c) Show also $D(E)$ at $E \geq E_3$ is the same as $D_{3D}(E)$ where D_{3D} represents the density of states in 3D without the quantum well (still per unit area in the (x,y)-plane for a width w in the z direction)

d) Make a sketch showing the results of b) and c) in graphic form.

12.5 For the situation of Problem 12.4 impose a magnetic field B in the z direction. Show then that the allowed energies are discrete with values

$$E_{n,p} = \frac{\hbar^2 \pi^2 n^2}{2mw^2} + \hbar \omega_c \left(p + \frac{1}{2}\right),$$

where n, p are integers and $\omega_c = |eB/m|$ is the cyclotron frequency. Show also the two-dimensional density of states per spin (and per unit energy and area in (x,y)-plane) is

$$D(E) = \frac{eB}{h} \sum_p \delta\left[E - \hbar\omega\left(p + \frac{1}{2}\right)\right],$$

when

$$\frac{\hbar^2 \pi^2}{2mw^2} < E < \frac{4\hbar^2 \pi^2}{2mw^2}.$$

These results are applicable to a 2D Fermi gas, see, e.g., Shik [12.54, Chaps. 7] as well as 12.7.2 and 12.7.3.

Appendices

A Units

The choice of a system of units to use is sometimes regarded as an emotionally charged subject. Although there are many exceptions, experimental papers often use mksa (or SI) units, and theoretical papers may use Gaussian units (or perhaps a system in which several fundamental constants are set equal to one).

All theories of physics must be checked by comparison to experiment before they can be accepted. For this reason, it is convenient to express final equations in the mksa system. Of course, much of the older literature is still in Gaussian units, so one must have some familiarity with it. The main thing to do is to settle on a system of units and stick to it. Anyone who has reached the graduate level in physics can convert units whenever needed. It just may take a little longer than we wish to spend.

In this appendix, no description of the mksa system will be made. An adequate description can be found in practically any sophomore physics book.[1]

In solid-state physics, another unit system is often convenient. These units are called Hartree atomic units. Let e be the charge on the electron, and m be the mass of the electron. The easiest way to get the Hartree system of units is to start from the Gaussian (cgs) formulas, and let $|e|$ = Bohr radius of hydrogen = $|m|$ = 1. The results are summarized in Table A.1. The Hartree unit of energy is 27.2 eV. Expressing your answer in terms of the fundamental physical quantities shown in Table A.1 and then using Hartree atomic units leads to simple numerical answers for solid-state quantities. In such units, the solid-state quantities usually do not differ by too many orders of magnitude from one.

---

[1] Or see "Guide for Metric Practice," by Robert A. Nelson at http://www.physicstoday.org/guide/metric.html.

Table A.1. Fundamental physical quantities*

Quantity	Symbol	Expression / value in mksa units	Expression / value in Gaussian units	Value in Hartree units
Charge on electron	e	1.6×10^{-19} coulomb	4.80×10^{-10} esu	1
Mass of electron	m	0.91×10^{-30} kg	0.91×10^{-27} g	1
Planck's constant	$\hbar$	1.054×10^{-34} joule s	1.054×10^{-27} erg s	1
Compton wavelength of electron	λ_c	$2\pi(\hbar/mc)$ 2.43×10^{-12} m	$2\pi(\hbar/mc)$ 2.43×10^{-10} cm	$(2\pi)\frac{1}{137}$
Bohr radius of hydrogen	a_0	$4\pi\varepsilon_0\hbar^2/me^2$ 0.53×10^{-10} m	$\hbar^2/me^2$ 0.53×10^{-8} cm	1
Fine structure constant	α	$e^2/\hbar c$ $\frac{1}{137}$ (approx.)	$e^2/\hbar c$ $\frac{1}{137}$	$\frac{1}{137}$
Speed of light	c	3×10^8 m s^{-1}	3×10^{10} cm s^{-1}	137
Classical electron radius	r_0	$e^2/4\pi\varepsilon_0 mc^2$ 2.82×10^{-15} m	e^2/mc^2 2.82×10^{-13} cm	$(\frac{1}{137})^2$
Energy of ground state of hydrogen (1 Rydberg)	E_0	$e^4 m/32(\pi\varepsilon_0\hbar)^2$ 13.61† eV	$me^4/2\hbar^2$ 13.61† eV	$\frac{1}{2}$
Bohr magneton (calculated from above)	μ_B	$e\hbar/2m$ 0.927×10^{-23} amp meter^2	$e\hbar/2mc$ 0.927×10^{-20} erg gauss^{-1}	$\frac{1}{274}$
Cyclotron frequency (calculated from above)	ω_c, or ω_h	$(\mu_0 e/2m)(2H)$	$(e/2mc)(2H)$	$\frac{1}{274}(2H)$

* The values given are greatly rounded off from the standard values. The list of fundamental constants has been updated and published yearly in part B of the August issue of *Physics Today*. See, e.g., Peter J. Mohr and Barry N. Taylor, "The Fundamental Physical Constants," *Physics Today*, pp. BG6-BG13, August, 2003. Now see http://www.physicstoday.org/guide/fundcon.html.
† 1 eV = 1.6×10^{-12} erg = 1.6×10^{-19} joule.

B Normal Coordinates

The main purpose of this appendix is to review clearly how the normal coordinate transformation arises, and how it leads to a diagonalization of the Hamiltonian. Our development will be made for classical systems, but a similar development can be made for quantum systems. An interesting discussion of normal modes has been given by Starzak.[2] The use of normal coordinates is important for collective excitations such as encountered in the discussion of lattice vibrations.

We will assume that our mechanical system is described by the Hamiltonian

$$\mathcal{H} = \tfrac{1}{2}\sum_{i,j}(\dot{x}_i\dot{x}_j\delta_{ij} + v_{ij}x_ix_j). \tag{B.1}$$

In (B.1) the first term is the kinetic energy and the second term is the potential energy of interaction among the particles. We consider only the case that each particle has the same mass that has been set equal to one. In (B.1) it is also assumed that $v_{ij} = v_{ji}$; and that each of the v_{ij} is real. The coordinates x_i in (B.2) are measured from equilibrium that is assumed to be stable. For a system of N particles in three dimensions, one would need $3N$ x_i to describe the vibration of the system. The dot of $\dot{x}_i$ of course means differentiation with respect to time, $\dot{x}_i = dx_i/dt$.

The Hamiltonian (B.1) implies the following equation of motion for the mechanical system:

$$\sum_j(\delta_{ij}\ddot{x}_j + v_{ij}x_j) = 0. \tag{B.2}$$

The normal coordinate transformation is the transformation that takes us from the coordinates x_i to the normal coordinates. A normal coordinate describes the motion of the system in a normal mode. In a normal mode each of the coordinates vibrates with the same frequency. Seeking a normal mode solution is equivalent to seeking solutions of the form

$$x_j = ca_j e^{-i\omega t}. \tag{B.3}$$

In (B.3), c is a constant that is usually selected so that $\sum_j|x_j|^2 = 1$, and $|ca_j|$ is the amplitude of vibration of x_j in the mode with frequency ω. The different frequencies ω for the different normal modes are yet to be determined.

Equation (B.2) has solutions of the form (B.3) provided that

$$\sum_j(v_{ij}a_j - \omega^2\delta_{ij}a_j) = 0. \tag{B.4}$$

Equation (B.4) has nontrivial solutions for the a_j (i.e. solutions in which all the a_j do not vanish) provided that the determinant of the coefficient matrix of the a_j vanishes. This condition determines the different frequencies corresponding to the different normal modes of the mechanical system. If V is the matrix whose

---

[2] See Starzak [A.25 Chap. 5].

elements are given by v_{ij} (in the usual notation), then the eigenvalues of V are ω^2, determined by (B.4). V is a real symmetric matrix; hence it is Hermitian; hence its eigenvalues must be real.

Let us suppose that the eigenvalues ω^2 determined by (B.4) are denoted by Ω_k. There will be the same number of eigenvalues as there are coordinates x_i. Let a_{jk} be the value of a_j, which has a normalization determined by (B.7), when the system is in the mode corresponding to the kth eigenvalue Ω_k. In this situation we can write

$$\sum_j v_{ij} a_{jk} = \Omega_k \sum_j \delta_{ij} a_{jk} . \tag{B.5}$$

Let A stand for the matrix with elements a_{jk} and Ω be the matrix with elements $\Omega_{lk} = \Omega_k \delta_{lk}$. Since $\Omega_k \sum_j \delta_{ij} a_{jk} = \Omega_k a_{ik} = a_{ik} \Omega_k = \sum_l a_{il} \Omega_k \delta_{lk} = \sum_l a_{il} \Omega_{lk}$, we can write (B.5) in matrix notation as

$$VA = A\Omega . \tag{B.6}$$

It can be shown [2] that the matrix A that is constructed from the eigenvectors is an orthogonal matrix, so that

$$A\tilde{A} = \tilde{A}A = I . \tag{B.7}$$

$\tilde{A}$ means the transpose of A. Combining (B.6) and (B.7) we have

$$\tilde{A}VA = \Omega . \tag{B.8}$$

This equation shows how V is diagonalized by the use of the matrix that is constructed from the eigenvectors.

We still must indicate how the new eigenvectors are related to the old coordinates. If a column matrix a is constructed from the a_j as defined by (B.3), then the eigenvectors E (also a column vector, each element of which is an eigenvector) are defined by

$$E = \tilde{A}a , \tag{B.9a}$$

or

$$a = AE . \tag{B.9b}$$

That (B.9) does define the eigenvectors is easy to see because substituting (B.9b) into the Hamiltonian reduces the Hamiltonian to diagonal form. The kinetic energy is already diagonal, so we need consider only the potential energy

$$\sum v_{ij} a_i a_j = \tilde{a} V a = \widetilde{EA} V A E = \tilde{E} \Omega E$$
$$= \sum_{k,j} (\tilde{E})_j \Omega_{jk} E_k = \sum_{j,k} (\tilde{E})_j \Omega_k \delta_{jk} E_k$$
$$= \sum_j (\tilde{E})_j \Omega_j E_j = \sum_{j,k} \omega_j^2 (\tilde{E}_i) E_i \delta_{jk} ,$$

which tells us that the substitution reduces V to diagonal form. For our purposes, the essential thing is to notice that a substitution of the form (B.9) reduces the Hamiltonian to a much simpler form.

An example should clarify these ideas. Suppose the eigenvalue condition yielded

$$\det\begin{pmatrix} 1-\omega^2 & 2 \\ 2 & 3-\omega^2 \end{pmatrix} = 0. \tag{B.10}$$

This implies the two eigenvalues

$$\omega_1^2 = 2+\sqrt{5} \tag{B.11a}$$

$$\omega_2^2 = 2-\sqrt{5}. \tag{B.11b}$$

Equation (B.4) for each of the eigenvalues gives for

$$\omega = \omega_1^2 : a_1 = \frac{2a_2}{1+\sqrt{5}}, \tag{B.12a}$$

and for

$$\omega = \omega_2^2 : a_1 = \frac{2a_2}{1-\sqrt{5}}. \tag{B.12b}$$

From (B.12) we then obtain the matrix A

$$\widetilde{A} = \begin{pmatrix} \dfrac{2N_1}{1+\sqrt{5}}, & N_1 \\ \dfrac{2N_2}{1-\sqrt{5}}, & N_2 \end{pmatrix}, \tag{B.13}$$

where

$$(N_1)^{-1} = \left[\frac{4}{(\sqrt{5}+1)^2} + 1 \right]^{1/2}, \tag{B.14a}$$

and

$$(N_2)^{-1} = \left[\frac{4}{(\sqrt{5}-1)^2} + 1 \right]^{1/2}. \tag{B.14b}$$

The normal coordinates of this system are given by

$$E = \begin{pmatrix} E_1 \\ E_2 \end{pmatrix} = \begin{pmatrix} \dfrac{2N_1}{1+\sqrt{5}}, & N_1 \\ \dfrac{2N_2}{1-\sqrt{5}}, & N_2 \end{pmatrix} \begin{pmatrix} a_1 \\ a_2 \end{pmatrix}. \tag{B.15}$$

Problems

B.1 Show that (B.13) satisfies (B.7)

B.2 Show for A defined by (B.13) that

$$\tilde{A}\begin{pmatrix} 1 & 2 \\ 2 & 3 \end{pmatrix} A = \begin{pmatrix} 2+\sqrt{5}, & 0 \\ 0, & 2-\sqrt{5} \end{pmatrix}.$$

This result checks (B.8).

C Derivations of Bloch's Theorem

Bloch's theorem concerns itself with the classifications of eigenfunctions and eigenvalues of Schrödinger-like equations with a periodic potential. It applies equally well to electrons or lattice vibrations. In fact, Bloch's theorem holds for any wave going through a periodic structure. We start with a simple one-dimensional derivation.

C.1 Simple One-Dimensional Derivation[3-5]

This derivation is particularly applicable to the Kronig–Penney model. We will write the Schrödinger wave equation as

$$\frac{d^2\psi(x)}{dx^2} + U(x)\psi(x) = 0, \qquad (C.1)$$

where $U(x)$ is periodic with period a, i.e.,

$$U(x+na) = U(x), \qquad (C.2)$$

with n an integer. Equation (C.1) is a second-order differential equation, so that there are two linearly independent solutions ψ_1 and ψ_2:

$$\psi_1'' + U\psi_1 = 0, \qquad (C.3)$$

$$\psi_2'' + U\psi_2 = 0. \qquad (C.4)$$

---

[3] See Ashcroft and Mermin [A.3].
[4] See Jones [A.10].
[5] See Dekker [A.4].

From (C.3) and (C.4) we can write

$$\psi_2\psi_1'' + U\psi_2\psi_1 = 0,$$

$$\psi_1\psi_2'' + U\psi_1\psi_2 = 0.$$

Subtracting these last two equations, we obtain

$$\psi_2\psi_1'' - \psi_1\psi_2'' = 0. \tag{C.5}$$

This last equation is equivalent to writing

$$\frac{dW}{dx} = 0, \tag{C.6}$$

where

$$W = \begin{vmatrix} \psi_1 & \psi_2 \\ \psi_1' & \psi_2' \end{vmatrix} \tag{C.7}$$

is called the *Wronskian*. For linearly independent solutions, the Wronskian is a constant not equal to zero.

It is easy to prove one result from the periodicity of the potential. By dummy variable change $(x) \to (x + a)$ in (C.1) we can write

$$\frac{d^2\psi(x+a)}{dx^2} + U(x+a)\psi(x+a) = 0.$$

The periodicity of the potential implies

$$\frac{d^2\psi(x+a)}{dx^2} + U(x)\psi(x+a) = 0. \tag{C.8}$$

Equations (C.1) and (C.8) imply that if $\psi(x)$ is a solution, then so is $\psi(x + a)$. Since there are only two linearly independent solutions ψ_1 and ψ_2, we can write

$$\psi_1(x+a) = A\psi_1(x) + B\psi_2(x) \tag{C.9}$$

$$\psi_2(x+a) = C\psi_1(x) + D\psi_2(x). \tag{C.10}$$

The Wronskian W is a constant $\neq 0$, so $W(x + a) = W(x)$, and we can write

$$\begin{vmatrix} A\psi_1 + B\psi_2 & C\psi_1 + D\psi_2 \\ A\psi_1' + B\psi_2' & C\psi_1' + D\psi_2' \end{vmatrix} = \begin{vmatrix} \psi_1 & \psi_2 \\ \psi_1' & \psi_2' \end{vmatrix} \begin{vmatrix} A & C \\ B & D \end{vmatrix} = \begin{vmatrix} \psi_1 & \psi_2 \\ \psi_1' & \psi_2' \end{vmatrix},$$

or

$$\begin{vmatrix} A & C \\ B & D \end{vmatrix} = 1,$$

or
$$AD - BC = 1. \tag{C.11}$$

We can now prove that it is possible to choose solutions $\psi(x)$ so that
$$\psi(x+a) = \Delta \psi(x), \tag{C.12}$$

where Δ is a constant $\neq 0$. We want $\psi(x)$ to be a solution so that
$$\psi(x) = \alpha \psi_1(x) + \beta \psi_2(x), \tag{C.13a}$$

or
$$\psi(x+a) = \alpha \psi_1(x+a) + \beta \psi_2(x+a). \tag{C.13b}$$

Using (C.9), (C.10), (C.12), and (C.13), we can write
$$\begin{aligned}\psi(x+a) &= (\alpha A + \beta C)\psi_1(x) + (\alpha B + \beta D)\psi_2(x) \\ &= \Delta \alpha \psi_1(x) + \Delta \beta \psi_2(x).\end{aligned} \tag{C.14}$$

In other words, we have a solution of the form (C.12), provided that
$$\alpha A + \beta C = \Delta \alpha,$$

and
$$\alpha B + \beta D = \Delta \beta.$$

For nontrivial solutions for α and β, we must have
$$\begin{vmatrix} A - \Delta & C \\ B & D - \Delta \end{vmatrix} = 0. \tag{C.15}$$

Equation (C.15) is equivalent to, using (C.11),
$$\Delta + \Delta^{-1} = A + D. \tag{C.16}$$

If we let Δ_+ and Δ_- be the eigenvalues of the matrix $\begin{pmatrix} A & C \\ B & D \end{pmatrix}$ and use the fact that the trace of a matrix is the sum of the eigenvalues, then we readily find from (C.16) and the trace condition
$$\Delta_+ + (\Delta_+)^{-1} = A + D,$$
$$\Delta_- + (\Delta_-)^{-1} = A + D, \tag{C.17}$$

and
$$\Delta_+ + \Delta_- = A + D.$$

Equations (C.17) imply that we can write

$$\Delta_+ = (\Delta_-)^{-1}. \tag{C.18}$$

If we set

$$\Delta_+ = e^b, \tag{C.19}$$

and

$$\Delta_- = e^{-b}, \tag{C.20}$$

the above implies that we can find linearly independent solutions ψ_i^1 that satisfy

$$\psi_1^1(x+a) = e^b \psi_1^1(x), \tag{C.21}$$

and

$$\psi_2^1(x+a) = e^{-b} \psi_2^1(x). \tag{C.22}$$

Real b is ruled out for finite wave functions (as $x \to \pm \infty$), so we can write $b = ika$, where k is real. Dropping the superscripts, we can write

$$\psi(x+a) = e^{\pm ika} \psi(x). \tag{C.23}$$

Finally, we note that if

$$\psi(x) = e^{ikx} u(x), \tag{C.24}$$

where

$$u(x+a) = u(x), \tag{C.25}$$

then (C.23) is satisfied. (C.23) or (C.24), and (C.25) are different forms of Bloch's theorem.

C.2 Simple Derivation in Three Dimensions

Let

$$\mathcal{H}\psi(x_1 \cdots x_N) = E\psi(x_1 \cdots x_N) \tag{C.26}$$

be the usual Schrödinger wave equation. Let T_l be a translation operator that translates the lattice by $l_1 a_1 + l_2 a_2 + l_3 a_3$, where the l_i are integers and the a_i are the primitive translation vectors of the lattice.

Since the Hamiltonian is invariant with respect to translations by T_l, we have

$$[\mathcal{H}, T_l] = 0, \tag{C.27}$$

and

$$[T_l, T_l'] = 0. \tag{C.28}$$

Now we know that we can always find simultaneous eigenfunctions of commuting observables. Observables are represented by Hermitian operators. The T_l are unitary. Fortunately, the same theorem applies to them (we shall not prove this here). Thus we can write

$$\mathcal{H}\psi_{E,l} = E\psi_{E,l}, \tag{C.29}$$

$$T_l \psi_{E,l} = t_l \psi_{E,l}. \tag{C.30}$$

Now certainly we can find a vector k such that

$$t_l = e^{i k \cdot l}. \tag{C.31}$$

Further

$$\int_{\text{all space}} |\psi(r)|^2 \, d\tau = \int |\psi(r+l)|^2 \, d\tau = |t_l|^2 \int |\psi(r)|^2 \, d\tau,$$

so that

$$|t_l|^2 = 1. \tag{C.32}$$

This implies that k must be a vector over the real field.

We thus arrive at Bloch's theorem

$$T_l \psi(r) = \psi(r+l) = e^{i k \cdot l} \psi(r). \tag{C.33}$$

The theorem says we can always choose the eigenfunctions to satisfy (C.33). It does not say the eigenfunction must be of this form. If periodic boundary conditions are applied, the usual restrictions on the k are obtained.

C.3 Derivation of Bloch's Theorem by Group Theory

The derivation here is relatively easy once the appropriate group theoretic knowledge is acquired. We have already discussed in Chaps. 1 and 7 the needed results from group theory. We simply collect together here the needed facts to establish Bloch's theorem.

1. It is clear that the group of the T_l is abelian (i.e. all the T_l commute).

2. In an abelian group each element forms a class by itself. Therefore the number of classes is O(G), the order of the group.

3. The number of irreducible representations (of dimension n_i) is the number of classes.

4. $\sum n_i^2 = O(G)$ and thus by above

$$n_1^2 + n_2^2 + \cdots + n_{0(G)}^2 = 0(G).$$

This can be satisfied only if each $n_i = 1$. Thus the dimensions of the irreducible representations of the T_l are all one.

5. In general

$$T_l \psi_i^k = \sum_j A_{ij}^{l,k} \psi_j^k ,$$

where the $A_{ij}^{l,k}$ are the matrix elements of the T_l for the kth representation and the sum over j goes over the dimensionality of the kth representation. The ψ_i^k are the basis functions selected as eigenfunctions of $\mathcal{H}$ (which is possible since $[\mathcal{H}, T^l] = 0$). In our case the sum over j is not necessary and so

$$T^l \psi^k = A^{l,k} \psi^k .$$

As before, the $A^{l,k}$ can be chosen to be $e^{il\cdot k}$. Also in one dimension we could use the fact that $\{T^l\}$ is a cyclic group so that the $A^{l,k}$ are automatically the roots of one.

D Density Matrices and Thermodynamics

A few results will be collected here. The proofs of these results can be found in any of several books on statistical mechanics.

If $\psi^i(x, t)$ is the wave function of system (in an ensemble of N systems where $1 \leq i \leq N$) and if $|n\rangle$ is a complete orthonormal set, then

$$\left| \psi^i(x,t) \right\rangle = \sum_n c_n^i(t) |n\rangle.$$

The density matrix is defined by

$$\rho_{nm} = \frac{1}{N} \sum_{i=1}^N c_m^{i*}(t) c_m^i(t) \equiv \overline{c_m^* c_n} .$$

It has the following properties:

$$Tr(\rho) \equiv \sum_n \rho_{nn} = 1,$$

the ensemble average (denoted by a bar) of the quantum-mechanical expectation value of an operator A is

$$\left\langle \overline{A} \right\rangle \equiv Tr(\rho A),$$

and the equation of motion of the density operator ρ is given by

$$-i\hbar \frac{\partial \rho}{\partial t} = [\rho, H],$$

where the density operator is defined in such a way that $\langle n|\rho|m\rangle \equiv \rho_{nm}$. For a canonical ensemble in equilibrium

$$\rho = \exp\left(\frac{F-H}{kT}\right).$$

Thus we can readily link the idea of a density matrix to thermodynamics and hence to measurable quantities. For example, the internal energy for a system in equilibrium is given by

$$U = \langle \overline{H} \rangle = Tr\left[H \exp\left(\frac{F-H}{kT}\right)\right] = \frac{Tr[H\exp(-H/kT)]}{Tr[\exp(-H/kT)]}.$$

Alternatively, the internal energy can be calculated from the free energy F where for a system in equilibrium,

$$F = -kT \ln Tr[\exp(-H/kT)].$$

It is fairly common to leave the bar off $\langle \overline{A} \rangle$ so long as the meaning is clear. For further properties and references see Patterson [A.19], see also Huang [A.8].

E Time-Dependent Perturbation Theory

A common problem in solid-state physics (as in other areas of physics) is to find the transition rate between energy levels of a system induced by a small time-dependent perturbation. More precisely, we want to be able to calculate the time development of a system described by a Hamiltonian that has a small time-dependent part. This is a standard problem in quantum mechanics and is solved by the time-dependent perturbation theory. However, since there are many different aspects of time-dependent perturbation theory, it seems appropriate to give a brief review without derivations. For further details any good quantum mechanics book such as Merzbacher[6] can be consulted.

---

[6] See Merzbacher [A.15 Chap. 18].

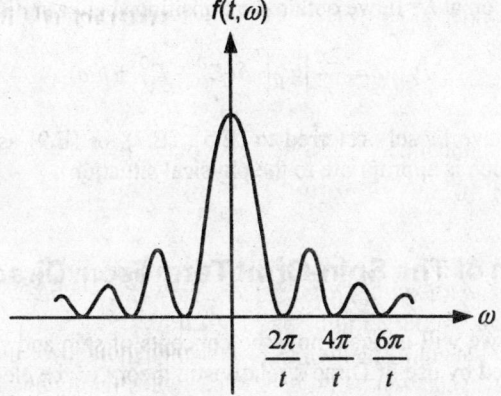

Fig. E.1. $f(t, \omega)$ versus ω. The area under the curve is $2\pi t$

Let

$$\mathcal{H}(t) = \mathcal{H}^0 + V(t), \tag{E.1}$$

$$\mathcal{H}^0 |l\rangle = E_l^0 |l\rangle, \tag{E.2}$$

$$V_{kl}(t) = \langle k | V(t) | l \rangle, \tag{E.3}$$

$$\omega_{kl} = \frac{E_k^0 - E_l^0}{\hbar}. \tag{E.4}$$

In first order in V, for V turned on at $t = 0$ and constant otherwise, the probability per unit time of a discrete $i \to f$ transition for $t > 0$ is

$$P_{i \to f} \cong \frac{2\pi}{\hbar} |V_{fi}|^2 \delta(E_i^0 - E_f^0). \tag{E.5}$$

In deriving (E.5) we have assumed that the $f(t, \omega)$ in Fig. E.1 can be replaced by a Dirac delta function via the equation

$$\lim_{t \to \infty} \frac{1 - \cos(\omega_{if} t)}{(\hbar \omega_{if})^2} = \frac{\pi t}{\hbar} \delta(E_i^0 - E_f^0) = \frac{f(t, \omega)}{2\hbar^2}. \tag{E.6}$$

If we have transitions to a group of states with final density of states $p_f(E_f)$, a similar calculation gives

$$P_{i \to f} = \frac{2\pi}{\hbar} |V_{fi}|^2 p_f(E_f). \tag{E.7}$$

In the same approximation, if we deal with periodic perturbations represented by

$$V(t) = g e^{i\omega t} + g^\dagger e^{-i\omega t}, \tag{E.8}$$

which are turned on at $t = 0$, we obtain for transitions between discrete states

$$P_{i \to f} = \frac{2\pi}{\hbar} |g_{fi}|^2 \delta(E_i^0 - E_f^0 \pm \hbar\omega). \tag{E.9}$$

In the text, we have loosely referred to (E.5), (E.7), or (E.9) as the Golden rule (according to which is appropriate to the physical situation).

F Derivation of The Spin-Orbit Term From Dirac's Equation

In this appendix we will indicate how the concepts of spin and spin-orbit interaction are introduced by use of Dirac's relativistic theory of the electron. For further details, any good quantum mechanics text such as that of Merzbacher[7], or Schiff[8] can be consulted. We will discuss Dirac's equation only for fields described by a potential V. For this situation, Dirac's equation can be written

$$[c(\alpha \cdot p) + m_0 c^2 \beta + V]\psi = E\psi. \tag{F.1}$$

In (F.1), c is the speed of light, α and β are 4×4 matrices defined below, p is the momentum operator, m_0 is the rest mass of the electron, ψ is a four-component column matrix (each element of this matrix may be a function of the spatial position of the electron), and E is the total energy of the electron (including the rest mass energy that is $m_0 c^2$). The α matrices are defined by

$$\alpha = \begin{pmatrix} 0 & \sigma \\ \sigma & 0 \end{pmatrix}, \tag{F.2}$$

where the three components of σ are the 2×2 Pauli spin matrices. The definition of β is

$$\beta = \begin{pmatrix} I & 0 \\ 0 & -I \end{pmatrix}, \tag{F.3}$$

where I is a 2×2 unit matrix.

For solid-state purposes we are not concerned with the fully relativistic equation (F.1), but rather we are concerned with the relativistic corrections that (F.1) predicts should be made to the nonrelativistic Schrödinger equation. That is, we want to consider the Dirac equation for the electron in the small velocity limit. More precisely, we will consider the limit of (F.1) when

$$\varepsilon \equiv \frac{(E - m_0 c^2) - V}{2 m_0 c^2} \ll 1, \tag{F.4}$$

---

[7] See Merzbacher [A.15 Chap. 23].
[8] See Schiff [A.23].

and we want results that are valid to first order in ε, i.e. first-order corrections to the completely nonrelativistic limit. To do this, it is convenient to make the following definitions:

$$E = E' + m_0 c^2, \tag{F.5}$$

and

$$\psi = \begin{pmatrix} \chi \\ \phi \end{pmatrix}, \tag{F.6}$$

where both χ and ϕ are two-component wave functions.

If we substitute (F.5) and (F.6) into (F.1), we obtain an equation for both χ and ϕ. We can combine these two equations into a single equation for χ in which ϕ does not appear. We can then use the small velocity limit (F.4) together with several properties of the Pauli spin matrices to obtain the Schrödinger equation with relativistic corrections

$$E'\chi = \left[\frac{p^2}{2m_0} - \frac{p^4}{8m_0^3 c^2} + V - \frac{\hbar^2}{4m_0^2 c^2}\nabla V \cdot \nabla + \frac{\hbar^2}{4m_0^2 c^2}\sigma \cdot ((\nabla V) \times p) \right] \cdot \chi. \tag{F.7}$$

This is the form that is appropriate to use in solid-state physics calculations. The term

$$\frac{\hbar^2}{4m_0^2 c^2}\sigma \cdot [(\nabla V) \times p] \tag{F.8}$$

is called the spin-orbit term. This term is often used by itself as a first-order correction to the nonrelativistic Schrödinger equation. The spin-orbit correction is often applied in band-structure calculations at certain points in the Brillouin zone where bands come together. In the case in which the potential is spherically symmetric (which is important for atomic potentials but not crystalline potentials), the spin-orbit term can be cast into the more familiar form

$$\frac{\hbar^2}{2m_0^2 c^2}\frac{1}{r}\frac{dV}{dr}\mathbf{L}\cdot\mathbf{S}, \tag{F.9}$$

where $\mathbf{L}$ is the orbital angular momentum operator and $\mathbf{S}$ is the spin operator (in units of $\hbar$).

It is also interesting to see how Dirac's theory works out in the (completely) nonrelativistic limit when an external magnetic field $\mathbf{B}$ is present. In this case the magnetic moment of the electron is introduced by the term involving $\mathbf{S}\cdot\mathbf{B}$. This term automatically appears from the nonrelativistic limit of Dirac's equation. In addition, the correct ratio of magnetic moment to spin angular momentum is obtained in this way.

G The Second Quantization Notation for Fermions and Bosons

When the second quantization notation is used in a nonrelativistic context it is simply a notation in which we express the wave functions in occupation-number space and the operators as operators on occupation number space. It is of course of great utility in considering the many-body problem. In this formalism, the symmetry or antisymmetry of the wave functions is automatically built into the formalism. In relativistic physics, annihilation and creation operators (which are the basic operators of the second quantization notation) have physical meaning. However, we will apply the second quantization notation only in nonrelativistic situations. No derivations will be made in this section. (The appropriate results will just be concisely written down.) There are many good treatments of the second quantization or occupation number formalism. One of the most accessible is by Mattuck.[9]

G.1 Bose Particles

For Bose particles we deal with b_i and $b_i^\dagger$ operators (or other letters where convenient): $b_i^\dagger$ *creates* a Bose particle in the state i; b_i *annihilates* a Bose particle in the state f. The b_i operators obey the following commutation relations:

$$[b_i, b_j] \equiv b_i b_j - b_j b_i = 0,$$
$$[b_i^\dagger, b_j^\dagger] = 0,$$
$$[b_i, b_j^\dagger] = \delta_{ij}.$$

The occupation number operator whose eigenvalues are the number of particles in state i is

$$n_i = b_i^\dagger b_i,$$

and

$$n_i + 1 = b_i b_i^\dagger.$$

The effect of these operators acting on different occupation number kets is

$$b_i |n_1, \ldots, n_i, \ldots\rangle = \sqrt{n_i} |n_1, \ldots, n_i - 1, \ldots\rangle,$$
$$b_i^\dagger |n_1, \ldots, n_i, \ldots\rangle = \sqrt{n_i + 1} |n_1, \ldots, n_i + 1, \ldots\rangle,$$

where $|n_1, \ldots, n_i, \ldots\rangle$ means the ket appropriate to the state with n_1 particles in state 1, n_2 particles in state 2, and so on.

---

[9] See Mattuck [A.14].

The matrix elements of these operators are given by

$$\langle n_i - 1 | b_i | n_i \rangle = \sqrt{n_i},$$
$$\langle n_i | b_i^\dagger | n_i - 1 \rangle = \sqrt{n_i}.$$

In this notation, any one-particle operator

$$f_{op}^{(1)} = \sum_l f^{(1)}(r_l)$$

can be written in the form

$$f_{op}^{(1)} = \sum_{i,k} \langle i | f^{(1)} | k \rangle b_i^\dagger b_k,$$

and the $|k\rangle$ are any complete set of one-particle eigenstates.

In a similar fashion any two-particle operator

$$f_{op}^{(2)} = \sum_{l,m} f^{(2)}(r_l - r_m)$$

can be written in the form

$$f_{op}^{(2)} = \sum_{i,k,l,m} \langle i(1)k(2) | f^{(2)} | l(1)m(2) \rangle b_i^\dagger b_k^\dagger b_m b_l.$$

Operators that create or destroy base particles at a given point in space (rather than in a given state) are given by

$$\psi(r) = \sum_\alpha u_\alpha(r) b_\alpha,$$
$$\psi^\dagger(r) = \sum_\alpha u_\alpha^*(r) b_\alpha^\dagger,$$

where $u_\alpha(r)$ is the single-particle wave function corresponding to state α. In general, r would refer to *both* space and *spin* variables. These operators obey the commutation relation

$$[\psi(r), \psi^\dagger(r)] = \delta(r - r').$$

G.2 Fermi Particles

For Fermi particles, we deal with a_i and $a_i^\dagger$ operators (or other letters where convenient): $a_i^\dagger$ creates a fermion in the state i; a_i annihilates a fermion in the state i. The a_i operators obey the following anticommutation relations:

$$\{a_i, a_j\} \equiv a_i a_j + a_j a_i = 0,$$
$$\{a_i^\dagger, a_j^\dagger\} = 0,$$
$$\{a_i, a_j^\dagger\} = \delta_{ij}.$$

The occupation number operator whose eigenvalues are the number of particles in state i is

$$n_i = a_i^\dagger a_i ,$$

and

$$1 - n_i = a_i a_i^\dagger .$$

Note that $(n_i)^2 = n_i$, so that the only possible eigenvalues of n_i are 0 and 1 (the Pauli principle is built in!).

The matrix elements of these operators are defined by

$$\langle \cdots n_i = 0 \cdots | a_i | \cdots n_i = 1 \cdots \rangle = (-)^{\Sigma(1,i-1)} ,$$

and

$$\langle \cdots n_i = 1 \cdots | a_i^\dagger | \cdots n_i = 0 \cdots \rangle = (-)^{\Sigma(1,i-1)} ,$$

where $\Sigma(1, i-1)$ equals the sum of the occupation numbers of the states from 1 to $i-1$.

In this notation, any one-particle operator can be written in the form

$$f_0^{(1)} = \sum_{i,j} \langle i | f^{(1)} | j \rangle a_i^\dagger a_j ,$$

where the $|j\rangle$ are any complete set of one-particle eigenstates. In a similar fashion, any two-particle operator can be written in the form

$$f_{op}^{(2)} = \sum_{i,j,k,l} \langle i(1) j(2) | f^{(2)} | k(1) l(2) \rangle a_j^\dagger a_i^\dagger a_k a_l .$$

Operators that create or destroy Fermi particles at a given point in space (rather than in a given state) are given by

$$\psi(r) = \sum_\alpha u_\alpha(r) a_\alpha ,$$

where $u_\alpha(r)$ is the single-particle wave function corresponding to state α, and

$$\psi^\dagger(r) = \sum_\alpha u_\alpha^*(r) a_\alpha^\dagger .$$

These operators obey the anticommutation relations

$$\{\psi(r), \psi^\dagger(r')\} = \delta(r - r') .$$

The operators also allow a convenient way of writing *Slater* determinants, e.g.,

$$a_\alpha^\dagger a_\beta^\dagger |0\rangle \leftrightarrow \frac{1}{\sqrt{2}} \begin{vmatrix} u_\alpha(1) & u_\alpha(2) \\ u_\beta(1) & u_\beta(2) \end{vmatrix} ;$$

$|0\rangle$ is known as the *vacuum* ket.

The easiest way to see that the second quantization notation is consistent is to show that matrix elements in the second quantization notation have the same values as corresponding matrix elements in the old notation. This demonstration will not be done here.

H The Many-Body Problem

Richard P. Feynman is famous for many things, among which is the invention, in effect, of a new quantum mechanics. Or maybe we should say of a new way of looking at quantum mechanics. His way involves taking a process going from A to B and looking at all possible paths. He then sums the amplitude of the all paths from A to B to find, by the square, the probability of the process.

Related to this is a diagram that defines a process and that contains by implication all the paths, as calculated by appropriate integrals. Going further, one looks at all processes of a certain class, and sums up all diagrams (if possible) belonging to this class. Ideally (but seldom actually) one eventually treats all classes, and hence arrives at an exact description of the interaction.

Thus, in principle, there is not so much to treating interactions by the use of Feynman diagrams. The devil is in the details, however. Certain sums may well be infinite—although hopefully disposable by renormalization. Usually doing a nontrivial calculation of this type is a great technical feat.

We have found that a common way we use Feynman diagrams is to help us understand what we mean by a given approximation. We will note below, for example, that the Hartree approximation involves summing a certain class of diagrams, while the Hartree–Fock approximation involves summing these diagrams along with another class. We believe, the diagrams give us a very precise idea of what these approximations do.

Similarly, the diagram expansion can be a useful way to understand why a perturbation expansion does not work in explaining superconductivity, as well as a way to fix it (the Nambu formalism).

The practical use of diagrams, and diagram summation, may involve great practical skill, but it seems that the great utility of the diagram approach is in clearly stating, and in keeping track of, what we are doing in a given approximation.

One should not think that an expertise in the technicalities of Feynman diagrams solves all problems. Diagrams have to be summed and integrals still have to be done. For some aspects of many-electron physics, density functional theory (DFT) has become the standard approach. Diagrams are usually not used at the beginning of DFT, but even here they may often be helpful in discussing some aspects.

DFT was discussed in Chap. 3, and we briefly review it here, because of its great practical importance in the many-electron problem of solid-state physics. In the beginning of DFT Hohenberg and Kohn showed that the N-electron Schrödinger wave equation in three dimensions could be recast. They showed that an equation for the electron density in three dimensions would suffice to determine ground-state properties. The Hohenberg–Kohn formulation may be regarded as a generalization of the Thomas–Fermi approximation. Then came the famous Kohn–Sham equations that reduced the Hohenberg–Kohn formulation to the problem of noninteracting electrons in an effective potential (somewhat analogous to the Hartree equations, for example). However, part of the potential, the exchange correlation part could only be approximately evaluated, e.g. in the local density approximation (LDA) – which assumed a locally homogeneous electron gas. A problem with DFT-LDA is that it is not necessarily clear what the size of the errors are, however, the DFT is certainly a good way to calculate, ab initio, certain ground-state properties of finite electronic systems, such as the ionization energies of atoms. It is also very useful for computing the electronic ground-state properties of periodic solids, such as cohesion and stability. Excited states, as well as approximations for the exchange correlation term in N-electron systems continue to give problems. For a nice brief summary of DFT see Mattsson [A.13].

For quantum electrodynamics, a brief and useful graphical summary can be found at: http://www2slac.stanford.edu/vvc/theory/feynman.html. We now present a brief summary of the use of diagrams in many-body physics.

In some ways, trying to do solid-state physics without Feynman diagrams is a little like doing electricity and magnetism (EM) without resorting to drawing Faraday's lines of electric and magnetic fields. However, just as field lines have limitations in describing EM interactions, so do diagrams for discussing the many-body problem [A.1]. The use of diagrams can certainly augment one's understanding.

The distinction between quasi- or dressed particles and collective excitations is important and perhaps is made clearer from a diagrammatic point of view. Both are 'particles' and are also elementary energy excitations. But after all a polaron (a quasi-particle) is not the same kind of beast as a magnon (a collective excitation). Not everybody makes this distinction. Some call all 'particles' quasiparticles. Bogolons are particles of another type, as are excitons (see below for definitions of both). All are elementary excitations and particles, but not really collective excitations or dressed particles in the usual sense.

H.1 Propagators

These are the basic quantities. Their representation is given in the next section. The single-particle propagator is a sum of probability amplitudes for all the ways of going from r_1, t_1 to r_2, t_2 (adding a particle at 1 and taking out at 2).

The two-particle propagator is the sum of the probability amplitudes for all the ways two particles can enter a system, undergo interactions and emerge again.

H.2 Green Functions

Propagators are represented by Green functions. There are both advanced and retarded propagators. Advanced propagators can describe particles traveling backward in time, i.e. holes. The use of Fourier transforms of time-dependent propagators led to simpler algebraic equations. For a retarded propagator the free propagator is:

$$G_0^+(k,\omega) = \frac{1}{\omega - \varepsilon_k + i\delta}.$$ (H.1)

For quasiparticles, the real part of the pole of the Fourier transform of the single-particle propagator gives the energy, and the imaginary part gives the width of the energy level. For collective excitations, one has a similar statement, except that two-particle propagators are needed.

H.3 Feynman Diagrams

Rules for drawing diagrams are found in Economu [A.5 pp. 251-252], Pines [A.22 pp. 49-50] and Schrieffer [A.24 pp. 127-128]. Also, see Mattuck [A.14 p. 165]. There is a one-to-one correspondence between terms in the perturbation expansion of the Green functions and diagrammatic representation. Green functions can also be calculated from a hierarchy of differential equations and an appropriate decoupling scheme. Such approximate decoupling schemes are always equivalent to a partial sum of diagrams.

H.4 Definitions

Here we remind you of some examples. A more complete list is found in Chap. 4.

Quasiparticle – A real particle with a cloud of surrounding disturbed particles with an effective mass and a lifetime. In the usual case it is a dressed fermion. Examples are listed below.

Electrons in a solid – These will be dressed electrons. They can be dressed by interaction with the static lattice, other electrons or interactions with the vibrating lattice. It is represented by a straight line with an arrow to the right ⟶ if time goes that way

Holes in a solid – One can view the ground state of a collection of electrons as a vacuum. A hole is then what results when an electron is removed from a normally occupied state. It is represented by a straight line with an arrow to the left ⟵ .

Polaron – An electron moving through a polarizable medium surrounded by its polarization cloud of virtual phonons.

Photon – Quanta of electromagnetic radiation (e.g. light) – it is represented by a wavy line ∿∿ .

Collective Excitation – These are elementary energy excitations that involve wave-like motion of all the particles in the systems. Examples are listed below.

Phonon – Quanta of normal mode vibration of a lattice of ions. Also often represented by wavy line.

Magnon – Quanta of low-energy collective excitations in the spins, or quanta of waves in the spins.

Plasmon – Quanta of energy excitation in the density of electrons in an interacting electron gas (viewing, e.g., the positive ions as a uniform background of charge).

Other Elementary Energy Excitations – Excited energy levels of many-particle systems.

Bogolon – Linear combinations of electrons in a state $+k$ with 'up' spin and $-k$ with 'down' spin. Elementary excitations in a superconductor.

Exciton – Bound electron–hole pairs.

Some examples of interactions represented by vertices (time going to the right):

An electron emitting a phonon

A hole emitting a phonon.

Diagrams are built out of vertices with conservation of momentum satisfied at the vertices. For example

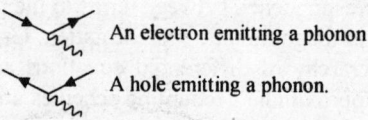

represents a coulomb interaction with time going up.

H.5 Diagrams and the Hartree and Hartree–Fock Approximations

In order to make these concepts clearer it is perhaps better to discuss an example that we have already worked out without diagrams. Here, starting from the Hamiltonian we will discuss briefly how to construct diagrams, then explain how to associate single-particle Green functions with the diagrams and how to do the partial sums representing these approximations. For details, the references must be consulted.

In the second quantization notation, a Hamiltonian for interacting electrons

$$\mathcal{H} = \sum_i V(i) + \frac{1}{2}\sum_{i,j} V(ij), \quad (H.2)$$

with one- and two-body terms can be written as

$$\mathcal{H} = \sum_{i,j} \langle i(1)|V(1)|j(1)\rangle a_i^\dagger a_j + \frac{1}{2}\sum_{ijkl} \langle i(1)j(2)|V(1,2)|k(1)j(2)\rangle a_j^\dagger a_i^\dagger a_k a_l, \quad (H.3)$$

where

$$\langle i(1)|V(1)|j(1)\rangle = \int \phi_i^*(r_1)V(r_1)\phi_j(r_1)d^3r_i , \qquad (H.4)$$

and

$$\langle i(1)j(2)|V(1,2)|k(1)l(2)\rangle = \int \phi_i^*(r_1)\phi_j^*(r_2)V(1,2)\phi_k(r_1)\phi_l(r_2)d^3r_1 d^3r_2 , \quad (H.5)$$

and the annihilation and creation operators have the usual properties

$$a_i a_j^\dagger + a_j^\dagger a_i = \delta_{ij} ,$$
$$a_i a_j + a_j a_i = 0.$$

We now consider the Hartree approximation. We assume, following Mattuck [A.14] that the interactions between electrons is mostly given by the forward scattering processes where the interacting electrons have no momentum change in the interaction. We want to get an approximation for the single-particle propagator that includes interactions. In first order the only possible process is given by a bubble diagram where the hole line joins on itself. One thinks of the particle in state k knocking a particle out of and into a state l instantaneously. Since this can happen any number of times, we get the following partial sum for diagrams representing the single-particle propagator. The first diagram on the right-hand side represents the free propagator where nothing happens (Mattuck [A.14 p. 89][10]).

Using the "dictionary" given by Mattuck [A.14 p. 86], we substitute propagators for diagrams and get

$$G^+(k,\omega) = \frac{1}{\omega - \varepsilon_k - \sum_{l(\text{occ.})} V_{klkl} + i\delta} . \qquad (H.6)$$

Since the poles give the elementary energy excitations we have

$$\varepsilon_k' = \varepsilon_k + \sum_{l(\text{occ.})} V_{klkl} , \qquad (H.7)$$

---

[10] Reproduced with permission from Mattuck RD, *A Guide to Feynman Diagrams in the Many-Body Problem*, 2nd edn, (4.67) p. 89, Dover Publications, Inc., 1992.

which is exactly the same as the Hartree approximation (see (3.21)) since

$$\sum_l V_{klkl} = \int d^3 r_2 \phi_k^*(r_2)\phi_k(r_2)\sum_l \int \phi_l^*(r_1)\phi_l(r_1)V(1,2)d^3 r_1 . \quad (H.8)$$

It is actually very simple to go from here to the Hartree–Fock approximation – all we have to do is to include the exchange terms in the interactions. These are the "open-oyster" diagrams

where a particle not only strikes a particle in l and creates an instantaneous hole, but is exchanged with it. Doing the partial sum of forward scattering and exchange scattering one has (Mattuck [A.14 p. 91][11]):

Associating propagators with the terms in the diagram gives

$$G^+(k,\omega) = \frac{1}{\omega - \varepsilon_k - \sum_{l(\text{occ.})}(V_{klkl} - V_{lkkl}) + i\delta}. \quad (H.9)$$

From this we identify the elementary energy excitations as

$$\varepsilon'_k = \varepsilon_k + \sum_{l(\text{occ.})}(V_{klkl} - V_{lkkl}), \quad (H.10)$$

which is just what we got for the Hartree–Fock approximation (see (3.50)).

The random-phase approximation [A.14] can also be obtained by a partial summation of diagrams, and it is equivalent to the Lindhard theory of screening.

---

[11] Reproduced with permission from Mattuck RD, *A Guide to Feynman Diagrams in the Many-Body Problem*, 2nd edn, (4.76) p. 91, Dover Publications, Inc., 1992.

H.6 The Dyson Equation

This is the starting point for many approximations both diagrammatic, and algebraic. Dyson's equation can be regarded as a generalization of the partial sum technique used in the Hartree and Hartree–Fock approximations. It is exact. To state Dyson's equation we need a couple of definitions. The self-energy part of a diagram is a diagram that has no incoming or outgoing parts and can be inserted into a particle line. The bubbles of the Hartree method are an example. An irreducible or proper self-energy part is a part that cannot be further reduced into unconnected self-energy parts. It is common to define

as the sum over all proper self-energy parts. Then one can sum over all repetitions of sigma ($\Sigma k, \omega$) to get

Dyson's equation yields an exact expression for the propagator,

$$G(k,\omega) = \frac{1}{\omega - \varepsilon_k - \Sigma_{l(\text{occ.})}(k,\omega) + i\delta_k}, \tag{H.11}$$

since all diagrams are either proper diagrams or their repetition.

In the Hartree approximation

and in the Hartree–Fock approximation

Although the Dyson equation is in principle exact, one still has to evaluate sigma, and this is in general not possible except in some approximation.

We cannot go into more detail here. We have given accurate results for the high and low-density electron gas in Chap. 2. In general, the ideas of Feynman diagrams and the many-body problem merit a book of their own. We have found the book by Mattuck [A.14] to be particularly useful, but note the list of references at the end of this section. We have used some ideas about diagrams when we discussed superconductivity.

Bibliography

Chapter 1

1.1. Anderson PW, *Science* **177**, 393-396 (1972).
1.2. Bacon GE, *Neutron Diffraction*, Clarendon Press, Oxford, 2nd edn (1962).
1.3. Bradley CJ and Cracknell AP, The Mathematical Theory of Symmetry in Solids: Representation Theory for Point Groups and Space Groups, Clarendon Press, Oxford (1972).
1.4. Brown PJ and Forsyth JB, *The Crystal Structure of Solids*, Edward Arnold, London, 1975 (Chap. 3).
1.5. Buerger MJ, *Elementary Crystallography*, John Wiley and Sons, New York, 1956.
1.6. Daw MS, "Model of metallic cohesion: The embedded-atom method," *Phys Rev B* **39**, 7441-7452 (1989).
1.7. de Gennes PG and Prost J, *The Physics of Liquid Crystals*, Clarendon Press, Oxford, 2nd edn (1993).
1.8. Evjan HM, *Physical Review*, **39**, 675 (1932)
1.9. Ghatak AK and Kothari LS, *An Introduction to Lattice Dynamics*, Addison Wesley Publishing Co., Reading, Mass. 1972 p165ff.
1.10. Herzfield CM and Meijer PHE, "Group Theory and Crystal Field Theory," *Solid State Physics, Advances in Research and Applications* **12**, 1-91 (1961).
1.11. Horton GK, "Ideal Rare-Gas Crystals," *American Journal of Physics*, **36**(2), 93 (1968).
1.12. Janot C, *Quasicrystals, A Primer*, Clarendon Press, Oxford (1992).
1.13. Kittel C, *Introduction to Solid State Physics*, John Wiley and Sons, New York, Seventh Edition, 1996, Chaps. 1-3, 19.
1.14. Koster GF, "Space Groups and Their Representations," *Solid State Physics, Advances in Research and Applications* **5**, 174-256 (1957).
1.15. Levine D and Steinhardt PJ, *Phys Rev Lett* **53**, 2477-83 (1984).
1.16. Maradudin AA, Montrol EW, and Weiss GW, *Theory of Lattice Dynamics in the Harmonic Approximation*, Academic Press, N.Y. 1963 p 245.
1.17. Moffatt WG, Pearsall GW, and Wulff J, *The Structure and Properties of Materials, Vol. 1*, John Wiley and Sons, Inc., New York, 1964.
1.18. Pauling L, *Nature of the Chemical Bond*, 2nd edn., Cornell University Press, Ithaca, 1945.
1.19. Phillips FC, *An Introduction to Crystallography*, John Wiley and Sons, Inc., New York, 4th edn., 1971.
1.20. Pollock Daniel D, "Physical Properties of Materials for Engineers," CRC Press, Boca Raton, 1993.
1.21. Shechtman D, Blech I, Gratias D, and Cahn JW, *Phys Rev Lett* **53**, 1951-3 (1984).

1.22. Steinhardt PJ and Ostlund S, *The Physics of Quasicrystals*, World Scientific Publishing, Singapore, 1987.
1.23. Streetman BG, *Solid State Electronic Devices*, Prentice Hall, Englewood Cliffs, NJ, 4th edn. (1995).
1.24. Tinkham M, *Group Theory and Quantum Mechanics*, McGraw-Hill Book Company, New York, 1964.
1.25. Tosi MP, "Cohesion of Ionic Solids in the Born Model," *Solid State Physics, Advances in Research and Applications* **16**, 1-120 (1964).
1.26. Tran HT and Perdew JP, "How Metals Bind: The deformable jellium model with correlated electrons," *Am J Phys* **71**, 1048-1061 (2003).
1.27. Webb MB and Lagally MG, "Elastic Scattering of Low Energy Electrons from Surfaces," *Solid State Physics, Advances in Research and Applications* **28**, 301-405 (1973).
1.28. West Anthony R, *Solid State Chemistry and its Properties*, John Wiley and Sons, New York, 1984.
1.29. Wigner EP and Seitz F, "Qualitative Analysis of the Cohesion in Metals," *Solid State Physics, Advances in Research and Applications* **1**, 97-126 (1955).
1.30. Wyckoff RWG, "Crystal Structures," Vols 1-5, John Wiley and Sons, New York 1963-1968.

Chapter 2

2.1. Anderson PW, *Science* **177**, 393-396 (1972).
2.2. Bak TA (ed), *Phonons and Phonon Interactions*, W A Benjamin, New York, 1964.
2.3. Bilz H and Kress W, *Phonon Dispersion Relations in Insulators*, Springer, 1979
2.4. Blackman M, "The Specific Heat of Solids," *Encyclopedia of Physics*, Vol. VII, Part 1, Crystal Physics 1, Springer Verlag, Berlin, 1955, p. 325.
2.5. Born M and Huang K, *Dynamical Theory of Crystal Lattices*, Oxford University Press, New York, 1954.
2.6. Brockhouse BN and Stewart AT, *Rev. Mod Phys* **30**, 236 (1958).
2.7. Brown FC, *The Physics of Solids*, W A Benjamin, Inc., New York, 1967, Chap. 5.
2.8. Choquard P, *The Anharmonic Crystal*, W A Benjamin, Inc., New York (1967).
2.9. Cochran W, "Interpretation of Phonon Dispersion Curves," in Proceedings of the International Conference on Lattice Dynamics, Copenhagen, 1963, Pergamon Press, New York, 1965.
2.10. Cochran W, "Lattice Vibrations," *Reports on Progress in Physics*, Vol. XXVI, The Institute of Physics and the Physical Society, London, 1963, p. 1. See also Cochran W, *The Dynamics of Atoms in Crystals*, Edward Arnold, London, 1973.
2.11. deLauney J, "The Theory of Specific Heats and Lattice Vibrations," *Solid State Physics: Advances in Research and Applications* **2**, 220-303 (1956).
2.12. Dick BG Jr, and Overhauser AW, *Physical Review* **112**, 90 (1958).
2.13. Dorner B, Burkel E, Illini T, and Peisl J, *Z für Physik* **69**, 179-183 (1989)
2.14. Dove MT, *Structure and Dynamics*, Oxford University Press, 2003.
2.15. Elliott RJ and Dawber DG, *Proc. Roy. Soc.* **A223**, 222 (1963)
2.16. Ghatak K and Kothari LS, *An Introduction to Lattice Dynamics*, Addison-Wesley Publ. Co., Reading, MA, 1972, Chap. 4.

2.17. Grosso G and Paravicini GP, *Solid State Physics, Academic Press*, New York, 2000, Chaps. VIII and IX.
2.18. Huntington HB, "The Elastic Constants of Crystals," *Solid State Physics: Advances in Research and Applications* **7**, 214-351 (1958).
2.19. Jensen HH, "Introductory Lectures on the Free Phonon Field," in *Phonons and Phonon Interactions*, Bak TA (ed), W. A. Benjamin, New York, 1964.
2.20. Jones W and March NA, *Theoretical Solid State Physics*, John Wiley and Sons, (1973), Vol I, Chap. 3.
2.21. Joshi SK and Rajagopal AK, "Lattice Dynamics of Metals," *Solid State Physics: Advances in Research and Applications* **22**, 159-312 (1968).
2.22. Kunc K, Balkanski M, and Nusimovici MA, *Phys Stat Sol B* **71**, 341; **72**, 229, 249 (1975).
2.23. Lehman GW, Wolfram T, and DeWames RE, *Physical Review*, **128**(4), 1593 (1962).
2.24. Leibfried G and Ludwig W, "Theory of Anharmonic Effects in Crystals," *Solid State Physics: Advances in Research and Applications* **12**, 276-444 (1961).
2.25. Lifshitz M and Kosevich AM, "The Dynamics of a Crystal Lattice with Defects," *Reports on Progress in Physics*, Vol. XXIX, Part 1, The Institute of Physics and the Physical Society, London, 1966, p. 217.
2.26. Maradudin A, Montroll EW, and Weiss GH, "Theory of Lattice Dynamics in the Harmonic Approximation," *Solid State Physics: Advances in Research and Applications*, Supplement 3 (1963).
2.27. Messiah A, *Quantum Mechanics*, North Holland Publishing Company, Amsterdam, 1961, Vol. 1, p 69.
2.28. Montroll EW, *J. Chem. Phys.* **10**, 218 (1942), 11, 481 (1943).
2.29. Schaefer G, *Journal of Physics and Chemistry of Solids* **12**, 233 (1960).
2.30. Scottish Universities Summer School, *Phonon in Perfect Lattices and in Lattices with Point Imperfections*, 1965, Plenum Press, New York, 1960.
2.31. Shull CG and Wollan EO, "Application of Neutron Diffraction to Solid State Problems," *Solid State Physics, Advances in Research and Applications* **2**, 137 (1956).
2.32. Srivastava GP, *The Physics of Phonons*, Adam Hilger, Bristol, 1990.
2.33. Strauch D, Pavone P, Meyer AP, Karch K, Sterner H, Schmid A, Pleti Th, Bauer R, Schmitt M, "Festorkorperproblem," *Advances in Solid State Physics* **37**, 99-124 (1998), Helbig R (ed), Braunschweig/Weisbaden: Vieweg.
2.34. Toya T, "Lattice Dynamics of Lead," in Proceedings of the International Conference on Lattice Dynamics, Copenhagen, 1963, Pergamon Press, New York, 1965.
2.35. Van Hove L, *Phys Rev* **89**, 1189 (1953).
2.36. Vogelgesang R et al, *Phys Rev* **B54**, 3989 (1996).
2.37. Wallis RF (ed), Proceedings of the International Conference on Lattice Dynamics, Copenhagen, 1963, Pergamon Press, New York, 1965.
2.38. Ziman JM, *Electrons and Phonons*, Oxford, Clarendon Press (1962).
2.39. 1962 Brandeis University Summer Institute Lectures in Theoretical Physics, Vol. 2, W. A. Benjamin, New York, 1963.

Chapter 3

3.1. Altman SL, *Band Theory of Solids*, Clarendon Press, Oxford, 1994. See also Singleton J, *Band Theory and Electronic Structure*, Oxford University Press, 2001.
3.2. Aryasetiawan F and Gunnarson D, *Rep Prog Phys* **61**, 237 (1998).
3.3. Austin BJ et al, *Phys Rev* **127**, 276 (1962)
3.4. Berman R, *Thermal Conduction in Solid*, Clarendon Press, Oxford, 1976, p. 125.
3.5. Blount EI, "Formalisms of Band Theory," *Solid State Physics, Advances in Research and Applications*, **13**, 305-373 (1962).
3.6. Blount EI, *Lectures in Theoretical Physics*, Vol. V, Interscience Publishers, New York, 1963, p422ff.
3.7. Bouckaert LP, Smoluchowski R, and Wigner E, *Physical Review*, **50**, 58 (1936).
3.8. Callaway J and March NH, "Density Functional Methods; Theory and Applications," *Solid State Physics, Advances in Research and Applications*, **38**, 135-221 (1984).
3.9. Ceperley DM and Alder BJ, *Phys Rev Lett* **45**, 566 (1980).
3.10. Chelikowsky JR and Louie SG (eds), *Quantum Theory of Real Materials*, Kluwer Academic Publishers, Dordrecht, 1996.
3.11. Cohen ML, *Physics Today* **33**, 40-44 (1979)
3.12. Cohen ML and Chelikowsky JR, *Electronic Structure and Optical Properties of Semiconductors*, Springer-Verlag, Berlin, 2nd edn. (1989).
3.13. Cohen ML and Heine V, "The Fitting of Pseudopotentials to Experimental Data and their Subsequent Application," *Solid State Physics, Advances in Research and Applications*, **24**, 37-248 (1970).
3.14. Cohen M and Heine V, *Phys Rev* **122**, 1821 (1961).
3.15. Cusack NE, *The Physics of Disordered Matter*, Adam Hilger, Bristol, 1987, see especially Chaps. 7 and 9.
3.16. Dimmock JO, "The Calculation of Electronic Energy Bands by the Augmented Plane Wave Method," *Solid State Physics, Advances in Research and Applications*, **26**, 103-274 (1971).
3.17. Fermi E, *Nuovo Cimento* **2**, 157 (1934)
3.18. Friedman B, *Principles and Techniques of Applied Mathematics*, John Wiley and Sons, New York, 1956.
3.19. Harrison WA, *Pseudopotentials in the theory of Metals*, W A Benjamin, Inc., New York, 1966.
3.20. Heine V, "The Pseudopotential Concept," *Solid State Physics, Advances in Research and Applications*, **24**, 1-36 (1970).
3.21. Herring C, *Phys Rev* **57**, 1169 (1940).
3.22. Herring C, *Phys Rev* **58**, 132 (1940).
3.23. Hohenberg PC and Kohn W, "Inhomogeneous Electron Gas," *Phys Rev*, **136**, B804-871 (1964).
3.24. Izynmov YA, *Advances in Physics* **14**(56), 569 (1965).
3.25. Jones RO and Gunnaisson O, "The Density Functional Formalism, its Applications and Prospects," *Rev Modern Phys*, **61**, 689-746 (1989).
3.26. Jones W and March NH, *Theoretical Solid State Physics*, Vol. 1 and 2, John Wiley and Sons, 1973.
3.27. Kohn W, "Electronic Structure of Matter—Wave Functions and Density Functionals," *Rev Modern Phys*, **71**, 1253-1266 (1999).

3.28. Kohn W and Sham LJ, "Self Consistent Equations Including Exchange and Correlation Effects," *Phys Rev*, **140**, A1133-1138 (1965).
3.29. Kohn W and Sham LJ, *Phys Rev* **145**, 561 (1966).
3.30. Kronig and Penny, *Proceedings of the Royal Society* (London), **A130**, 499 (1931).
3.31. Landau L, *Soviet Physics* JETP, **3**, 920 (1956).
3.32. Loucks TL, *Phys Rev Lett*, **14**, 693 (1965).
3.33. Löwdin PO, *Advances in Physics* **5**, 1 (1956)
3.34. Marder MP, *Condensed Matter Physics*, John Wiley and Sons, Inc., New York, 2000.
3.35. Mattuck RD, *A Guide to Feynman Diagrams in the Many-Body Problem*, McGraw-Hill Book Company, New York, 2nd Ed., 1976. See particularly Chap. 4.
3.36. Negele JW and Orland H, *Quantum Many Particle Systems*, Addison-Wesley Publishing Company, Redwood City, California (1988).
3.37. Nemoshkalenko VV and Antonov VN, *Computational Methods in Solid State Physics*, Gordon and Breach Science Publishers, The Netherlands, 1998.
3.38. Parr RG and Yang W, Oxford Univ. Press, New York, 1989.
3.39. Pewdew JP and Zunger A, *Phys Rev* **B23**, 5048 (1981).
3.40. Phillips JC and Kleinman L, *Phys Rev* **116**, 287-294 (1959).
3.41. Pines D, *The Many-Body Problem*, W A Benjamin, New York, 1961.
3.42. Raimes S, *The Wave Mechanics of Electrons in Metals*, North-Holland Publishing Company, Amsterdam (1961).
3.43. Reitz JR, "Methods of the One-Electron Theory of Solids," *Solid State Physics, Advances in Research and Applications*, **1**, 1-95 (1955).
3.44. Schlüter M and Sham LJ, "Density Functional Techniques," *Physics Today*, Feb. 1982, pp. 36-43.
3.45. Singh DJ, *Plane Waves, Pseudopotentials, and the APW Method*, Kluwer Academic Publishers, Boston (1994).
3.46. Singleton J, *Band Theory and Electronic Properties of Solids*, Oxford University Press (2001).
3.47. Slater JC [88, 89, 90].
3.48. Slater JC, "The Current State of Solid-State and Molecular Theory," *International Journal of Quantum Chemistry*, **I**, 37-102 (1967).
3.49. Slater JC and Koster GF, *Phys Rev* **95**, 1167 (1954).
3.50. Slater JC and Koster GF, *Phys Rev* **96**, 1208 (1954).
3.51. Smith N, "Science with soft x-rays," *Physics Today* **54**(1), 29-54 (2001).
3.52. Spicer WE, *Phys Rev* **112**, p114ff (1958).
3.53. Stern EA, "Rigid-Band Model of Alloys," *Phys Rev*, **157**(3), 544 (1967).
3.54. Thouless DJ, *The Quantum Mechanics of Many-Body Systems*, Academic Press, New York, 1961.
3.55. Tran HT and Pewdew JP, "How metals bind: The deformable-jellium model with correlated electrons," *Am. J. Phys.* **71**(10), 1048-1061 (2003).
3.56. Wannier GH, "The Structure of Electronic Excitation Levels in Insulating Crystals," *Phys Rev* **52**, 191-197, (1937).
3.57. Wigner EP and Seitz F, "Qualitative Analysis of the Cohesion in Metals," *Solid State Physics, Advances in Research and Applications*, **1**, 97-126 (1955).
3.58. Woodruff TO, "The Orthogonalized Plane-Wave Method," *Solid State Physics, Advances in Research and Applications*, **4**, 367-411 (1957).
3.59. Ziman JM, "The Calculation of Bloch Functions," *Solid State Physics, Advances in Research and Applications*, **26**, 1-101 (1971).

Chapter 4

4.1. Anderson HL (ed), *A Physicists Desk Reference*, 2nd edn, Article 20: Frederikse HPR, p. 310, AIP Press, New York, 1989.
4.2. Appel J, "Polarons," *Solid State Physics, Advances in Research and Applications*, **21**, 193-391 (1968). A comprehensive treatment.
4.3. Arajs S, *American Journal of Physics*, **37** (7), 752 (1969).
4.4. Bergmann DJ, *Physics Reports* **43**, 377 (1978).
4.5. Brockhouse BN, *Rev Modern Physics* **67**, 735-751 (1995).
4.6. Callaway J, "Model for Lattice Thermal Conductivity at Low Temperatures," *Physical Review*, **113**, 1046 (1959).
4.7. Feynman RP, Statistical Mechanics, Addison-Wesley Publ. Co., Reading MA, 1972, Chap. 8.
4.8. Fisher ME and Langer JS, "Resistive Anomalies at Magnetic Critical Points," *Physical Review Letters*, **20**(13), 665 (1968).
4.9. Garnett M, *Philos. Trans. R. Soc.* (London), **203**, 385 (1904).
4.10. Geiger Jr. FE and Cunningham FG, "Ambipolar Diffusion in Semiconductors," *American Journal of Physics*, **32**, 336 (1964).
4.11. Halperin BI and Hohenberg PC, "Scaling Laws for Dynamical Critical Phenomena," *Physical Review*, **177**(2), 952 (1969).
4.12. Holland MG, "Phonon Scattering in Semiconductors from Thermal Conductivity Studies," *Physical Review*, **134**, A471 (1964).
4.13. Howarth DJ and Sondheimer EH, Proc. Roy. Soc. **A219**, 53 (1953)
4.14. Jan JP, "Galvanomagnetic and Thermomagnetic Effects in Metals," *Solid State Physics, Advances in Research and Applications*, **5**, 1-96 (1957).
4.15. Kadanoff LP, "Transport Coefficients Near Critical Points," *Comments on Solid State Physics*, **1**(1), 5 (1968).
4.16. Katsnelson AA, Stepanyuk VS, Szász AI, and Farberovich DV, *Computational Methods in Condensed Matter: Electronic Structure*, American Institute of Physics, 1992.
4.17. Kawasaki K, "On the Behavior of Thermal Conductivity Near the Magnetic Transition Point," *Progress in Theoretical Physics (Kyoto)*, **29**(6), 801 (1963).
4.18. Klemens PG, "Thermal Conductivity and Lattice Vibration Modes," *Solid State Physics, Advances in Research and Applications*, **7**, 1-98 (1958).
4.19. Kohn W, *Physical Review*, **126**, 1693 (1962).
4.20. Kohn W, "Nobel Lecture: Electronic Structure of Matter–Wave Functions and density Functionals," *Rev. Modern Phys.* **71**, 1253-1266 (1998)
4.21. Kondo J, "Resistance Minimum in Dilute Magnetic Alloys," *Progress in Theoretical Physics (Kyoto)*, **32**, 37 (1964).
4.22. Kothari LS and Singwi KS, "Interaction of Thermal Neutrons with Solids," *Solid State Physics, Advances in Research and Applications*, **8**, 109-190 (1959).
4.23. Kuper CG and Whitfield GD, *POLARONS AND EXCITONS*, Plenum Press, New York, 1962. There are lucid articles by Fröhlich, Pines, and others here, as well as a chapter by F. C. Brown on experimental aspects of the polaron.
4.24. Langer JS and Vosko SH, *Journal of Physics and Chemistry of Solids*, **12**, 196 (1960).

4.25. MacDonald DKC, "Electrical Conductivity of Metals and Alloys at Low Temperatures," *Encyclopedia of Physics*, Vol. XIV, Low Temperature Physics I, Springer-Verlag, Berlin, 1956, p. 137.

4.26. Madelung O, *Introduction to Solid State Theory*, Springer-Verlag, 1978, pp. 153-155, 183-187, 370-373. A relatively simple and clear exposition of both the large and small polaron.

4.27. Mahan GD, *Many Particle Physics*, Plenum Press, New York, 1981, Chaps. 1 and 6. Green's functions and diagrams will be found here.

4.28. Mattuck RD, *A Guide to Feynman Diagrams in the Many-Body Problem*, McGraw-Hill Book Company, New York, 1967.

4.29. McMillan WL and Rowell JM, *Physical Review Letters*, **14** (4), 108 (1965).

4.30. Mendelssohn K and Rosenberg HM, "The Thermal Conductivity of Metals a Low Temperatures," *Solid State Physics, Advances in Research and Applications*, **12**, 223-274 (1961).

4.31. Mott NF, *Metal-Insulator Transitions*, Taylor and Francis, London, 1990 (2nd edn).

4.32. Olsen JL, *Electron Transport in Metals*, Interscience, New York, 1962.

4.33. Patterson JD, "Modern Study of Solids," *Am. J. Phys.* **32**, 269-278 (1964).

4.34. Patterson JD, "Error Analysis and Equations for the Thermal Conductivity of Composites," *Thermal Conductivity* **18**, Ashworth T and Smith DR (eds), Plenum Press, New York, 1985, pp 733-742.

4.35. Pines D, "Electron Interactions in Metals," *Solid State Physics, Advances in Research and Applications*, **1**, 373-450 (1955).

4.36. Reynolds JA and Hough JM, *Proc. Roy. Soc.* (London), **B70**, 769-775 (1957).

4.37. Sham LJ and Ziman JM, "The Electron-Phonon Interaction," *Solid State Physics, Advances in Research and Applications*, **15**, 223-298 (1963).

4.38. Stratton JA, *Electromagnetic Theory*, McGraw Hill, 1941, p. 211ff.

4.39. Ziman JM, *Electrons and Phonons*, Oxford, London, 1962, Chap. 5 and later chapters (esp. p. 497)

Chapter 5

5.1. Alexander, W. and Street A, *Metals in the Service of Man*, 7th edn. Middlesex, England: Penguin, 1979.

5.2. Blatt FJ, *Physics of Electronic Conduction in Solids*, McGraw-Hill (1968).

5.3. Borg RJ and Dienes GJ, *An Introduction to Solid State Diffusion*, Academic Press, San Diego, 1988, p 148-151.

5.4. Cottrell A, *Introduction to the Modern Theory of Metals*, the Institute of Metals, London, 1988.

5.5. Cracknell AP and Wong KC, *The Fermi Surface: Its Concept, Determination, and Use in the Physics of Metals*, Clarendon Press, Oxford, 1973.

5.6. Duke CB, "Tunneling in Solids," in Supplement 10, Solid State Physics, Advances in Research and Applications (1969).

5.7. Fiks VB, *Sov Phys Solid State*, **1**, 14 (1959).

5.8. Fisk Z et al, "The Physics and Chemistry of Heavy Fermions," *Proc Natl Acad Sci USA* **92**, 6663-6667 (1995).

5.9. Gantmakher VF, "Radio Frequency Size Effect in Metals," *Progress in Low Temperature Physics*, Vol. V, Gorter CJ (ed), North-Holland Publishing Company, Amsterdam, 1967, p. 181.
5.10. Harrison WA, *Applied Quantum Mechanics*, World Scientific, Singapore, 2000, Chap. 21.
5.11. Harrison WA and Webb MB (eds), *The Fermi Surface*, John Wiley and Sons, New York, 1960.
5.12. Huang K, Statistical Mechanics, John Wiley and Sons, 2nd edn, 1987, pp 247-255.
5.13. Huntington HB and Grove AR, *J Phys Chem Solids*, **20**, 76 (1961).
5.14. Kahn AH and Frederikse HPR, "Oscillatory Behavior of Magnetic Susceptibility and Electronic Conductivity," *Solid State Physics, Advances in Research and Applications* **9**, 259-291 (1959).
5.15. Kittel C and Kroemer H, *Thermal Physics*, W. H. Freeman and Company, San Francisco, 2nd edn., 1980, Chap. 11.
5.16. Langenberg DN, "Resource Letter OEPM-1 on the Ordinary Electronic Properties of Metals," *American Journal of Physics*, **36** (9), 777 (1968).
5.17. Lax B and Mavroides JG, "Cyclotron Resonance," *Solid State Physics, Advances in Research and Applications* **11**, 261-400 (1960).
5.18. Lloyd JR, "Electromigration in integrated circuit conductors", *J Phys D: Appl Phys* **32**, R109-R118 (1999).
5.19. Mackintosh AR, *Sci. Am.* **209**, 110, (1963).
5.20. Onsager L, Phil. Mag. **93**, 1006-1008 (1952).
5.21. Overhauser AW, "Charge Density Wave," *Solid State Physics Source Book*, Parker SP (ed), McGraw-Hill Book Co., 1987, pp 142-143.
5.22. Overhauser AW, "Spin-Density Wave," *Solid State Physics Source Book*, Parker SP (ed), McGraw-Hill Book Co., 1987, pp. 143-145
5.23. Peierls R, *More Surprises in Theoretical Physics*, Princeton University Press, NJ, 1991, p29.
5.24. Pippard AB, "The Dynamics of Conduction Electrons," *Low Temperature Physics*, deWitt C, Dreyfus B, and deGennes PG (eds), Gordon and Breach, New York, 1962. Also Pippard AB, *Magnetoresistance in Metals*, Cambridge University Press, 1988.
5.25. Radousky HB, *Magnetism in Heavy Fermion Systems*, World Scientific, Singapore, 2000.
5.26. Shapiro SL and Teukolsky SA, *Black Holes, White Dwarfs and Neutron Stars: The Physics of Compact Objects*, John Wiley and Sons, Inc., New York, 1983.
5.27. Shoenberg D, *Magnetic Oscillations in Metals*, Cambridge University Press, 1984.
5.28. Sorbello RS, "Theory of Electromigration," *Solid State Physics, Advances in Research and Applications* **51**, 159-231 (1997).
5.29. Stark RW and Falicov LM, "Magnetic Breakdown in Metals," *Progress in Low Temperature Physics*, Vol. V, Gorter CJ (ed), North-Holland Publishing Company, Amsterdam, 1967, p. 235.
5.30. Stewart GR, "Heavy-Fermion Systems," *Rev Modern Physics*, **56**, 755-787 (1984).
5.31. Thorne RE, "Charge Density Wave Conductors," *Physics Today*, pp 42-47 (May 1996).
5.32. Wigner E and Huntington HB, *J Chem Phys* **3**, 764-770 (1935).
5.33. Wilson AH, *The Theory of Metals*, Cambridge, 1954.

5.34. Zak J, "Dynamics of Electrons in Solids in External Fields," I, *Physical Review*, **168**(3), 686 (1968); II, *Physical Review*, **177**(3), 1151 (1969). Also "The kq-Representation in the Dynamics of Metals" *Solid State Physics, Advances in Research and Applications* **27**, 1-62 (1972).
5.35. Ziman JM, *Principles of Theory of Solids*, 2nd edn., Cambridge, 1972.
5.36. Ziman JM, *Electrons in Metals, A Short Guide to the Fermi Surface*, Taylor and Francis, London, 1963.

Chapter 6

6.1. Alferov ZI, "Nobel Lecture: The Double Heterostructure Concept and its application in Physics, Electronics, and Technology," *Rev. Modern Phys.* **73**(3), 767-782 (2001).
6.2. Ashcroft NW and Mermin ND, *Solid State Physics*, Holt, Rinehart and Winston, New York, 1976, Chapters 28 and 29.
6.3. Bardeen J, "Surface States and Rectification at a Metal-Semiconductor Contact", *Physical Review*, **71**, 717-727 (1947).
6.4. Blakemore JS, *Solid State Physics*, Second Edition, W. B. Saunders Co., Philadelphia, 1974.
6.5. Boer KW, Survey of Semiconductor Physics, Electrons and Other Particles in Bulk Semiconductors, Van Nostrand Reinhold, New York, 1990.
6.6. Bube R, *Electronics in Solids*, Academic Press, Inc., New York, 1992 3rd edn.
6.7. Chen A and Sher A, *Semiconductor Alloys*, Plenum Press, New York, 1995.
6.8. Cohen ML and Chelikowsky JR, *Electronic Structure and Optical Properties of Semiconductors*, Springer-Verlag, Berlin, 2nd edn, 1989.
6.9. Conwell E and Weisskopf VF, *Physical Review*, **77**, 388 (1950).
6.10. Dalven R, *Introduction to Applied Solid State Physics*, Plenum Press, New York, 1980. See also second edition, 1990.
6.11. Dresselhaus G, Kip AF and Kittel C, *Phys Rev* **98**, 368 (1955).
6.12. Einspruch NG, "Ultrasonic Effects in Semiconductors," *Solid State Physics, Advances in Research and Applications* **17**, 217-268 (1965).
6.13. Fan HY, "Valence Semiconductors, Ge and Si," *Solid State Physics, Advances in Research and Applications* **1**, 283-265 (1955).
6.14. Fraser DA, *The Physics of Semiconductor Devices*, Clarendon Press, Oxford, 4th edition, 1986.
6.15. Handler P, "Resource Letter Scr-1 on Semiconductors," *American Journal of Physics*, **32** (5), 329 (1964).
6.16. Kane EO, *J. Phys. Chem. Solids* **1**, 249 (1957).
6.17. Kittel C, *Introduction to Solid State Physics*, Seventh Edition, John Wiley and Sons, New York, 1996, Chap. 8.
6.18. Kohn W, "Shallow Impurity States in Si and Ge," *Solid State Physics, Advances in Research and Applications* **5**, 257-320 (1957).
6.19. Kroemer H, "Nobel Lecture: Quasielectronic Fields and Band Offsets: Teaching Electrons New Tricks," *Rev. Modern Phys.* **73**(3), 783-793 (2001).
6.20. Li M-F, "Modern Semiconductor Quantum Physics," *World Scientific*, Singapore, 1994.

6.21. Long D, *Energy Bands in Semiconductors*, Interscience Publishers, New York, 1968.
6.22. Ludwig GW and Woodbury HH, "Electron Spin Resonance in Semiconductors," *Solid State Physics, Advances in Research and Applications* 13, 223-304 (1962).
6.23. McKelvey JP, *Solid State and Semiconductor Physics*, Harper and Row Publishers, New York, 1966.
6.24. Merzbacher E, *Quantum Mechanics*, 2nd edn., John Wiley & Sons, Inc., New York, 1970, Chap. 2.
6.25. Moss TS (ed), *Handbook on Semiconductors*, Vol. 1, Landberg PT (ed), North Holland/Elsevier (1992), Amsterdam (There are additional volumes).
6.26. Nakamura S, Pearton S, Fasol G, *The Blue Laser Diode: The Complete Story*, Springer-Verlag, New York, 2000.
6.27. Ovshinsky SR, "Reversible Electrical Switching Phenomena in Disordered Structures," *Physical Review Letters*, 21, 1450 (1968).
6.28. Pankove JI and Moustaka TD (eds), "Gallium Nitride I," *Semiconductors and Semimetals*, Vol. 50, Academic Press, New York, 1997.
6.29. Pantiledes ST (editor), *Deep Centers in Semiconductors*, Gordon and Breach Publishers, Yverdon, Switzerland, 1992.
6.30. Patterson JD, "Narrow Gap Semiconductors," *Condensed Matter News* 3 (1), 4-11 (1994).
6.31. Perkowitz S, *Optical Charaterization of Semiconductors*, Academic Press, San Diego, 1993.
6.32. Ridley BK, *Quantum Processes in Semiconductors*, Clarendon Press, Oxford, 1988.
6.33. Sapoval B and Hermann C, *Physics of Semiconductors*, Springer-Verlag, New York, 1995.
6.34. Seeger K, *Semiconductor Physics*, Springer-Verlag, Berlin, 4th edn, 1989.
6.35. Seitz F, *Physical Review*, 73, 549 (1948).
6.36. Shockley W, *Electrons and Holes in Semiconductors*, D. Van Nostrand, New York, 1950.
6.37. Slater JC, *Quantum Theory of Molecules and Solids*, Vol. III, *Insulators, Semiconductors, and Metals*, McGraw-Hill Book Company, New York, 1967.
6.38. Smith RA, *Wave Mechanics of Crystalline Solids*, John Wiley and Sons, 1961, section 8.8 and appendix 1.
6.39. Smith RA (ed), *Semiconductors*, Proceedings of the International School of Physics, "Enrico Fermi" Course XXII, Academic Press, New York, 1963.
6.40. Streetman BG, *Solid State Electronic Devices*, 2nd ed. Prentice Hall, Englewood Cliffs, N.J., 1980. Also see the third edition (1990).
6.41. Sze SM, *Physics of Semiconductor Devices*, 2nd edn, Wiley, New York, 1981.
6.42. Sze SM (ed), *Modern Semiconductor Device Physics*, John Wiley and Sons, Inc., New York, 1998.
6.43. Willardson RK, and Weber ER, "Gallium Nitride II," *Semiconductors and Semimetals*, Vol. 57, Academic Press, 1998.
6.44. Yu PY and Cardona M, *Fundamentals of Semiconductors*, Springer Verlag, Berlin, 1996.

Chapter 7

7.1. Anderson PW, "Theory of Magnetic Exchange Interactions: Exchange in Insulators and Semiconductors," *Solid State Physics, Advances in Research and Applications*, **14**, 99-214 (1963).

7.2. Ashcroft NW and Mermin ND, *Solid State Physics*, Holt, Rinehart and Winston, New York, 1976, Chaps. 31, 32 and 33.

7.3. Auld BA, "Magnetostatic and Magnetoelastic Wave Propagation in Solids", *Applied Solid State Science*, Vol. 2, Wolfe R and Kriessman CJ (eds), Academic Press 1971.

7.4. Baibich MN, Broto JM, Fert A, Nguyen Van Dau F, Petroff F, Eitenne P, Creuzet G, Friederich A, and Chazelas J, *Phys. Rev. Lett.*, **61**, 2472 (1988).

7.5. Bennett C, "Quantum Information and Computation," *Physics Today*, October 1995, pp. 24-30.

7.6. Bertram HN, *Theory of Magnetic Recording*, Cambridge University Press, 1994, Chap. 2.

7.7. Bitko D et.al., *J Research of NIST*, **102**(2), 207-211 (1997)).

7.8. Blackman JA and Tagüeña J, *Disorder in Condensed Matter Physics, A Volume in Honour of Roger Elliott*, Clarendon Press, Oxford, 1991.

7.9. Blundell S, *Magnetism in Condensed Matter*, Oxford University Press, 2001.

7.10. Charap SH and Boyd EL, *Physical Review*, **133**, A811 (1964).

7.11. Chikazumi S, *Physics of Ferromagnetism*, (Translation editor, Graham CD), Oxford at Clarendon Press, 1977.

7.12. Chowdhury D, *Spin Glasses and Other Frustrated Systems*, Princeton University Press, 1986.

7.13. Cooper B, "Magnetic Properties of Rare Earth Metals," *Solid State Physics, Advances in Research and Applications*, **21**, 393-490 (1968).

7.14. Cracknell AP and Vaughn RA, *Magnetism in Solids Some Current Topics*, Scottish Universities Summer School, 1981.

7.15. Craik D, *Magnetism Principles and Applications*, John Wiley and Sons, 1995.

7.16. Cullity BD, *Introduction to Magnetic Materials*, Addison-Wesley, Reading, Mass., 1972.

7.17. Damon R and Eshbach J, *J Phys Chem Solids*, **19**, 308 (1961).

7.18. Dyson FJ, *Physical Review*, **102**, 1217 (1956).

7.19. Elliott RJ, *Magnetic Properties of Rare Earth Metals*, Plenum Press, London, 1972.

7.20. Fetter AL and Walecka JD, *Theoretical Mechanics of Particles and Continua*, McGraw-Hill, pp. 399-402, 1980.

7.21. Fisher ME, "The Theory of Equilibrium Critical Phenomena," *Reports on Progress in Physics*, XXX(II), 615 (1967).

7.22. Fischer KH and Hertz JA, *Spin Glasses*, Cambridge University Press, 1991.

7.23. Fontcuberta J, "Colossal Magnetoresistance," *Physics World*, February 1999, pp. 33-38.

7.24. Gibbs MRJ (ed), *Modern Trends in Magnetostriction Study and Application*, Kluwer Academic Publishers, Dordrecht, 2000.

7.25. Gilbert W, *De Magnete* (originally published in 1600), Translated by P. Fleury Mottelay, Dover, New York (1958).

7.26. Griffiths RB, *Physical Review*, **136**(2), 437 (1964).

7.27. Heitler W, *Elementary Wave Mechanics*, Oxford, 1956, 2nd edn, Chap. IX.
7.28. Heller P, "Experimental Investigations of Critical Phenomena," *Reports on Progress in Physics*, XXX(II), 731 (1967).
7.29. Herbst JF, *Rev Modern Physics* **63**(4), 819-898 (1991).
7.30. Herring C, *Exhange Interactions among Itinerant Electrons in Magnetism*, Rado GT and Suhl H (eds), Academic Press, New York, 1966.
7.31. Herzfield CM and Meijer HE, "Group Theory and Crystal Field Theory," *Solid State Physics, Advances in Research and Applications*, **12**, 1-91 (1961).
7.32. Huang K, *Statistical Mechanics*, 2nd edn, John Wiley and Sons, New York, 1987.
7.33. Ibach H and Luth H, *Solid State Physics*, Springer-Verlag, Berlin, 1991, p 152.
7.34. Julliere M, *Phys Lett* **54A**, 225 (1975).
7.35. Kadanoff LP et al, *Reviews of Modern Physics*, **39** (2), 395 (1967).
7.36. Kasuya T, *Progress in Theoretical Physics (Kyoto)*, **16**, 45 and 58 (1956).
7.37. Keffer F, "Spin Waves," *Encyclopedia of Physics*, Vol. XVIII, Part 2, Ferromagnetism, Springer-Verlag, Berlin, 1966.
7.38. Kittel C, "Magnons," *Low Temperature Physics*, DeWitt C, Dreyfus B, and deGennes PG (eds), Gordon and Breach, New York, 1962.
7.39. Kittel C, *Introduction to Solid State Physics*, 7th edn, John Wiley and Sons, New York, 1996, Chapters 14, 15, and 16.
7.40. Kosterlitz JM and Thouless DJ, *J Phys C* **6**, 1181 (1973).
7.41. Kouwenhoven L and Glazman L, *Physics World*, pp. 33-38, Jan. 2001.
7.42. Langer JS and S. H. Vosko, *J Phys Chem Solids* **12**, 196 (1960).
7.43. Levy RA and Hasegawa R, *Amorphous Magnetism II*, Plenum Press, New York, 1977.
7.44. Malozemoff AP and Slonczewski JC, *Magnetic Domain Walls in Bubble Materials*, Academic Press, New York, 1979.
7.45. Manenkov AA and Orbach R (eds), *Spin-Lattice Relaxation in Ionic Solids*, Harper and Row Publishers, New York, 1966.
7.46. Marshall W (ed), *Theory of Magnetism in Transition Metals*, Proceedings of the International School of Physics, "Enrico Fermi" Course XXXVII, Academic Press, New York, 1967.
7.47. Mathews J and Walker RL, *Mathematical Methods of Physics*, W. A. Benjamin, New York, 1967.
7.48. Mattis DC, The Theory of Magnetism I Statics and Dynamics, Springer-Verlag, 1988 and II Thermodynamics and Statistical Mechanics, Springer-Verlag 1985.
7.49. Mermin ND and Wagner H, *Physical Review Letters*, **17**(22), 1133 (1966).
7.50. Muller B and Reinhardt J, *Neural Networks, An Introduction*, Springer-Verlag, Berlin, 1990.
7.51. Pake GE, "Nuclear Magnetic Resonance," *Solid State Physics, Advances in Research and Applications*, **2**, 1-91 (1956).
7.52. Parkin S, *J App Phys* **85**, 5828 (1999).
7.53. Patterson JD, *Introduction to the Theory of Solid State Physics*, Addison-Wesley Publishing Co., Reading MA, 1971 p176ff.
7.54. Patterson JD et al, *Journal of Applied Physics*, **39** (3), 1629 (1968), and references cited therein.
7.55. Prinz GA, Science Vol. 282, 27 Nov. 1998, p 1660.

7.56. Rado GT and Suhl H (eds), Vol. II Part A, *Statistical Models, Magnetic Symmetry, Hyperfine Interactions, and Metals*, Academic Press, New York, 1965. Vol. IV, *Exchange Interactions among Itinerant Electrons* by Conyers Herring, Academic Press, New York, 1966.
7.57. Ruderman MA and Kittel C, *Physical Review*, 96, 99 (1954).[1]
7.58. Salamon MB and Jaime M, "The Physics of Manganites Structure and Transport," *Rev. Modern Physics*, 73, 583-628 (2001).
7.59. Schrieffer JR, "The Kondo Effect–The Link Between Magnetic and NonMagnetic Impurities in Metals?", *Journal of Applied Physics*, 38(3), 1143 (1967).
7.60. Slichter CP, *Principles of Magnetic Resonance*, Harper and Row, Evanston,1963.
7.61. Slonczewski JC, *Phys Rev* B39, 6995 (1989).
7.62. Tyalblikov SV, *Methods in the Quantum Theory of Magnetism*, Plenum Press, New York, 1967.
7.63. Van Vleck JH, *The Theory of Electric and Magnetic Susceptibilities*, Oxford University Press, 1932.
7.64. Von der Lage FC and Bethe HA, *Phys Rev* 71, 612 (1947).
7.65. Walker LR, *Phys Rev* 105, 309 (1957).
7.66. Waller I, *Z Physik*, 79, 370 (1932).
7.67. Weinberg S, *The Quantum Theory of Fields, Vol I Modern Applications*, Cambridge University Press, 1996, pp. 332-352.
7.68. White RM, *Quantum Theory of Magnetism*, McGraw Hill, New York, 1970.
7.69. Wohlfarth EP, *Magnetism Vol. III*, Rado GT and Suhl H, (eds), Academic Press, New York, 1963. Wohlfarth EP, *Rev Mod Phys.* 25, 211 (1953). Wohlfarth EP, *Handbook of Magnetic Materials*, Elsevier, several volumes (~1993).
7.70. Wojtowicz PJ, *Journal of Applied Physics*, 35, 991 (1964).
7.71. Yosida K, *Physical Review*, 106, 893 (1957).[1]
7.72. Yosida K, *Theory of Magnetism*, Springer, Berlin, 1998.
7.73. Zutic I et al, "Spintronics: Fundamentals and applications," *Rev Mod Phys* 76, 325 (2004).

Chapter 8

8.1. Allen PB and Mitrovic B, "Theory of Superconductivity T_c," *Solid State Physics, Advances in Research and Applications*, 37, 2-92 (1982).
8.2. Anderson PW, "The Josephson Effect and Quantum Coherence Measurements in Superconductors and Superfluids," *Progress in Low Temperature Physics*, Vol. V, Gorter CJ (ed), North-Holland Publishing Company, Amsterdam, 1967, p. 1. See also Annett JF, *Superconductivity, Superfluids, and Condensates*, Oxford University Press, 2004.
8.3. Annett JF, *Superconductivity, Superfluids, and Condensates*, Oxford University Press, 2004.

---

[1] These papers deal with the indirect interaction of nuclei by their interaction with the conduction electrons and of the related indirect interaction of ions with atomic magnetic moments by their interaction with the conduction electrons. In the first case the hyperfine interaction is important and in the second the exchange interaction is important.

8.4.	Bardeen J, "Superconductivity," *1962 Cargese Lectures in Theoretical Physics*, Levy M (ed), W. A. Benjamin, New York, 1963.
8.5.	Bardeen J and Schrieffer JR, "Recent Developments in Superconductivity," *Progress in Low Temperature Physics*, Vol. III, Gorter CJ (ed), North-Holland Publishing Company, Amsterdam, 1961, p. 170.
8.6.	Bardeen J, Cooper LN, and Schreiffer JR, "Theory of Superconductivity," *Phys Rev* **108**, 1175-1204 (1957).
8.7.	Beyers R and Shaw TM, "The Structure of $Y_1Ba_2Cu_3O_{7-\delta}$ and its Derivatives," *Solid State Physics, Advances in Research and Applications*, **42**, 135-212 (1989).
8.8.	Bogoliubov NN (ed), *The Theory of Superconductivity*, Gordon and Breoch, New York, 1962.
8.9.	Burns G, *High Temperature Superconductivity an Introduction*, Academic Press, Inc., Boston, 1992.
8.10.	Cooper L, "Bound Electron Pairs in a Degenerate Fermi Gas," *Phys Rev* **104**, 1187-1190 (1956).
8.11.	Dalven R, *Introduction to Applied Solid State Physics*, Plenum Press, New York, 1990, 2nd edn, Ch. 8.
8.12.	deGennes PG, *Superconductivity of Metals and Alloys*, W. A. Benjamin, New York, 1966.
8.13.	Feynman RP, Leighton RB, and Sands M, *The Feynman Lectures on Physics*, Vol. III, Addison-Wesley Publishing Co., 1965.
8.14.	Giaever I, "Electron Tunneling and Superconductivity," *Rev Modern Phys*, **46**(2), 245-250 (1974).
8.15.	Goodman BB, "Type II Superconductors," *Reports on Progress in Physics*, Vol. XXIX, Part 11, The Institute of Physics and The Physical Society, London, 1966, p. 445.
8.16.	Hass KC, "Electronics Structure of Copper-Oxide Superconductors," *Solid State Physics, Advances in Research and Applications*, **42**, 213-270 (1989).
8.17.	Jones W and March NH, *Theoretical Solid State Physics*, Vol. 2, Dover, 1985, p895ff and p1151ff.
8.18.	Josephson B, "The Discovery of Tunneling Supercurrents," *Rev Modern Phys*, **46**(2), 251-254 (1974).
8.19.	Kittel C, [60], Chap. 8.
8.20.	Kuper CG, *An Introduction to the Theory of Superconductivity*, Clarendon Press, Oxford, 1968.
8.21.	Mahan G, *Many-Particle Physics*, Plenum Press, New York, 1981, Ch. 9.
8.22.	Marder, *Condensed Matter Physics*, John Wiley and Sons, 2000, p. 819.
8.23.	Mattuck RD, *A Guide to Feynman Diagrams in the Many-Body Problem*, 2nd edn, Dover, New York, 1992, Chap. 15.
8.24.	Parker WH, Taylor BN and Langenberg DN, *Physical Review Letters*, **18** (8), 287 (1967).
8.25.	Parks RD (ed), *Superconductivity*, Vols. 1 and 2, Marcel Dekker, New York, 1969.
8.26.	Rickaysen G, *Theory of Superconductivity*, Interscience, New York, 1965.
8.27.	Saint-James D, Thomas EJ, and Sarma G, Type II Superconductivity, Pergamon, Oxford, 1969.
8.28.	Scalapino DJ, "The Theory of Josephson Tunneling," *Tunneling Phenomena in Solids*, Burstein E and Lundquist S (eds), Plenum Press, New York, 1969.

8.29. Schafroth MR, "Theoretical Aspects of Superconductivity," *Solid State Physics, Advances in Research and Applications*, **10**, 293-498 (1960).
8.30. Schrieffer JR, *Theory of Superconductivity*, W. A. Benjamin, New York, 1964.
8.31. Silver AH and Zimmerman JE, *Phys Rev* **157**, 317 (1967).
8.32. Tinkham M, Introduction to Superconductivity, McGraw-Hill, New York, 2nd edn, 1996.
8.33. Tinkham M and Lobb CJ, "Physical Properties of the New Superconductors," *Solid State Physics, Advances in Research and Applications*, **42**, 91-134 (1989).

Chapter 9

9.1. Bauer S, Gerhard-Multhaupt R, and Sessler GM, "Ferroelectrets: Soft Electroactive Foams for Transducers," *Physics Today* **57**, 39-43 (Feb. 2004).
9.2. Böttcher CJF, *Theory of Electric Polarization*, Elsevier Publishing Company, New York, 1952.
9.3. Brown WF Jr, "Dielectrics," *Encyclopedia of Physics*, Vol. XVII, Flügge S (ed), Springer-Verlag, Berlin, 1956.
9.4. Devonshire AF, "Some Recent Work on Ferroelectrics," *Reports on Progress in Physics*, Vol. XXVII, The Institute of Physics and The Physical Society, London, 1964, p. 1.
9.5. Elliot RJ and Gibson AF, *An Introduction to Solid State Physics and its Applications*, Harper and Row 1974 p277ff.
9.6. Fatuzzo E and Merz WJ, *Ferroelectricity*, John Wiley and Sons, New York, 1967.
9.7. Forsbergh PW Jr., "Piezoelectricity, Electrostriction, and Ferroelectricity," *Encyclopedia of Physics*, Vol. XVII, Flügge S (ed), Springer Verlag, Berlin, 1956.
9.8. Fröhlich H, *Theory of Dielectrics—Dielectric Constant and Dielectric Loss*, Oxford University Press, New York, 1949.
9.9. Gutmann F, *Rev Modern Phys* **70**, 457 (1948).
9.10. Jona F and Shirane G, *Ferroelectric Crystals*, Pergamon Press, New York, 1962.
9.11. Kanzig W, "Ferroelectrics and Antiferroelectrics," *Solid State Physics, Advances in Research and Applications* **4**, 1-97 (1957).
9.12. Lines ME and Glass AM, Principles and Applications of Ferroelectrics and Related Materials, Oxford, 1977.
9.13. Moss TS, *Optical Properties of Semi-Conductors*, Butterworth and Company Pubs., London, 1959.
9.14. Pines D, "Electron Interaction in Metals," *Solid State Physics, Advances in Research and Applications* **1**, 373-450 (1955).
9.15. Platzman PM and Wolff PA, *Waves and Interactions in Solid State Plasmas*, Academic Press, New York, 1973, Chaps. VI and VII.
9.16. Smyth CP, *Dielectric Behavior and Structure*, McGraw-Hill Book Company, New York, 1955.
9.17. Samara GA and Peercy PS, "The Study of Soft-Mode Transitions at High Pressure," *Solid State Physics, Advances in Research and Applications* **36**, 1-118 (1981).
9.18. Steele MC and Vural B, *Wave Interactions in Solid State Plasmas*, McGraw-Hill Book Company, New York, 1969.

9.19. Tonks L and Langmuir I, *Phys Rev* **33**, 195-210 (1929).
9.20. Uehling EA, "Theories of Ferroelectricity in KH_2PO_4," in *Lectures in Theoretical Physics*, Vol. V, Briton WE, Downs BW, and Downs J (eds), Interscience Publishers, New York, 1963.
9.21. Zheludev IS, "Ferroelectricity and Symmetry," *Solid State Physics, Advances in Research and Applications* **26**, 429-464 (1971).

Chapter 10

10.1. Born M and Wolf E, Principles of Optics, 2nd (Revised) edn, MacMillan, 1964, especially Optics of Metal (Chap. XIII) and Optics of Crystals (Chap. XIV).
10.2. Born M and Huang K, *Dynamical Theory of Crystal Lattices*, Oxford at the Clarenden Press, 1954, see especially the optical effects (Chap. VII).
10.3. Brown FC, "Ultraviolet Spectroscopy of Solids with the Use of Synchrotron Radiation," *Solid State Physics, Advances in Research and Applications* **29**, 1-73 (1974).
10.4. Bube RH, *Photoconductivity of Solids*, John Wiley and Sons, New York, 1960.
10.5. Caldwell DJ, "Some Observations of the Faraday Effect," *Proc Natl Acad Sci* **56**, 1391-1398 (1966).
10.6. Callaway J, "Optical Absorption in an Electric-Field," *Physical Review*, **130**(2), 549 (1963).
10.7. Cochran W, *The Dynamics of Atoms in Crystals*, Edward Arnold, London, 1973, p 90.
10.8. Cohen MH, *Phil Mag* **3**, 762 (1958)
10.9. Cohen ML and Chelikowsky JR, *Electronic Structure and Optical Properties of Semiconductors*, 2nd edn, Springer-Verlag, Berlin, 1989.
10.10. Dexter DL, "Theory of the Optical Properties of Imperfections in Nonmetals," *Solid State Physics, Advances in Research and Applications* **6**, 353-411 (1958).
10.11. Elliott RJ and Gibson AF, *An Introduction to Solid State Physics*, Macmillan, 1974, Chap. 6, 7.
10.12. Fox M, *Optical Properties of Solids*, Oxford University Press, 2002.
10.13. Frova A, Handler P, Germano FA, and Aspnes DE, "Electro-Absorption Effects at the Band Edges of Silicon and Germanium," *Physical Review*, **145**(2), 575 (1966).
10.14. Givens MP, "Optical Properties of Metals," *Solid State Physics, Advances in Research and Applications* **6**, 313-352 (1958).
10.15. Gobeli GW and Fan HY, *Phys Rev* **119**(2), 613-620, (1960).
10.16. Greenaway DL and Harbeke G, *Optical Properties and Band Structures of Semiconductors*, Pergamon Press, Oxford, 1968.
10.17. Hagen E and Rubens H, *Ann. d. Physik* (4) **11**, 873 (1903).
10.18. Kane EO, *J Phys Chem Solids* **12**, 181 (1959).
10.19. Knox RS, "Theory of Excitons," *Solid State Physics, Advances in Research and Applications*, Supplement 5, 1963.
10.20. Lyddane RH, Sachs RG, and Teller E, *Phys Rev* **59**, 673 (1941).
10.21. Moss TS, *Optical Properties of Semiconductors*, Butterworth, London, 1961.
10.22. Pankove JL, *Optical Processes in Semiconductors*, Dover, New York, 1975.

10.23. Phillips JC, "The Fundamental Optical Spectra of Solids," *Solid State Physics, Advances in Research and Applications* **18**, 55-164 (1966).
10.24. Stern F, "Elementary Theory of the Optical Properties of Solids," *Solid State Physics, Advances in Research and Applications* **15**, 299-408 (1963).
10.25. Tauc J (ed), *The Optical Properties of Solids*, Proceedings of the International School of Physics, "Enrico Fermi" Course XXXIV, Academic Press, New York, 1966.
10.26. Tauc J, *The Optical Properties of Semiconductors*, Academic Press, New York, 1966.
10.27. Yu PY and Cordona M, *Fundamentals of Semiconductors*, Springer-Verlag, Berlin, 1996, chapters 6,7,8.

Chapter 11

11.1. Bastard G, *Wave Mechanics Applied to Semiconductor Heterostructures*, Halsted (1988)
11.2. Borg RJ and Dienes GJ, *An Introduction to Solid State Diffusion*, Academic Press, San Diego, 1988.
11.3. Bube RH, "Imperfection Ionization Energies in CdS-Type Materials by Photo-electronic Techniques," *Solid State Physics: Advances in Research and Applications* **11**, 223-260 (1960).
11.4. Chelikowsky JR and Louie SG, *Quantum Theory of Real Materials*, Kluwer Academic Publishers, Dordrecht, 1996.
11.5. Compton WD and Rabin H, "F-Aggregate Centers in Alkali Halide Crystals," *Solid State Physics: Advances in Research and Applications* **16**, 121-226 (1964).
11.6. Cottrell AH, *Dislocations and Plastic Flow in Crystals*, Oxford University Press, New York, 1953.
11.7. Crawford JH Jr. and Slifkin LM, *Point Defects in Solids, Vol. 1 General and Ionic Crystals, Vol. 2 Defects in Semiconductors*, Plenum Press, New York, Vol. 1 (1972), Vol. 2 (1975).
11.8. Davison SG and Steslicka M, *Basic Theory of Surface States*, Clarendon Press, Oxford, 1992, p. 155.
11.9. deWit R, "The Continuum Theory of Stationary Dislocations," *Solid State Physics: Advances in Research and Applications* **10**, 249-292 (1960).
11.10. Dexter DL, "Theory of Optical Properties of Imperfections in Nonmetals," *Solid State Physics: Advances in Research and Applications* **6**, 353-411 (1958).
11.11. Eshelby JD, "The Continuum Theory of Lattice Defects," *Solid State Physics: Advances in Research and Applications* **3**, 79-144 (1956).
11.12. Fowler WB (ed), *Physics of Color Centers*, Academic Press, New York, 1968.
11.13. Gilman JJ and Johnston WG, "Dislocations in Lithium Fluoride Crystals," *Solid State Physics: Advances in Research and Applications* **13**, 147-222 (1962).
11.14. Gundry PM and Tompkins FC, "Surface Potentials," in *Experimental Methods of Catalysis Research*, Anderson RB (ed), Academic Press, New York, 1968, pp. 100-168.

11.15. Gourary BS and Adrian FJ, "Wave Functions for Electron-Excess Color Centers in Alkali Halide Crystals," *Solid State Physics: Advances in Research and Applications* **10**, 127-247 (1960).
11.16. Henderson B, *Defects in Crystalline Solids*, Crane, Russak and Company, Inc., New York, 1972.
11.17. Kohn W, "Shallow Impurity States in Silicon and Germanium," *Solid State Physics: Advances in Research and Applications* **5**, 257-320 (1957).
11.18. Kröger FA and Vink HJ, "Relations between the Concentrations of Imperfections in Crystalline Solids," *Solid State Physics: Advances in Research and Applications* **3**, 307-435 (1956).
11.19. Lehoczky SL et al, NASA CR-101598, "Advanced Methods for Preparation and Characterization of Infrared Detector Materials, Part I," July 5, 1981.
11.20. Li W and Patterson JD, "Deep Defects in Narrow-Gap Semiconductors," *Phys Rev* **50**, 14903-14910 (1994).
11.21. Li W and Patterson JD, "Electronic and Formation Energies for Deep Defects in Narrow-Gap Semiconductors," *Phys Rev* **53**, 15622-15630 (1996), and references cited therein.
11.22. Luttinger JM and Kohn W, "Motion of Electrons and Holes in Perturbed Periodic Fields," *Phys Rev* **99**, 869-883 (1955).
11.23. Madelung O, *Introduction to Solid-State Theory*, Springer-Verlag, Berlin (1978), Chaps. 2 and 9.
11.24. Markham JJ, "F-Centers in the Alkali Halides," *Solid State Physics: Advances in Research and Applications*, Supplement 8 (1966).
11.25. Mitin V, Kochelap VA, and Stroscio MA, *Quantum Heterostructures*, Cambridge University Press, 1999.
11.26. Pantelides ST, *Deep Centers in Semiconductors*, 2nd edn, Gordon and Breach, Yverdon, Switzerland, (1992).
11.27. Sarid D, *Exploring Scanning Probe Microscopy with Mathematica*, John Wiley and Sons, Inc., 1997, Chap. 11.
11.28. Schulman JH and Compton WD, *Color Centers in Solids*, The Macmillan Company, New York, 1962.
11.29. Seitz F and Koehler JS, "Displacement of Atoms During Irradiation," *Solid State Physics: Advances in Research and Applications* **2**, 305-448 (1956).
11.30. Stoneham HM, *Theory of Defects in Solids*, Oxford, 1973.
11.31. Wallis RF (ed), *Localized Excitations in Solids*, Plenum Press, New York, 1968.
11.32. West AR, *Solid State Chemistry and its Applications*, John Wiley and Sons, New York, 1984.
11.33. Zanquill A, *Physics at Surfaces*, Cambridge University Press, 1988, p. 293.

Chapter 12

12.1. Barnham K and Vvendensky D, *Low Dimensional Semiconductor Structures*, Cambridge University Press, Cambridge, 2001.
12.2. Bastard G, *Wave Mechanics Applied to Semiconductor Heterostructures*, Halsted Press, New York, 1988.

12.3. Blakemore JS, *Solid State Physics*, 2nd edn, W. B. Saunders Company, Philadelphia, 1974, p. 168.
12.4. Brown TL, LeMay, HE Jr., and Bursten BE, *Chemistry The Central Science*, 6th edn, Prentice Hall, Englewood Cliff, NJ 07632, 1994.
12.5. Bullis WM, Seiler DG, and Diebold AC (eds), *Semiconductor Characterization—Present Status and Future Needs*, AIP Press, Woodbury, New York, 1996.
12.6. Butcher P, March NH, and Tosi MP, *Physics of Low-Dimensional Semiconductor Structures*, Plenum Press, New York, 1993.
12.7. Callen HB, *Thermodynamics and an introduction to Thermostatistics*, John Wiley and Sons, New York, 1985, p339ff.
12.8. Capasso F and Datta S, "Quantum Electron Devices," *Physics Today* **43**, 74-82 (1990).
12.9. Capasso F, Gmachl C, Siveo D, and Cho A, "Quantum Cascade Lasers," *Physics Today* **55**, 34-40 (May 2002).
12.10. Cargill GS, "Structure of Metallic Alloy Glasses," *Solid State Physics, Advances in Research and Applications* **30**, 227-320 (1975).
12.11. Chaikin PM and Lubensky TC, *Principles of Condensed Matter Physics*, Cambridge University Press, 1995.
12.12. Chen CT and Ho KM, Chap. 20 "Metal Surface Reconstructions" in Chelikowsky JR and Louie SG, *Quantum Theory of Real Materials*, Kluwer Academic Publishers, Dordrecht, 1996.
12.13. Davies JH and Long AR (eds), *Nanostructures*, Scottish Universities Summer School and Institute of Physics, Bristol and Philadelphia, 1992.
12.14. Davison SG and Steslika M, *Basic Theory of Surface States*, Clarendon Press, Oxford, 1992.
12.15. deGennes PG and Prost J, *The Physics of Liquid Crystals*, Clarendon Press, Oxford, 2nd edn (1993).
12.16. Doi M and Edwards SF, *The Theory of Polymer Dynamics*, Oxford University, Oxford, 1986.
12.17. Dresselhaus MS, Dresselhaus G, and Avouris P, *Carbon Nanotubes*, Springer-Verlag, 2000.
12.18. Esaki L and Tsu R, *IBM J Res Devel* **14**, 61 (1970).
12.19. Fergason JL, *The Scientific American*, 74 (Aug. 1964).
12.20. Fisher KH and Hertz JA, *Spin Glasses*, Cambridge University Press, 1991, see especially p. 55, pp. 346-353.
12.21. Gaponenko SV, *Optical Properties of Semiconductor Nanocrystals*, Cambridge University Press, 1998.
12.22. Girvin S, "Spin and Isospin: Exotic Order in Quantum Hall Ferromagnets," *Physics Today*, 39-45 (June 2000).
12.23. Grahn HT (ed), *Semiconductor Superlattices–Growth and Electronic Properties*, World Scientific, Singapore, 1998.
12.24. Halperin BI, "The Quantized Hall Effect", *Scientific American*, April 1986, 52-60.
12.25. Hebard A, "Superconductivity in Doped Fullerenes," *Physics Today*, 26-32 (November, 1992).
12.26. Herbst JF, "$R_2Fe_{14}B$ Materials, Intrinsic Properties and Technological Aspects," *Rev Modern Physics* **63**, 819-898 (1991).
12.27. Isihara A, *Condensed Matter Physics*, Oxford University Press, New York, 1991.
12.28. Jacak L, Hawrylak P, Wojs A, *Quantum Dots*, Springer, Berlin 1998.

12.29. Jain JK, "The Composite Fermion: A Quantum Particle and its Quantum Fluid", *Physics Today*, April 2000, 39-45.
12.30. Jones RAL, *Soft Condensed Matter*, Oxford University Press, 2002.
12.31. Joyce BA, *Rep Prog Physics* **48**, 1637 (1985).
12.32. Kastner M, "Artificial Atoms," *Physics Today* **46**(1), 24-31 (Jan. 1993).
12.33. Kelly MJ, Low Dimensional Semiconductors–Materials, Physics, Technology, Devices, Clarendon Press, Oxford, 1995.
12.34. Kivelson S, Lee D-H and Zhang S-C, "Global phase diagram in the quantum Hall effect", *Phys Rev B* **46**, 2223-2238 (1992).
12.35. Kivelson S, Lee D-H and Zhang S-C, "Electrons in Flatland", *Scientific American*, March 1996, 86-91.
12.36. Levy RA and Hasegana R (eds), *Amorphous Magnetism II*, Plenum Press, New York, 1977.
12.37. Lockwood DJ and Pinzuk A (eds), *Optical Phenomena in Semiconductor Structures of Reduced Dimensions*, Kluwer Academic Publishers, Dordrecht, 1993.
12.38. Laughlin RB, "Quantized Hall conductivity in two dimensions", *Phys Rev B* **23**, 5632-5633 (1981).
12.39. Laughlin RB, "Anomalous quantum Hall effect: An incompressible quantum fluid with fractionally charged excitations," *Phys Rev Lett* **50**, 1395-1398 (1983).
12.40. Laughlin RB, *Phys Rev Lett* **80**, 2677 (1988).
12.41. Laughlin RB, *Rev Mod Phys* **71**, 863-874 (1999).
12.42. Lee DH, *Phys Rev Lett* **80** 2677 (1988).
12.43. Lee DH, "Anyon superconductivity and the fractional quantum Hall effect", *International Journal of Modern Physics B* **5**, 1695 (1991).
12.44. Lu ZP et al, *Phys Rev Lett* **92**, 245503 (2004).
12.45. Lyssenko VG, et al, "Direct Measurement of the Spatial Displacement of Bloch-Oscillating Electrons in Semiconductor Superlattices," *Phys Rev Lett* **79**, 301 (1997).
12.46. Mendez EE and Bastard G, "Wannier-Stark Ladders and Bloch Oscillations in Superlattices," *Physics Today* **46**(6), 39-42 (June, 1993).
12.47. Mitin VV, Kochelap VA, and Stroscio MA, *Quantum Heterostructures*, Cambridge University Press, 1999.
12.48. Mott NF, *Metal-Insulator Transitions*, Taylor and Francis, London, 1990, 2nd edn, See especially pp. 50-54.
12.49. Perkowitz S, *Optical Characterization of Semiconductors*, Academic Press, New York, 1993.
12.50. Poon W, McLeish T, and Donald A, "Soft Condensed Matter: Where Physics meets Biology," *Physics Education* **37**(1), 25-33 (2002).
12.51. Prange RE and Girvin SM (eds), *The Quantum Hall Effect*, 2nd edn, Springer New York, 1990.
12.52. Prutton M, *Introduction to Surface Physics*, Clarendon Press, Oxford, 1994.
12.53. Schab K et al, *Nature* **404**, 974 (2000).
12.54. Shik A, *Quantum Wells, Physics and Electronics of Two-Dimensional Systems*, World Scientific, Singapore, 1997.
12.55. Shklovskii BI and Efros AL, *Electronic Properties of Doped Semiconductors*, Springer-Verlag, Berlin, 1984.
12.56. Stormer HL, *Rev Mod Phys* **71**, 875-889 (1999).
12.57. Strobl G, *The Physics of Polymers*, Springer-Verlag, Berlin, 2nd edn, 1997.

12.58. Tarton R, *The Quantum Dot*, Oxford Press, New York, 1995.
12.59. Tsui DC, *Rev Modern Phys* **71**, 891-895 (1999).
12.60. Tsui DC, Stormer HL, and Gossard AC, "Two-dimensional magnetotransport in the extreme quantum limit", *Phys Rev Lett* **48**, 1559-1562 (1982).
12.61. Vasko FT and Kuznetsov AM, *Electronic States and Optical Transitions in Semiconductor Heterostructures*, Springer, Berlin 1993.
12.62. von Klitzing K, Dorda G, and Pepper M, "New Method for high-accuracy determination of the fine-structure constant based on quantum Hall resistance", *Phys Rev Lett* **45**, 1545-1547 (1980).
12.63. von Klitzing K, "The quantized Hall effect", *Rev Modern Phys* **58**, 519-531 (1986).
12.64. Wannier GH, *Phys Rev* **117**, 432 (1969).
12.65. Weisbuch C and Vinter B, *Quantum Semiconductor Structures*, Academic Press, Inc., Boston, 1991.
12.66. Wilczek F, "Anyons," *Scientific American*, May 1991, 58-65.
12.67. Zallen R, *The Physics of Amorphous Solids*, John Wiley, New York, 1983.
12.68. Zargwill A, *Physics at Surfaces*, Cambridge University Press, New York, 1988.
12.69. Zhang SC, "The Chern-Simons-Landau-Ginzburg theory of the fractional quantum Hall effect", *International Journal of Modern Physics B* **6**, 25, 1992.

Additional References on nanophysics, especially nanomagnetism (some of this material also relates to Chap. 7, see section 7.5.1 on spintronics). Thanks to D. J. Sellmyer, Univ. of Nebraska-Lincoln, for this list.

12.70. Hadjipanayis GC and Prinz GA (eds), "Science and Technology of Nanostructured Magnetic Materials," *NATO Proceedings*, Kluwer, Dordrecht (1991)
12.71. Hadjipanayis GC and Siegel RW (eds), *Nanophase Materials: Synthesis - Properties - Applications*, Kluwer, Dordrecht (1994)
12.72. Hernando A (Ed.), "Nanomagnetism," *NATO Proceedings*, Kluwer, Dordrecht (1992)
12.73. Jena P, Khanna SN, and Rao BK (eds), "Cluster and Nanostructure Interfaces," *Proceedings of International Symposium*, World Scientific, Singapore (2000)
12.74. Maekawa S and Shinjo T, *Spin Dependent Transport in Magnetic Nanostructures*, Taylor & Francis, London (2002).
12.75. Nalwa HS (ed), *Magnetic Nanostructures*, American Scientific Publishers, Los Angeles (2001)
12.76. Nedkov I and Ausloos M (eds), *Nano-Crystalline and Thin Film Magnetic Oxides*, Kluwer, Dordrecht, (1999).
12.77. Shi D, Aktas B, Pust L, and Mikallov F (eds), *Nanostructured Magnetic Materials and Their Applications*, Springer, Berlin (2003)
12.78. Wang ZL, Liu Y, and Zhang Z (eds), *Handbook of Nanophase and Nanostructured Materials*, Kluwer, Dordrecht (2002)
12.79. Zhang J et al (eds), *Self-Assembled Nanostructures*, Kluwer, Dordrecht (2002)

Appendices

A.1. Anderson PW, "Brainwashed by Feynman?," *Physics Today*, 53(2), 11-12, (Feb. 2000).
A.2. Anderson PW, *Concepts in Solids*, W. A. Benjamin, New York, 1963.
A.3. Ashcroft NW and Mermin ND, *Solid State Physics*, Holt, Rhinhart, and Wilson, New York, 1976, pp. 133-141.
A.4. Dekker AJ, *Solid State Physics*, Prentice-Hall, Inc., Englewood Cliffs, NJ, 1957, pp. 240-242.
A.5. Economou EN, *Green's Functions in Quantum Physics*, Springer, Berlin 1990.
A.6. Enz CP, "A Course on Many-Body Theory Applied to Solid-State Physics", *World Scientific*, Singapore, 1992.
A.7. Fradkin E, *Field Theories of Condensed Matter Systems*, Addison-Wesley Publishing Co., Redwood City, CA, 1991.
A.8. Huang K, *Statistical Mechanics*, 2nd edn., John Wiley and Sons, New York, 1987, pp. 174-178.
A.9. Huang K, *Quantum Field Theory From Operators to Path Integrals*, John Wiley and Sons, Inc., New York, 1998.
A.10. Jones H, *The Theory of Brillouin Zones and Electronic States in Crystals*, North-Holland Pub. Co., Amsterdam, 1960, Chap. 1.
A.11. Levy M (ed), *1962 Cargese Lectures in Theoretical Physics*, W. A. Benjamin, Inc., New York 1963.
A.12. Mahan GD, *Many-Particle Physics*, Plenum, New York, 1981.
A.13. Mattsson AE, "In Pursuit of the "Divine" Functional," *Science* **298**, 759-760 (25 October 2002).
A.14. Mattuck RD, *A Guide to Feynman Diagrams in the Many-Body Problem*, 2nd edn, Dover edition, New York 1992.
A.15. Merzbacher E, *Quantum Mechanics*, 2nd edn., John Wiley and Sons, Inc., New York, 1970.
A.16. Mills R, *Propagators for Many-particle Systems*, Gordon and Breach Science Publishers, New York, 1969.
A.17. Negele JW and Henri Orland, *Quantum Many-Particle Systems*, Addison-Wesley Publishing Co., Redwood City, CA, 1988.
A.18. Nozieres P, *Theory of Interacting Fermi Systems*, W. A. Benjamin, Inc., New York 1964, see especially pp. 155-167 for rules about Feynman diagrams.
A.19. Patterson JD, *American Journal of Physics*, **30**, 894 (1962).
A.20. Phillips P, *Advanced Solid State Physics*, Westview Press, Boulder. CO, 2003.
A.21. Pines D, *The Many-Body Problem*, W. A. Benjamin, New York 1961.
A.22. Pines D, *Elementary Excitation in Solids*, W. A. Benjamin, New York, 1963.
A.23. Schiff LI, *Quantum Mechanics* 3rd edn, McGraw-Hill Book Company, New York, 1968.
A.24. Schrieffer JR, *Theory of Superconductivity*, W. A. Benjamin, Inc., New York 1964.
A.25. Starzak ME, *Mathematical Methods in Chemistry and Physics*, Plenum Press, New York, 1989, Chap. 5.
A.26. Van Hove L, Hugenholtz NM, and Howland LP, *Quantum Theory of Many-Particle Systems*, W. A. Benjamin, Inc., New York 1961.
A.27. Zagoskin AM, *Quantum Theory of Many-Body Systems*, Springer, Berlin 1998.

Subject References

Solid state, of necessity, draws on many other disciplines. Suggested background reading is listed in this bibliography.

Mechanics

1. Fetter AL and Walecka JD, *Theoretical Mechanics of Particles and Continua*, McGraw-Hill Book Co., New York, 1980. Advanced
2. Goldstein H, *Classical Mechanics*, 2nd edn, Addison-Wesley Publishing Co., Reading, MA 1980. Advanced
3. Marion JB and Thornton ST, *Classical Dynamics of Particles and Systems*, Saunders College Publ. Co., Fort Worth, 1995. Intermediate

Electricity

4. Jackson JD, *Classical Electrodynamics*, John Wiley and Sons, 2nd edn, New York, 1975. Advanced
5. Reitz JD, Milford FJ, and Christy RW, *Foundations of Electromagnetic Theory*, Addison-Wesley Publishing Co., Reading, MA, 1993. Intermediate

Optics

6. Guenther RD, *Modern Optics*, John Wiley and Sons, New York, 1990. Intermediate
7. Klein MV and Furtak TE, *Optics*, 2nd edn, John Wiley and Sons, New York, 1986. Intermediate

Thermodynamics

8. Espinosa TP, *Introduction to Thermophysics*, W. C. Brown, Dubuque, IA 1994. Intermediate
9. Callen HB, *Thermodynamics and an Introduction to Thermostatics*, John Wiley and Sons, New York, 1985. Intermediate to Advanced

Statistical Mechanics

10. Kittel C and Kroemer H, *Thermal Physics*, 2nd edn, W. H. Freeman and Co., San Francisco, 1980. Intermediate
11. Huang K, *Statistical Physics*, 2nd edn, John Wiley and Sons, New York, 1987. Advanced

Critical Phenomena

12. Binney JJ, Dowrick NJ, Fisher AJ, and Newman MEJ, *The Theory of Critical Phenomena*, Clarendon Press, Oxford, 1992. Advanced

Crystal Growth

13. Tiller WA, *The Science of Crystallization-Macroscopic Phenomena and Defect Generation*, Cambridge U. Press, 1991 and *The Science of Crystallization-Microscopic Interfacial Phenomena*, Cambridge University Press, Cambridge, 1991. Advanced

Modern Physics

14. Born M, *Atomic Physics*, 7th edn, Hafner Publishing Company, New York, 1962. Intermediate
15. Eisberg R and Resnick R, *Quantum Physics of Atoms, Molecules, Solids, Nuclei, and Particles*, 2nd edn, John Wiley and Sons, 1985. Intermediate

Quantum Mechanics

16. Bjorken JD and Drell SD, *Relativistic Quantum Mechanics*, McGraw-Hill, New York, 1964 and *Relativistic Quantum Fields*, McGraw-Hill, New York, 1965. Advanced
17. Mattuck RD, *A Guide to Feynman Diagrams in the Many-body Problem*, 2nd edn, McGraw-Hill Book Company, New York, 1976. Intermediate to Advanced
18. Merzbacher E, *Quantum Mechanics*, 2nd edn, John Wiley, New York, 1970. Intermediate to Advanced
19. Park D, *Introduction to the Quantum Mechanics*, 3rd edn, McGraw-Hill, Inc., New York, 1992. Intermediate and very readable.

Math Physics

20. Arfken G, *Mathematical Methods for Physicists*, 3rd edn, Academic Press, Orlando, 1980. Intermediate

Solid State

21. Ashcroft NW and Mermin ND, *Solid State Physics*, Holt Reiehart and Winston, New York, 1976. Intermediate to Advanced
22. Jones W and March NH, *Theoretical Solid State Physics*, Vol. 1, *Perfect Lattices in Equilibrium*, Vol. 2, *Non-equilibrium and Disorder*, John Wiley and Sons, London, 1973 (also available in a Dover edition). Advanced
23. Kittel C, *Introduction to Solid State Physics*, 7th edn, John Wiley and Sons, Inc., New York, 1996. Intermediate
24. Parker SP, Editor in Chief, *Solid State Physics Source Book*, McGraw-Hill Book Co., New York, 1987. Intermediate
25. Ziman JM, *Principles of the Theory of Solids*, Second Edition, Cambridge University Press, Cambridge 1972. Advanced

Condensed Matter

26. Chaikin PM and Lubensky TC, *Condensed Matter Physics*, Cambridge University Press, Cambridge, 1995. Advanced
27. Isihara A, *Condensed Matter Physics*, Oxford University Press, Oxford, 1991. Advanced

Computational Physics

28. Koonin SE, *Computational Physics*, Benjamin/Cummings, Menlo Park, CA, 1986. Intermediate to Advanced
29. Press WH, Flannery BP, Teukolsky SA, and Vetterling WT, *Numerical Recipes-The Art of Scientific Computing*, Cambridge University Press, Cambridge, 1986. Advanced

Problems

30. Goldsmid HJ (ed), *Problems in Solid State Physics*, Academic Press, New York, 1968. Intermediate

General Comprehensive Reference

31. Seitz F, Turnbull D, Ehrenreich H (and others depending upon volume), *Solid State Physics, Advances in Research and Applications*, Academic Press, New York, a continuing series at research level.

Applied Physics

32. Dalven R, *Introduction to Applied Solid State Physics*, 2nd edn, Plenum, New York, 1990. Intermediate
33. Fraser DA, *The Physics of Semiconductor Devices*, 4th edn, Oxford University Press, Oxford, 1986. Intermediate
34. Kroemer H, *Quantum Mechanics for Engineering, Materials Science, and Applied Physics*, Prentice-Hall, Englewood Cliffs, NJ, 1994. Intermediate
35. Sze SM, *Semiconductor Devices*, Physics and Technology, John Wiley and Sons, 2nd edn, New York, 1985. Advanced

Rocks

36. Gueguen Y and Palciauskas V, *Introduction to the Physics of Rocks*, Princeton University Press, Princeton, 1994. Intermediate

History of Solid State Physics

37. Seitz F, *On the Frontier-My Life in Science*, AIP Press, New York, 1994. Descriptive

38. Hoddeson L, Braun E, Teichmann J, and Weart S (eds), *Out of the Crystal Maze–Chapters from the History of Solid State Physics*, Oxford University Press, Oxford, 1992. Descriptive plus technical

The Internet

39. (http://xxx.lanl.gov/lanl/), This gets to arXiv which is an e-print source in several fields including physics. It is presently owned by Cornell University.
40. (http://online.itp.ucsb/online/), Institute of Theoretical Physics at the University of California, Santa Barbara, programs and conferences available on line.

Periodic Table

When thinking about solids it is often useful to have a good tabulation of atomic properties handy. The Welch periodic chart of the atoms by Hubbard and Meggers is often useful as a reference tool.

Further Reading

The following mostly older books have also been useful in the preparation of this book, and hence the student may wish to consult some of them from time to time.

41. Anderson PW, *Concepts in Solids*, W. A. Benjamin, New York, 1963. Emphasizes modern and quantum ideas of solids.
42. Bates LF, *Modern Magnetism*, Cambridge University Press, New York, 1961. An experimental point of view.
43. Billington DS, and Crawford JH Jr., *Radiation Damage in* Solids, Princeton University Press; Princeton, New Jersey, 1961. Describes a means for introducing defects in solids.
44. Bloembergen N, *Nuclear Magnetic Relaxation*, W. A. Benjamin, New York, 1961. A reprint volume with a pleasant mixture of theory and experiment.
45. Bloembergen N, *Nonlinear Optics*, W. A. Benjamin, New York, 1965. Describes the types of optics one needs with high intensity laser beams.
46. Born M, and Huang K, *Dynamical Theory of Crystal Lattices*, Oxford University Press, New York, 1954. Useful for the study of lattice vibrations.
47. Brillouin L, *Wave Propagation in Periodic Structures*, McGraw-Hill Book Company, New York, 1946. Gives a unifying treatment of the properties of different kinds of waves in periodic media.
48. Brout R, *Phase Transitions,* W. A. Benjamin, New York, 1965. A very advanced treatment of freezing, ferromagnetism, and superconductivity.
49. Brown FC, *The Physics of Solids—Ionic Crystals, Lattice Vibrations, and Imperfections*, W. A. Benjamin, New York, 1967. A textbook with an unusual emphasis on ionic crystals. The book has a particularly complete chapter on color centers.
50. Buerger MJ, *Elementary Crystallography,* John Wiley and Sons, New York, 1956. A very complete and elementary account of the symmetry properties of solids.

51. Choquard P, *The Anharmonic Crystal*, W. A. Benjamin, New York, 1967. This book is intended mainly for theoreticians, except for a chapter on thermal properties. The book should convince you that there are still many things to do in the field of lattice dynamics.
52. Debye P, *Polar Molecules*, The Chemical Catalog Company, 1929, reprinted by Dover Publications, New York. Among other things this book should aid the student in understanding the concept of the dielectric constant.
53. Dekker AJ, Solid *State Physics*, Prentice-Hall, Engelwood Cliffs, New Jersey, 1957. Has many elementary topics and treats them well.
54. Frauenfelder H, *The Mossbauer Effect*, W. A. Benjamin, New York, 1962. A good example of relationships between solid state and nuclear physics.
55. Grosso G. and Paravicini GP, *Solid State Physics*, Academic Press, 2000, modern.
56. Harrison WA, *Pseudopotentials in the Theory of Metals*, W. A. Benjamin, New York, 1966. The first book-length review of pseudopotentials.
57. Holden A, *The Nature of Solids*, Columbia University Press, New York, 1965. A greatly simplified view of solids. May be quite useful for beginners.
58. Jones H, *The Theory of Brillouin Zones and Electronic States in Crystals*, North-Holland Publishing Company, Amsterdam, 1960. Uses group theory to indicate how the symmetry of crystals determines in large measure the electronic band structure.
59. Kittel C, *Introduction to Solid State Physics*, John Wiley and Sons, New York. All editions have some differences and can be useful. The latest is listed in [23]. The standard introductory text in the field.
60. Kittel C, *Quantum Theory of Solids*, John Wiley and Sons, New York, 1963. Gives a good picture of how the techniques of field theory have been applied to solids. Most of the material is on a high level.
61. Knox RS, and Gold A, *Symmetry in the Solid State*, W. A. Benjamin, New York, 1964. Group theory is vital for solid state physics, and this is a good review and reprint volume.
62. Lieb EH, and Mattis DC, *Mathematical Physics in One Dimension*, Academic Press, New York, 1966. This book is a collection of reprints with an introductory text. Because of mathematical simplicity, many topics in solid state physics can best be introduced in one dimension. This book offers many examples of one dimensional calculations which are of interest to solid state physics.
63. Loucks T, *Augmented Plane Wave Method*, W. A. Benjamin, New York, 1967. With the use of the digital computer, the APW method developed by J. C. Slater in 1937 has been found to be a practical and useful technique for doing electronic band structure calculations. This lecture note and reprint volume is by the man who developed a relativistic generalization of the APW method.
64. *Magnetism and Magnetic Materials Digest*, A Survey of Technical Literature of the Preceding Year, Academic Press, New York. This is a useful continuing series put out by different editors in different years. Mention should also be made of the survey volumes of Bell Telephone Laboratories, called *Index to the Literature of Magnetism*.
65. *Materials, A Scientific American Book*, W. H. Freeman and Company, San Francisco, *1967*. A good, elementary, and modern view of many of the properties of solids. Written in the typical *Scientific American* style.

66. Mattis DC, *The Theory of Magnetism—An Introduction to the Study of Cooperative Phenomena,* Harper and Row Publishers, New York, 1965. A modern authoritative account of magnetism; advanced. See also The Theory of Magnetism I and II, Springer-Verlag, Berlin, 1988 (I), 1985 (II).
67. Mihaly L and Martin MC, *Solid State Physics—Problems and Solutions,* John Wiley, 1996
68. Morrish AH, *The Physical Principles of Magnetism,* John Wiley and Sons, New York, 1965. A rather complete and modern exposition of intermediate level topics in magnetism.
69. Moss TS, *Optical Properties of Semi-Conductors,* Butterworth and Company Publishers, London, 1959. A rather special treatise, but it gives a good picture of the power of optical measurements in determining the properties of solids.
70. Mott NF, and Gurney RW, *Electronic Processes in Ionic Crystals,* Oxford University Press, New York, 1948. A good introduction to the properties of the alkali halides.
71. Mott NF, and Jones H, *Theory of the Properties of Metals and Alloys,* Oxford University Press, New York, 1936. A classic presentation of the free-electron properties of metals and alloys.
72. Nozieres P, *Theory of Interacting Fermi Systems,* W. A. Benjamin, New York, 1964. A good account of Landau's ideas of quasi-particles. Very advanced, but helps to explain why "free-electron theory" seems to work for many metals. In general it discusses the many-body problem, which is a central problem of solid state physics.
73. Olsen JL, *Electron Transport in Metals,* Interscience Publishers, New York, 1962. A simple outline of theory and experiment.
74. Pake GE, *Paramagnetic Resonance,* W. A. Benjamin, New York, 1962. This book is particularly useful for the discussion of crystal field theory.
75. Peierls RE, *Quantum Theory of Solids,* Oxford University Press, New York, 1955. Very useful for physical insight into the basic nature of a wide variety of topics.
76. Pines D, *Elementary Excitations in Solids,* W. A. Benjamin, New York, 1963. The preface states that the course on which the book is based concerns itself with the "view of a solid as a system of interacting particles which, under suitable circumstances, behaves like a collection of nearly independent elementary excitations."
77. Rado, GT and Suhl H (eds), *Magnetism,* Vols. I, IIA, IIB, III, and IV, Academic Press, New York. Good summaries in various fields of magnetism which take one up to the level of current research.
78. Raimes S, *The Wave Mechanics of Electrons in Metals,* North-Holland Publishing Company, Amsterdam, 1961. Gives a fairly simple approach to the applications of quantum mechanics in atoms and metals.
79. Rice FO, and Teller E, *The Structure of Matter,* John Wiley and Sons, New York, 1949. A very simply written book; mostly words and no equations.
80. Schrieffer JR, *Theory of Superconductivity,* W. A. Benjamin, New York, 1964. An account of the Bardeen, Cooper, and Schrieffer theory of superconductivity, by one of the originators of the theory.
81. Schulman JH, and Compton WD, *Color Centers in Solids,* The Macmillan Company, New York, 1962. A nonmathematical account of color center research.

82. Seitz F, *The Modern Theory of Solids*, McGraw-Hill Book Company, New York, 1940. This is still probably the most complete book on the properties of solids, but it may be out of date in certain sections.
83. Seitz F, and Turnbull D (eds) (these are the original editors, later volumes have other editors), *Solid State Physics Advances in Research and Applications*, Academic Press, New York. Several volumes; a continuing series. This series provides excellent detailed reviews of many topics.
84. Shive JN, *Physics of Solid State Electronics*, Charles E. Merrill Books, Columbus, Ohio, 1966. An undergraduate level presentation of some of the solid state topics of interest to electrical engineers.
85. Shockley W, *Electrons and Holes in Semiconductors*, D. van Nostrand Company, Princeton, New Jersey, 1950. An applied point of view.
86. Slater JC, *Quantum Theory of Matter*, McGraw-Hill Book Company, New York, 1951, also 2nd edn, 1968. Good for physical insight.
87. Slater JC, *Atomic Structure*, Vols. I, II, McGraw-Hill Book Company, New York, 1960.
88. Slater JC, *Quantum Theory of Molecules and Solids*, Vol. I, *Electronic Structure of Molecules*, McGraw-Hill Book Company, New York, 1963.
89. Slater JC, *Quantum Theory of Molecules and Solids*, Vol. II, *Symmetry and Energy Bands in Crystals*, McGraw-Hill Book Company, New York, 1965.
90. Slater JC, *Quantum Theory of Molecules and Solids*, Vol. 111, *Insulators, Semiconductors, and Metals*, McGraw-Hill Book Company, New York, 1967. The titles of these books [87 through 90] are self-descriptive. They are all good books. With the advent of computers, Slater's ideas have gained in prominence.
91. Slichter CP, *Principles of Magnetic Resonance with Examples from Solid State Physics*, Harper and Row Publishers, New York, 1963. This is a special topic but the book is very good and it has many transparent applications of quantum mechanics. Also, see 3rd edn, Springer-Verlag, Berlin, 1980.
92. Smart JS, *Effective Field Theories of Magnetism*, W. B. Saunders Company, Philadelphia, 1966. A good summary of Weiss field theory and its generalizations.
93. Smith RA, *Wave Mechanics of Crystalline Solids*, John Wiley and Sons, New York, 1961. Among other things, this book has some good sections on one-dimensional lattice vibrations.
94. Van Vleck JH, *Theory of Electric and Magnetic Susceptibilities*, Oxford University Press, New York, 1932. Old, but still very useful.
95. Wannier GH, *Elements of Solid State Theory*, Cambridge University Press, New York, 1959. Has novel points of view on many topics.
96. Weinreich G, *Solids: Elementary Theory for Advanced Students*, John Wiley and Sons, New York, 1965. The title is descriptive of the book. The preface states that the book's "purpose is to give the reader some feeling for what solid state physics is all about, rather than to cover any appreciable fraction" of the theory of solids.
97. Wilson AH, *The Theory of Metals*, Cambridge University Press, New York, 1954, 2nd edn. This book gives an excellent, detailed account of the quasi-free electron picture of metals and its application to transport properties.
98. Wood EA, *Crystals and Light*, D. van Nostrand Company, Princeton, New Jersey, 1964. An elementary viewpoint of this subject.

99. Ziman JM, *Electrons and Phonons,* Oxford University Press, New York, 1960. Has interesting treatments of electrons, phonons, their interactions, and applications to transport processes.
100. Ziman JM, *Electrons in Metals—A Short Guide to the Fermi Surface*, Taylor and Francis, London, 1963. Short, qualitative, and excellent.
101. Ziman JM, *Elements of Advanced Quantum Theory*, Cambridge University Press, New York, 1969. Excellent for gaining an understanding of the many-body techniques now in vogue in solid state physics.

Index

A

Abelian group 16, 656
Absolute zero temperature 133, 162
Absorption by excitons 543
Absorption coefficient 348, 463, 544, 551
Absorptivity 547
AC Josephson Effect 478, 482
Acceptors 293, 561, 590
Accidental degeneracy 199, 448
Acoustic mode 64, 91, 223, 524
Actinides 265
Adiabatic approximation 44, 46, 47
Ag 291
Al 95, 284, 290, 293, 451, 503, 657
Alfvén Waves 529
Alkali Halides 95, 561
Alkali metals 184, 216, 265, 271
Alkaline earth metals 271
Allowed and forbidden regions of energy 178
Alloys 278, 677
Amorphous chalcogenide semiconductors 637
Amorphous magnet 430, 639
Amorphous semiconductors 637
Amorphous solids 1
Anderson localization transition 637
Angle-resolved Photoemission 172
Angular momentum operators 81, 443
Anharmonic terms 44, 113
Anisotropy energy 421, 423, 441
Annihilation operator 79, 394

Anomalous skin effect 273, 274
Anticommutation relations 125, 495, 663
Antiferromagnetic resonance 439
Antiferromagnetism 365, 370, 383, 418
Anti-site defects 588
Anti-Stokes line 580, 581
Antisymmetric spatial wave function 376
Ar 2, 12
Asperomagnetic 430
Associative Law 14
Atomic Force Microscopy 611
Atomic form factor 35
Atomic number of the nucleus 42, 116
Atomic polyhedra 184
Atomic wave functions 179, 190, 207, 407
Attenuation 157, 273, 462, 527, 556
Au 271, 273, 291, 445, 454, 455
Au1-xSix 279
Auger Electron Spectroscopy 611
Augmented plane wave 173, 185
Avalanche mechanism 344
Average drift velocity 171
Axially symmetric bond bending 48
Axis of symmetry 18, 20, 423
Azbel-Kaner 274

B

B.C.S. theory 461
Band bending 334
Band ferromagnetism 406, 450

Band filling 554
Band structure 172
Base-centered orthorhombic cell 24
Basis vectors 17
$BaTiO_3$ 29, 30
$Ba_xLa_{2-x}CuO_{4-y}$ 502
Bcc lattice 38, 269
Becquerel 596
Bell Labs 345, 350
Beryllium 191
Binding forces 10, 41
Bipolar junction transistor 350
Bitter Patterns 429
Black hole 289
Bloch Ansatz 250
Bloch condition 174, 179, 186, 191, 207
Bloch frequency 623
Bloch $T^{3/2}$ law 215
Bloch Walls 428
Body-centered cubic cell 25
Body-centered cubic lattice 111, 181, 210, 269
Body-centered orthorhombic cell 24
Bogoliubov-Valatin transformation 494
Bogolons 214, 218, 495
Bohr magneton 356, 648
Boltzmann equation 244, 307, 629
Boltzmann gas constant 110
Boltzmann statistics 311, 603
Bond stretching 48
Born approximation 453
Born-Haber cycle 9
Born-Mayer Theory 6, 9
Born-Oppenheimer approximation 41, 44, 191, 225
Born-von Kárman or cyclic boundary conditions 51
Bose statistics 224
Bose-Einstein condensation 505, 635
Boson annihilation operators 89

Boson creation operators 89
Bosons 62, 401, 476, 635
Bragg and von Laue Diffraction 31
Bragg peaks 21, 22, 39, 96, 283
Bragg reflection 175, 176, 266
Brass 29, 290
Bravais lattice 13, 23, 177
Brillouin function 358
Brillouin scattering 96, 580
Brillouin zone 54, 85
Brillouin zone surface 177
Brownian motion 642
Buckyball 636
Built-in potential 329, 343, 347
Bulk negative conductivity 327
Burgers Vector 600

C

Ca 191, 271, 273
Canonical ensemble 61, 167, 356, 658
Canonical equations 59
Carbon 11, 12, 290, 609, 636
Carbon nanotubes 636
Carrier drift currents 342
$CaTiO_3$ 29
Cauchy principal value 548
CCD 350
Cd 293, 320
CdS 241, 293
CdSe 293, 295
CdTe 241, 322
$CeAl_2$ 284
$CeAl_3$ 284
$CeCu_2Si_2$ 284, 501, 503
Center of mass 50, 68, 476
Central forces 3, 6, 11, 106
Ceramic oxide 502
Chandrasekhar Limit 289, 290
Character 446, 449
Charge density oscillations 380
Charge density waves 282
Chemical potential 248, 493, 497, 619

Chemical vapor deposition 613
Chemically saturated units 2, 6
Classical Diatomic Lattices 64
Classical elastic isotropic continuum waves 93
Classical equipartition theorem 43, 93
Classical Heisenberg Ferromagnet 389
Classical specific heat 222
Clausius-Mossotti Equation 515, 520
Close packing 6
Close-packed crystal structures 3
Closure 14
Co 424
Coercive Force 420, 427
Coercivity 428, 639
Coherence length 467, 472
Cohesive energy 8, 9
Cold-field emission 601, 604
Color center 596, 597
Commutation relations 76, 79, 394, 488, 558, 662
Complex conjugate 117, 206, 465
Complex dielectric constant 511, 544
Complex index of refraction 510, 545
Complex refractive index 511
Composite 216, 217, 256, 257, 258, 259, 262, 263, 635
Compressibility 9
Concentration gradients 307, 308
Conductance 291, 616, 618, 620, 621, 629, 632, 634, 636
Conduction band 133, 216, 550, 560, 590, 614, 619, 627
Conservation law 83, 218, 225, 569
Constant energy surfaces 182, 266, 275
Constraint of normalization 117
Continuity equation 328, 340
Continuum frequencies 93

Controlled doping 293, 325
Cooper pairs 490
Core electrons 178, 192, 354
Core repulsion 6
Correlation length 416
Correlations 2, 115, 129, 135, 163, 193, 217, 242, 265, 639
Coulomb blockade model 616
Coulomb gauge 550
Coulomb potential energy 7, 42, 116, 243
Covalent bonds 11, 443, 590, 591
Covalent crystals 47
Cr 291
Creation operator 79, 407, 493, 662, 669
Critical exponents 415
Critical magnetic field 460, 502
Critical point 97, 416, 560, 578
Critical temperature 353, 367, 393, 460, 499, 502, 509
Crystal classes 13
Crystal field 356, 376, 384, 442
Crystal growth 326, 590, 600, 606, 610
Crystal Hamiltonian 42, 179
Crystal lattice with defects 70, 206
Crystal mathematics 85, 109
Crystal structure determination 30
Crystal symmetry operations 13, 22
Crystal systems 13, 23, 26
Crystalline anisotropy 375, 423
Crystalline potential 186, 192, 661
Crystalline solid 1, 2, 19, 22, 23, 640
Crystalline state 1
Crystalline symmetry 2, 13
Crystallography 1, 13, 21, 23
CsCl structure 29
Cu 9, 265, 270, 271, 290, 345, 452, 453, 454, 502
Cubic point group 18, 19
Cubic symmetry 205, 426
Curie constants 366

Curie law 357
Curie temperature 358, 360, 414, 639
Curie-Weiss behavior 518, 521
Cutoff frequency 38, 93, 172, 489
Cyclic group 16, 657
Cyclotron frequency 167, 215, 274, 309, 314, 527, 528, 583, 628
Cyclotron resonance 273, 302, 311, 314, 315

D

Damon-Eshbach wave solutions 404
DC Josephson Effect 477, 481
de Broglie wavelength 157, 175, 286, 563
de Haas-Schubnikov Effect 273
de Haas-van Alphen effect 165
Debye approximation 38, 91, 251
Debye density of states 97
Debye function 95
Debye temperature 94, 251, 500, 505
Debye-Huckel Theory 531, 535
Debye-Waller factor 37, 38
Deep defects 205, 561, 595
Defect 70, 72, 205, 301, 562, 587, 589, 595
Degenerate semiconductors 347, 555
deGennes factor 381
Degree of polymerization 641
Demagnetization field 388, 439
Density functional theory 137, 147, 191, 665
Density matrix 657, 658
Density of states 63, 93, 94, 97, 159, 321, 413, 474, 534
Density of states for magnons 403
Depletion width 331
Destructive interference 157
Diamagnetic 164, 354, 370

Diamond 11, 28, 29, 191, 295, 590, 636
Diatomic linear lattice 64, 65, 67
Dielectric constant 233, 258, 509, 512, 514, 591
Dielectric function 280, 531, 578
Dielectric screening 509
Differential conductivity 326
Diffusion 222, 284, 308, 332, 338, 342, 598
Diodes 615
Dipole moment 234, 238, 384, 514
Dipole-dipole interactions 404
Dirac delta function 85, 110, 253, 533, 659
Dirac delta function potential 149
Dirac Hamiltonian 193
Direct band gap 237, 554, 555
Direct optical transitions 551
Direct product group 16
Dodecahedron 22, 269
Domain wall 420, 427, 456
Donors 293, 298, 561, 589
Drift current density 308
Drift velocity 171, 294, 307, 317, 326, 528, 631
Drude theory 563
d-wave pairing 501, 502
Dynamical matrix 64, 91
Dyson equation 671

E

Edge dislocation 599
Effective magneton number 357
Effective mass 156, 274, 283, 294, 304
Effective mass theory 591, 622
Effusion cell 613
Einstein A and B coefficients 579, 580
Einstein relation 309
Einstein theory of specific heat of a crystal 112
Elastic continuum 41, 92, 95

Elastic restoring force 5
Elastically scattered 31
Electric current density 171, 246, 256, 284
Electric dipole interactions 47
Electric polarization 522
Electrical and thermal conductors 10
Electrical conductivity 171, 250, 273, 293, 296, 302, 307
Electrical current density 257, 308, 327
Electrical mobility 307
Electrical neutrality condition 300, 331
Electrical resistivity 41, 46, 247, 251, 265, 293, 459, 483, 485
Electrochemical deposition 613
Electromagnetic wave 32, 68, 215, 510, 527, 529, 543, 569
Electromechanical transducers 516
Electromigration 284
Electron correlations 137, 243, 252, 265
Electron Energy Loss Spectroscopy 611
Electron paramagnetic resonance 433
Electron spin resonance 433
Electron tunneling 463
Electron volt 6, 39, 136, 293, 324, 612
Electron-electron interaction 10, 136, 193, 203, 242, 489, 616
Electron-hole pairs 344, 350
Electronic conduction spin density 379
Electronic conductivity 11
Electronic configuration 2
Electronic flux of heat energy 246
Electronic mass 42
Electronic surface states 587
Electron-lattice interaction 280, 473, 504

Electron-phonon coupling 232, 476, 499, 564
Electron-phonon interaction 41, 158, 216, 225, 232, 238, 279, 459, 461, 482
Ellipsometry 611
Empty lattice 266
Energy Dispersive X-ray Spectroscopy 611
Energy tubes 277
Entropy 8, 62, 110, 420, 588, 642
Envelope Functions 591, 622
Epitaxial layers 622
Equilibrium properties 414
Equivalent one-electron problem 113
Euler-Lagrange 141, 425
EuO 430, 458
Eutectic 278
Exchange charge density 129
Exchange correlation energy 142, 143, 144, 148
Exchange coupled spin system 388
Exchange energy 144, 372, 410, 421, 423
Exchange enhancement factor 413
Exchange integral 363, 365, 373, 390, 421
Exchange operator 128, 132
Excitation and ionization of impurities 543
Excitation of lattice vibrations 544
Exciton absorption 306
Excitons 215, 218
Exhaustion region 300
Extended X-ray Absorption Spectroscopy 611
Eyjen counting scheme 39

F

Face-centered cubic cell 25
Face-centered cubic lattice 111, 181, 210
Factor group 16, 23

Faraday and Kerr effects 429
Faraday effect 582, 585
Fast Fourier Transform 613
f-band superconductivity 284
F-centers 596, 597
Fe 290, 425, 444
Fe_3O_4 430
Fermi energy 160, 253, 301, 489, 534
Fermi function 160, 248, 296, 301, 497, 534, 619
Fermi gas 287
Fermi hole 129, 133, 134
Fermi liquid 135, 217
Fermi momentum 136, 288
Fermi surface 244, 265, 268, 272
Fermi temperature 163, 535
Fermi wave vector 283
Fermi-Dirac distribution 136, 161, 244, 249, 252
Fermions 119, 124, 138, 635
Fermi-Thomas potential 540
Ferrimagnetism 365, 367, 418
Ferroelectric 12, 29, 509, 516
Ferroelectric crystals 516
Ferromagnetic 190, 353, 358, 360, 371, 383, 397, 405, 419, 423, 452, 482, 509
Ferromagnetic resonance 438
Feynman diagrams 667
f-fold axis of rotational symmetry 26
Field effect transistor 325, 350
Field Ion Microscopy 611
Finite phonon lifetimes 222
Five-fold symmetry 21
Fluctuating dipoles 12
Fluorescence 30, 579
Flux penetration 461, 462, 471
Fluxoid 472, 502
Fourier analysis 71, 87, 174
Fourier coefficient 178
Fourier transform infrared spectroscopy 613

Fowler-Nordheim equation 605
Free carrier absorption 544, 563
Free electrons 134, 161, 266, 563
Free energy 1, 99, 465, 658
Free ions 8
Free-electron gas 130, 134, 165
Frenkel excitons 215, 561
Friction drag 307
Friedel oscillation 134, 379, 527, 540
Fullerides 636
Fundamental absorption edge 554, 555
Fundamental symmetries 225

G

GaAs 241, 293, 295, 322, 324, 326, 327, 335, 345, 561, 562, 577, 614, 624, 625, 633
GaAs-AlAs 335, 345
Gamma functions 401
Gap parameter 494, 499, 506
Gas crystals 2
Gas discharge plasmas 525
Gauge symmetry 419
Ge 29, 191, 351
Generator 16
Gibbs free energy 518, 541
Ginzburg, Landau, Abrikosov, and Gor'kov theory 465
Ginzburg-Landau equation 465, 467
Glide plane symmetry 20
Golden rule of perturbation theory 219, 220, 221, 226, 228, 660
Goldstone excitations 419
Graded junction 328, 332
Grain boundary 590, 607
Grand canonical ensemble 298
Gravitational energy 289
Green functions 667, 668
Group element 15, 445, 446
Group multiplication 15, 39
Group theory 445, 656

Group velocity 55, 304, 529, 560, 623
Gruneisen Parameter 99, 102
Gunn effect 326, 327
Gyromagnetic ratio 360, 432, 433

H

Hall coefficient 291, 311, 528, 628
Hall effect 309, 627, 629, 631
Halogen 6
Harmonic approximation 44, 51
Harmonic oscillator 5, 50
Hartree approximation 115, 158, 174, 665, 669
Hartree atomic units 647
Hartree-Fock analysis 411
Hartree-Fock approximation 119, 137, 242, 665, 670
He 1, 209, 419, 459, 502
Heat transport by phonons 223
Heavy electron superconductors 459, 501
Heavy holes 294, 321, 324
Heisenberg Hamiltonian 353, 359, 371, 375, 377, 396
Heitler-London approximation 11, 372
Helicons 215, 527
Helimagnetism 418
Helmholtz free energy 61, 168, 356
Hermann-Mauguin 26
Heterostructures 614
Hexagonal symmetry 25
Hg 293, 320, 500, 503, 506
$Hg_{1-x}Cd_xTe$ 320, 322
HgTe 322
Higgs mode 419
Hilbert space 419
Hohenberg-Kohn Theorem 137, 147
Hole Conduction 305
Hole effective mass 306, 315
Holes 215, 667

Holstein-Primakoff transformation 394, 396
Homopolar bonds 11
Homostructures 609
Hopping conductivity 638, 639
Hubbard Hamiltonian 406
Hume-Rothery 270
Hund-Mulliken method 377
Hydrogen atom 12, 81, 371, 541, 591, 607, 608
Hydrogen bond 12, 516
Hydrogen Metal 271
Hydrogenic wave functions 594
Hyperfine interaction 442, 463, 685
Hyperfine splitting 479
Hysteresis loop 427, 431

I

Ice 12
Icosahedron 22
Ideal crystals 17
Imperfections 427, 561
Impurity mode 74, 108
Impurity states 205, 244, 638
Index of refraction 509, 545, 559, 582
Induced transition 579
Inelastic neutron diffraction 228
Infinite crystal 17, 55
Infinite one-dimensional periodic potential 148
Infrared Absorption 463
Infrared detector 293, 322, 606
Inhomogeneous Semiconductors 338
Injected minority carrier densities 343
Injection current 339
Inner product 16, 123, 199
InP 326
Interatomic potential 3
Interatomic spacing 3, 223, 243
Inter-band frequencies 509

Internal energy 8, 62, 101, 365, 658
Interpolation methods 173, 194
Interstitial atoms 589
Interstitial defects 599
Intra-band absorption 563
Ion core potential energies 130
Ionic conductivity 6, 11
Ionic crystals 6, 12, 509, 588, 597
Irreducible representation 13, 656
Isomorphic 16, 23, 39, 79, 322
Isoprene group 642
Isothermal compressibility 8, 100
Itinerant electron magnetism 137

J

Jahn-Teller effect 444
Jellium 137, 242, 265, 535
Joint density of states 553, 559
Jones zones 175
Josephson effects 463, 475, 479
Junction capacitance 332, 334

K

K 40, 191, 516
k space 159, 273
KH_2PO_4 516, 517, 521
Kinematic correlations 129
Kohn anomalies 540
Kohn Effect 527
Kohn-Sham Equations 141, 148
Kondo effect 282, 353, 453
Korteweg-de Vries equation 456
Kronecker delta 36, 108, 199, 221, 447
Kronig-Kramers equations 548
Kronig-Penney model 148, 652

L

Lagrange equations 53, 425
Lagrange multiplier 117, 140
Lagrangian 53, 59, 76, 146
Lagrangian mechanics 53, 146

Landau diamagnetism 165, 167, 168, 354, 601, 630
Landau levels 632, 633
Landau quasi-particles 158, 217
Landau theory 135, 415, 459
Landauer equation 620
Landau-Lifshitz equations 441
Lande g-factors 366
Lanthanides 265
Larmor frequency 437
Lasers 626
Latent heat 414, 517, 518
Lattice constant 9, 296, 527, 622
Lattice of point ions 37
Lattice thermal conductivity 223
Lattice vibrations 41, 42, 47, 55, 62, 64
Law of constancy of angle 13
Law of Dulong and Petit 63
Law of geometric progression 36
Law of Mass Action 298, 330
Law of Wiedeman and Franz 171, 264
LiF 6
Light holes 320, 321
Lindhard theory 531, 535
Line defects 589
Linear combination of atomic orbitals 178
Linear lattice 61, 64, 66, 75, 108
Linear metal 265, 279
Liquid crystals 1, 610, 640
Liquid nitrogen 322, 502
Liquidus branches 278
Local density approximation 137, 143, 192, 203, 666
Local density of states 612
London penetration depth 466
Longitudinal mode 64, 94
Longitudinal optic modes 322
Longitudinal plasma oscillations 525, 568
Lorentz-Lorenz Equation 515
Lorenz number 172

Low Energy Electron Diffraction 611
Low temperature magnon specific heat 402
Lower critical field 462, 472, 504
Luminescence 579
Lyddane-Sachs-Teller Relation 574

M

Madelung constant 7, 38, 39
Magnetic anisotropy 383
Magnetic charge 426
Magnetic domains 420
Magnetic flux 257, 460, 504
Magnetic hysteresis 421
Magnetic induction 32
Magnetic interactions 116
Magnetic moment 164, 353, 422, 685
Magnetic phase transition 414
Magnetic potential 387
Magnetic resonance 353, 432, 613
Magnetic specific heat 365, 415, 454
Magnetic structure 353, 358, 393, 418
Magnetic susceptibility 164, 273, 277, 354, 370, 379, 454, 457
Magnetization 164, 168, 354
Magnetoacoustic 274
Magnetoelectronics 450
Magnetoresistance 273, 451, 453
Magnetostatic energy 386, 421
Magnetostatic self energy 387
Magnetostatic Spin Waves 404
Magnetostriction 425, 426
Magnetostrictive energy 421
Magnon-magnon interactions 394
Magnons 215, 392, 397
Mass defect in a linear chain 72
Mass of the electron 46, 158, 647
Mass of the nucleus 42
Maxwell equations 32, 263, 387

Mean field theory 353, 359, 366, 416
Medium crystal field 443
Meissner effect 419, 466, 504
Membranes 640
Metal Oxide Semiconductor Field Effect Transistor 350
Metal Semiconductor Junctions 334
Metal-Barrier-Metal Tunneling 619
Metallic binding 10
Metallic densities 135, 193, 242
Mg 272, 501
MgB_2 459, 501
Microgravity 322, 606
Miller Indices 31
Minibands 622
Minority carrier concentrations 342, 348
Mobility gap 637
Models of Band Structure 319
Molecular Beam Epitaxy 613
Molecular crystals 2, 3, 215
Molecular field constant 360, 361, 405
Monatomic case 66
Monatomic Lattice 51
Monoclinic Symmetry 24
Monomer 641, 642, 643
Monovalent metal 130, 158
MOS transistors 350
MOSFET 337, 350, 633
Mott transition 173, 639
Mott-Wannier excitons 215

N

N 2, 502
N interacting atoms 59
Na 9, 56, 76, 155, 184, 271, 527, 596
NaCl 6, 9, 12, 28, 577, 596
$NaKC_4H_4O_6 \cdot 4H_2O$ (Rochelle salt) 516
Narrow gap insulator 293

Narrow gap semiconductor 293, 615, 625, 626
N-body problem 59
$Nd_2Fe_{14}B$ 431
Nearest neighbor repulsive interactions 8
Nearly free-electron approximation 173, 186, 201
Néel temperature 365, 368
Néel walls 428
Neutron diffraction 90, 95, 283, 368
Neutron star 289
Ni 188, 290, 359, 362, 405, 424
Noble metals 265, 271
Nonequilibrium statistical properties 218
Non-radiative (Auger) transitions 562
Normal coordinate transformation 4, 58, 77, 87, 649
Normal coordinates 51, 57, 76, 649
Normal mode 41, 49, 65, 649, 668
Normal or N-process 221
Normal subgroup 16
n-type semiconductor 334, 351
Nuclear coordinates 42
Nuclear magnetic resonance 433
Nuclear spin relaxation time 463, 464

O

O 335, 641
Occupation number space 79, 662
Octahedron 22
One-dimensional crystal 53, 207
One-dimensional harmonic oscillators 3
One-dimensional lattices 47
One-dimensional potential well 210, 211
One-electron Hamiltonian 354
One-electron models 148
One-particle operator 116, 126, 663

Optic mode 64, 91, 233, 520, 524
Optical absorption 555, 597
Optical magnons 399
Optical phenomena 543
Optical phonons 216, 544, 569, 580
Orbital angular momentum operator 356, 661
Order parameter 416, 454, 465
Order-disorder transition 521
Orthogonality constraints 142
Orthogonalized plane wave 173, 190, 203
Orthorhombic symmetry 24
Oscillating polarization 380
Oscillator strength 512, 553, 558, 585
Overlap catastrophe 375

P

Padé approximant 415
Pair tunneling 506
Parabolic bands 296, 322, 553
Paraelectric phase 517, 518
Parallelepiped 17, 22, 84, 159
Paramagnetic Curie temperature 361
Paramagnetic effects 355
Paramagnetic ions 358, 442
Paramagnetic resonance 597
Paramagnetic susceptibility 355, 413
Paramagnetism 166, 355, 365, 413
Particle tunneling 506, 507
Particle-in-a-box 154
Partition function 61, 99, 167, 415, 457
Passivation 335
Pauli paramagnetism 135, 355, 601
Pauli principle 120, 129, 168, 288
Pauli spin paramagnetism 164
Pauli susceptibility 413
Peierls transitions 279, 282
Peltier coefficient 254, 256

Index 713

Penetration depth 466, 471, 473, 502, 578
Perfect diamagnetism 506
Periodic boundary conditions 48, 51
Permalloy 428
Permittivity of free space 4, 234
Permutation operator 120
Perovskite 29, 517
Perpendicular twofold axis 26
Perturbation expansion 44, 667
Phase space 182, 245
Phase transition 203, 279, 359, 406, 414, 454, 517, 522
Phonon 41, 61, 62, 82, 556
Phonon absorption 229
Phonon current density 223
Phonon density of states 98, 228
Phonon emission 229
Phonon frequencies 222
Phonon radiation 222
Phonon-phonon interaction 41, 113, 219, 221
Phosphorescence 579
Photoconductivity 585, 597
Photoelectric effect 172, 347
Photoemission 172, 543, 578
Photoluminescent 579
Photon absorption 551
Photons 172, 563, 579
Photovoltaic effect 344, 347
Physical observables 77
Piezoelectric crystals 516
Pinned 208, 336, 439
Planar defects 589
Planck distribution 579
Plane polarized light 582, 611
Plane wave solution 132, 210
Plasma frequency 216, 265, 515, 527, 564, 583
Plasmons 136, 216, 527, 611
Platinum 291
pn-junction 331, 338, 344
Point defects 587, 597

Point group 13, 17, 22, 23, 26, 27, 39, 442
Point scatterers 32, 36
Point transformations 17
Poisson bracket relations 76
Polar crystals 216, 516
Polar solids 544
Polaritons 569
Polarization 234, 509, 514, 523, 563, 572, 583
Polarization catastrophe 520
Polarization vectors 89, 91, 111, 229, 238
Polarons 216, 232, 241
Polyhedron 22, 184
Polymers 1, 610, 640, 641
Polyvalent metals 265
Population inversion 626
Positive definite Hermitian operator 114
Positrons 213
Potential barrier 336, 338, 556, 605
Potential gradients 307
Primitive cells 17
Primitive translation 17, 23, 84, 269, 610, 655
Principal threefold axis 26
Projection operators 179
Propagators 666
Proper subgroup 15, 16
Pseudo binary alloys 320
Pseudo-Hamiltonian 200
Pseudopotential 178, 190, 194, 265, 318, 319, 500, 611
p-type semiconductor 335, 336, 337
p-wave pairing 501
Pyroelectric crystals 516

Q

Quantum conductance 609, 620
Quantum dot 609, 615, 616
Quantum electrodynamics 217, 233, 479, 666
Quantum Hall Effect 165, 632

Quantum mechanical inter-band tunneling 344
Quantum wells 325, 609, 626
Quantum wires 609, 615, 636
Quasi Periodic 21
Quasi-classical approximation 344
Quasicrystals 21
Quasi-electrons 135, 136, 217, 484
Quasi-free electron 113, 158, 509
Quasi-particles 135, 217, 666

R

Radiation damage 587
Radiative transitions 562, 626
Raman scattering 96, 580, 613
Rare earths 265, 378, 383, 405, 430, 443
Rayleigh-Ritz variational principle 114
Real orthogonal transformation 88
Real solids 6, 173, 243, 388
Reciprocal lattice 31, 36, 37, 84, 174, 221
Reciprocal lattice vectors 35, 268, 610
Reciprocal space 36, 176
Reducible representation 445
Reflection coefficient 545
Reflection High Energy Electron Diffraction 612
Reflection symmetry 18, 19
Reflectivity 203, 547, 564, 576
Refractive index 509
Registry 613
Regular polyhedron 22
Relativistic corrections 116, 192, 193, 204, 243, 660, 661
Relativistic dynamics 287
Relativistic effects 43
Relativistic pressure 288
Relaxation region 565
Relaxation time 171, 172, 250, 307
Relaxation time approximation 250, 253, 264

Remanence 420, 427
Renormalization 414, 416, 484, 665
Reptation 642
Repulsive force 3, 6
Resonance frequencies 313, 509
Resonant tunneling 614, 625
Rest mass 193, 287, 660
Restrahl frequency 68, 576
Restrahlen effect 510
Reverse bias breakdown 344
Reversible processes 8
Richardson-Dushmann equation 603
Riemann zeta functions 401
Rigid ion approximation 229, 230
RKKY interaction 378
Rochelle salt 516
Rotary reflection 18
Rotation inversion axis 26
Rotational operators 81
Rotational symmetry 18, 20, 81, 90, 188, 419
Rubber 641
Rushbrooke inequality 417

S

Saturation magnetization 405, 409, 420, 428
s-band 179, 184, 407
sc lattice 38, 210
Scaling laws 415
Scanning Auger microscope 612
Scanning Electron Microscopy 429, 612
Scanning Tunneling Microscopy 612
Scattered amplitude 40
Schoenflies 26, 27
Schottky and Frenkel defects 588
Schottky Barrier 334
Schottky emission 605
Screening 535
Screening parameter 533
Screw axis symmetry 19, 20

Screw dislocation 599, 600, 601
Second classical turning point 605
Second quantization 124, 662
Secondary Ion Mass Spectrometry 612
Second-order phase transitions 414
Secular equation 50
Selection rules 225, 228, 569
Self-consistent one-particle Hamiltonian 126
Semiconductor 293, 555, 568, 587, 590, 637
Seven crystal systems 23
Shallow defects 296, 589, 591
Shell structure 115
Shockley diode 339, 344
Shockley state 587
Short range forces 3
Si 29, 191, 315, 318
Similarity transformation 445
Simple cubic cell 25
Simple cubic lattice 181, 210, 392, 422
Simple monoclinic cell 24
Simple orthorhombic cell 24
Simple tetragonal cell 24
Single crystal 273, 325, 424, 590
Single domain 420
Single electron transistors 615
Single particle wave functions 141
Single-ion anisotropy 385, 386
Singlet state 373, 501
SiO_2 335, 633
Skin-depth 274
Slater determinant 120, 664
Slater-Koster model 205, 596
Slow neutron diffraction 228
$SmCo_5$ 431
Sn 265, 290
Soft condensed matter 1, 609, 640
Soft mode theory 517, 521
Soft x-ray emission 210
Soft x-ray emission spectra 168
Solar cell 294, 345

Solid state symmetry 17
Solitons 353, 456
Space degrees of freedom 43
Space groups 13, 23, 28
Specific heat of an electron gas 135, 161
Specific heat of an insulator 52, 97
Specific heat of linear lattice 61
Specific heat of spin waves 400
Specific heat of the one-dimensional crystal 61
Speed of light 32, 635, 660
Speromagnetic 430
Spherical harmonics 187
Spin 1/2 particle 43
Spin coordinate 119, 373
Spin degeneracy 602
Spin density waves 283
Spin deviation quantum number 397, 402
Spin diffusion length 451
Spin glass 353, 430, 454
Spin Hamiltonian 373
Spin polarization 450
Spin wave theory 364, 389, 413, 417
Spin-lattice interaction 222
Spin-orbit interaction 204, 443, 660
Spin-polarized transport 450
Spin-spin relaxation time 433, 437
Split-off band 320
Spontaneous emission 579
Spontaneous magnetism 359
Spontaneous polarization 517
Spontaneously broken symmetry 419, 505
Stark-Wannier Ladder 623
Steel 290, 639
Stereograms 26, 39
Stokes line 580
Stoner criterion 412
Stoner model 406, 450
Strained layer 622
Strong crystal field 443

Structure factor 35, 38, 177, 201
Subgroup 16, 23, 39
Substitutional atoms 70, 589
Superconducting metals 460
Superconducting wave function 468, 472
Superconductive state 459
Superconductivity 464
Superconductors 459
Superlattice 622, 623, 626
Surface defects 587
Surface reconstruction 612
Surface states 335, 345, 350, 636
s-wave pairing 501
Symmetry operations 17, 18, 23, 26, 192
Symmorphic 23

T

Tamm states 587
Tensor effective mass 158
Tetragonal Symmetry 24
Tetrahedron 22
Thermal conductivity 99, 169, 223, 246, 254
Thermal energy 172, 561
Thermal neutrons 30, 228
Thermal resistance 52, 222
Thermionic emission 135, 601
Thermodynamic fluctuations 359, 416
Thermodynamics of irreversible process 255
Thermoelectric power 255
Thomas-Fermi approximation 531
Thomas-Fermi-Dirac method 137
Three-dimensional lattice vibration 47, 85
Three-dimensional periodic potential 148
Threefold axis 20, 26
Ti 291

Tight binding approximation 113, 173, 178, 184, 190, 207
Total cohesive energy 8
Total exchange charge 129
Total reflection 568
Transistors 294, 328, 338, 350, 615, 636
Transition metals 265, 273, 639
Translation operator 80, 655
Translational symmetry 17
Transmission coefficient 605, 619
Transmission Electron Microscopy 429, 612
Transport coefficients 218, 246
Transverse and longitudinal acoustic modes 229
Trial wave function 115, 190, 243, 490, 493, 541
Triclinic Symmetry 23
Triglycine selenate 516
Triglycine sulfate 516
Trigonal Symmetry 25
Triplet state 373
Triplet superconductivity 284
Two-atom crystal 50
Two-body forces 3, 42
Two-dimensional defect 590
Two-fold axis of symmetry 26
Two-fold degeneracy 4, 150, 444
Two-particle operator 116, 121, 663
Type I superconductors 461, 471, 505
Type II superconductors 459, 471, 504

U

UAl_2 284
UBe_{13} 284, 501, 503
Ultrasonic absorption 274
Ultrasonic attenuation 463
Ultrasonic wave 274
Ultraviolet photoemission 578

Umklapp process 221, 232, 251, 483
Uncertainty principle 5
Unit cells 17
Unitary transformation 88, 207, 451
Unrestricted force constants approach 47
Upper critical field 462, 470, 502

V

Vacancies 285, 587, 588, 596, 598
Valence band 590
Valence crystalline binding 12
Valence crystals 11, 12, 371
van der Waals forces 4, 6
Van Hove singularities 559
Varactor 332
Variational principle 114
Variational procedure 140
Vector potential 165, 166, 465, 550
Velocity operator 195
Verdet constant 584
Vertical transitions 554
Vibrating dipoles 4
Virgin curve 427
Virtual crystal approximation 320
Virtual excited states 2
Virtual magnons 501
Virtual phonons 233, 237, 501, 502
Volume coefficient of thermal expansion α 100
Vortex region 462

W

W 291
Wall energy 424, 425
Wannier excitons 561
Wannier function 207, 407
Wave vector 55
Weak crystal fields 443
Weak superconductors 505
Weiss theory 359, 361, 406, 415
Whiskers 599
White dwarf 286
Wiedeman-Franz Law 169
Wigner-Seitz cell 184, 186, 231
Wigner-Seitz method 9, 184, 187
WKB approximation 557, 605, 620

X

X-ray photoemission 578
X-rays 30, 96, 168

Y

$YBa_2Cu_3O_7$ 502, 503

Z

Zeeman energy 388
Zener Breakdown 344, 556
Zero point energy 5, 62
Zincblende 295, 320, 323
Zn 290

图书在版编目（CIP）数据

固态物理学 = Solid – State Physics：Introduction to the Theory. 第 2 卷：英文／（美）帕特森（Patterson, J. D.）著．—影印本．—北京：世界图书出版公司北京公司，2011. 1
ISBN 978-7-5100-2976-9

Ⅰ.①固…　Ⅱ.①帕…　Ⅲ.①固体物理学—英文　Ⅳ.①O48

中国版本图书馆 CIP 数据核字（2010）第 212113 号

书　名：	Solid – State Physics：Introduction to the Theory Vol. 2
作　者：	James D. Patterson, Bernard C. Bailey
中译名：	固态物理学 第 2 卷
责任编辑：	高蓉　刘慧
出 版 者：	世界图书出版公司北京公司
印 刷 者：	三河国英印务有限公司
发　行：	世界图书出版公司北京公司（北京朝内大街 137 号　100010）
联系电话：	010-64021602，010-64015659
电子信箱：	kjb@ wpcbj. com. cn
开　本：	24 开
印　张：	15.5
版　次：	2011 年 01 月
版权登记：	图字：01-2010-1604
书　号：	978-7-5100-2976-9/O・842　　　定　价：　49.00 元